DUE DATE

Control of Crop
Productivity

Control of Crop Productivity

Edited by

C. J. Pearson

School of Agriculture
The University of Western Australia

ACADEMIC PRESS

(Harcourt Brace Jovanovich, Publishers)

Sydney   Orlando   San Diego   Petaluma   New York
London   Toronto   Montreal   Tokyo
1984

ACADEMIC PRESS AUSTRALIA
Centrecourt, 25–27 Paul Street North
North Ryde, N.S.W. 2113

United States Edition published by
ACADEMIC PRESS INC.
Orlando, Florida 32887

United Kingdom Edition published by
ACADEMIC PRESS, INC. (LONDON) LTD.
24/28 Oval Road, London NW1 7DX

Printed in Australia

National Library of Australia Cataloguing-in-Publication Data

Control of crop productivity.

  Includes bibliographies and index.
  ISBN 0 12 548280 9.

  1. Crop yields – Addresses, essays, lectures.
  2. Agricultural productivity – Addresses, essays,
  lectures. I. Pearson, C. J. (Craig John), 1946- .

631

Library of Congress Catalog Card Number: 83-71463

This book is dedicated to Professor Fred L. Milthorpe

Contents

12. Tropical Pastures

J. R. WILSON

13. Natural Grasslands

E. K. CHRISTIE

14. Fruit Crops

K. A. OLSSON, P. R. CARY AND D. W. TURNER

15. Protected Crops

S. W. BURRAGE AND P. NEWTON

16. Mangroves

B. F. CLOUGH

17. Trees and Forest Restoration

F. J. BURROWS

18. Modelling Environmental Effects on Crop Productivity

J. M. MORGAN

Contributors

Numbers in parentheses indicate the pages on which the authors' contributions begin.

D. ASPINALL (91), Department of Plant Physiology, Waite Agricultural Research Institute, The University of Adelaide, Glen Osmond, South Australia 5064, Australia

S. W. BURRAGE (239), Department of Horticulture, Wye College, Kent TN25 5AH, U.K.

F. J. BURROWS (269), School of Biological Sciences, Macquarie University, North Ryde, New South Wales 2113, Australia

P. R. CARY (219), Centre for Irrigation Research, Commonwealth Scientific and Industrial Research Organization, Griffith, New South Wales 2680, Australia

K. R. CHRISTIAN (111), Division of Plant Industry, Commonwealth Scientific and Industrial Research Organization, Canberra, Australian Capital Territory 2600, Australia

E. K. CHRISTIE (199), School of Australian Environmental Studies, Griffith University, Nathan, Queensland 4111, Australia

B. F. CLOUGH (253), Australian Institute of Marine Science, Private Mail Bag 3, Townsville, Queensland 4810, Australia

I. COWAN (13), Research School of Biological Sciences, Australian National University, Canberra, Australian Capital Territory 2600, Australia

J. L. DAVIDSON (111), Division of Plant Industry, Commonwealth Scientific and Industrial Research Organization, Canberra, Australian Capital Territory 2600, Australia

G. D. FARQUHAR (43), Research School of Biological Sciences, Australian National University, Canberra, Australian Capital Territory 2600, Australia

P. GOODWIN (127), Department of Agronomy and Horticultural Science, The University of Sydney, New South Wales 2006, Australia

A. J. HALL (141), Department of Ecology, Faculty of Agronomy, University of Buenos Aires, Buenos Aires, Argentina

D. R. KEMP (159), Agricultural Research and Veterinary Centre, Orange, New South Wales 2800, Australia

J. M. MORGAN (289), Agricultural Research Centre, Roadside Mail Box 944, Tamworth, New South Wales 2340, Australia

W. A. MUIRHEAD (73), Centre for Irrigation Research, Commonwealth Scientific and Industrial Research Organization, Griffith, New South Wales 2680, Australia

P. NEWTON (239), Department of Botany, University of Manchester, Manchester M13 9PL, U.K.

K. A. OLSSON (219), Irrigation Research Institute, Tatura, Victoria 3616, Australia

J. W. PATRICK (59), Department of Biological Sciences, The University of Newcastle, New South Wales 2308, Australia

C. J. PEARSON (1, 73, 141), School of Agriculture, The University of Western Australia, Nedlands 6009, Western Australia, Australia

M. A. SIDDIQUE (127), Department of Horticulture, Bangladesh Agricultural University, Mymensingh, Bangladesh

N. TERRY (43), Department of Plant and Soil Biology, University of California, Berkeley, California 94720, U.S.A.

N. THORPE (33), School of Biological Sciences, Macquarie University, North Ryde, New South Wales 2113, Australia

D. W. TURNER (219), Tropical Fruit Research Station, Alstonville, New South Wales 2477, Australia

J. R. WILSON (185), Division of Tropical Crops and Pastures, Commonwealth Scientific and Industrial Research Organization, St Lucia, Queensland 4067, Australia

Foreword

In many branches of science the development and transmission of ideas can be traced from one generation to another through the work of inspiring teachers and supervisors of research students. This book is dedicated to such a teacher, Professor F. L. Milthorpe, and written largely by his former research students. The book presents the state of the art over the whole range of crop physiology in which Fred Milthorpe has worked: metabolic processes, stomatal behaviour, leaf growth and productivity in relation to environmental factors. Previously unpublished data and new hypotheses and viewpoints on the processes which control crop productivity are discussed.

Fred Milthorpe belongs to F. G. Gregory's family of distinguished students which included Robert Brown, O. V. S. Heath, Helen Porter, F. J. Richards, W. W. Schwabe, D. C. Spanner and W. T. Williams at Imperial College, London. Over the past 30 years Fred Milthorpe has acquired a similar family of his own which includes M. Aluko, D. Aspinall, G. Baines, A. Bertus, E. Boerema, M. Borah, W. Bryant, M. Burgman, S. Burrage, F. Burrows, R. Burt, P. Cary, J. Castel, K. Christian, E. Christie, G. Cobrera, L. Coke, I. Cowan, E. Cox, A. Dakkak, D. DasGupta, J. Davidson, W. Dean, W. Farooqi, T. Flowers, B. Freeman, J. Gilmour, P. Goodwin, A. Hall, D. Headford, J. Hopkinson, M. Hossain, H. Idris, A. Kheiralla, E. J. M. Kirby, B. Lewis, I. McCorquodale, R. Mobayen, E. Mordue, J. M. Morgan, D. Morris, W. Muirhead, L. Mutton, P. Newton, T. O'Brien, K. Olsson, J. Patrick, C. Pearson, W. Peat, R. Pulver, A. Rainbow, A. ElRayah, E. Sabalvoro, E. Sadler, E. Sambo, C. R. Slack, J. Slater, R. Smart, N. Terry, N. Thorpe, S. Thrower, D. Turner, W. Whittington and R. Wills.

Milthorpe's studies led to the recognition, or more correctly, re-recognition, of the quantitative importance of stomata in controlling crop productivity. His contribution to our understanding of stomatal physiology includes many publications, some of which are listed in Chapter 3; his quantification, rather than mere description, of physiological processes, has encouraged modelling as a tool to summarize our knowledge and simulate crop productivity. Milthorpe's achievements in these fields and in education generally have been recognized by the J. G. Wood Memorial Lecture in plant physiology, the medal in Australian Agricultural Science and by election as Fellow of the Australian Institute of Agricultural Science, Fellow of the Institute of Biology and Fellow of the Royal Society of Arts.

Milthorpe's students, now distributed world-wide, are contributing to research in fields as diverse as plant biophysics and biochemistry (C. R. Slack in New Zealand and P. Goodwin in Australia), whole plant growth (for example, D. Aspinall, currently in Indonesia and E. J. M. Kirby at Cambridge), the agronomy and micrometeorology of pastures, crops, orchards and forests, and the design and management of glasshouses (P. Newton and S. Burrage in England). Some of his students, in turn, are helping to train another generation of physiologists; many have reached senior posts in universities and research stations. All bear the stamp of their teaching by F. L. Milthorpe at the University of Nottingham where he worked from 1954 to 1967 or subsequently at Macquarie University in Sydney from 1967 to 1982.

J. L. Monteith, F.R.S.
UNIVERSITY OF NOTTINGHAM
OCTOBER, 1983

Acknowledgements

This tribute to mark Professor F. L. Milthorpe's retirement was conceived by Jeff Moorby and the editor at Macquarie University, refined by a wider group at the Thirteenth International Botanical Congress, in Sydney, and worked on at The Universities of Sydney and Western Australia. Support from the following is acknowledged: J. Moorby and others, particularly the staff of Academic Press and my secretaries, Karen Chia and Diana Watson, and the contributors whose support, sometimes at short notice, is obvious.

CHAPTER **1**

Introduction

C. J. PEARSON

I. BACKGROUND

Crop productivity depends on a crop's development pattern and process physiology in response to its management and environment. Such an assertion will be accepted by almost all plant biologists, ecologists (although they may be upset by the distinction between management and "environment") and agriculturalists, yet each of us will place emphasis on the different components within this statement. I hope that by emphasizing the unity of plant growth and productivity this book will encourage you to look again at all the controls, and the interplay between them, in order to decide which interactions are more important in determining productivity in particular situations.

We must recognize that any one component of development or growth will not exert absolute control over productivity in all or even in most field situations. The belief that productivity results from interplay between a number of controls is supported by current crop yields (Fig. 1.1). These show no pattern between countries: high yields are associated with "good" interaction between genotype, management and environment but no one aspect

1

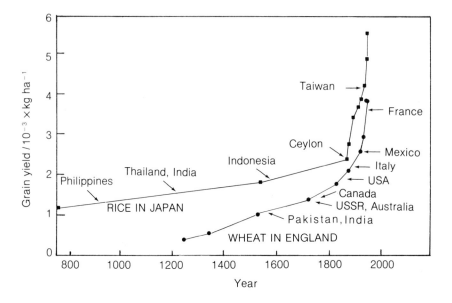

Fig. 1.1 Historical trends in grain yield of rice in Japan and of wheat in England, and (arrowed) yields of wheat and rice in several countries in 1968 (from Evans 1975b).

dominates, as shown by the wide range of yields of rice throughout the lowland tropics or the relatively low yields of wheat in highly mechanized, "scientific" countries such as Australia and Canada. It may be argued that the remarkable rate of increase in crop productivity which has occurred recently (Fig. 1.1) has taken place because farmers have increasingly managed this interplay between controls.

Emphasis on the unity of crop productivity contrasts with the current fashion of scientific research, with its orientation to work within narrow disciplines and its emphasis on process physiology. This book may contribute to shifting plant research towards a more balanced and integrated view of the control of crop productivity.

We can look at the interplay between controls over crop productivity at any level of biological organization. In this book we have arbitrarily made a division between "physiological processes" (Part I) and interactions at a higher biological level where we consider the development of specific crops (Part II).

In Part I we consider first the interplay between the two most important components of growth, carbon gain and water loss (Chapter 2). This raises the questions that recur throughout the book: what is the relative importance of, and interactions between, various processes in controlling productivity, and do plants adopt strategies to adjust or optimize processes in response to changing development, management or environment? Subsequent chapters develop these questions for each part of the carbon pathway: exchange (stomata, Chapter 3), fixation (Chapter 4) and translocation (Chapter 5); and for nitrogen exchange and distribution (Chapter 6).

In Part II we attempt to integrate these physiological controls with development and management of specific crops. I have defined "crop" in a wide, non-agricultural sense to include row crops (Chapters 7–10), sown and natural grasslands (Chapters 11–13), horticultural crops, mangroves and forests (Chapters 14–17). This draws attention to the fact that all plant communities are now managed; absence of interference is itself a management option that applies even to the clearing of rainforest by shifting hill-tribesmen of Sumatra. This wide, non-agricultural viewpoint compels us all to look at the unity, not the pigeon holes, within our biology.

This unitary approach complements existing texts which deal with single components of the crop system; for example, Milthorpe's contributions to aspects of development (Milthorpe, 1956; Dale and Milthorpe, 1983) and process physiology (Milthorpe and Moorby, 1979) and case studies of the physiological ecology of particular crops (Evans, 1975a; Norman et al., 1984).

II. CONTROLS OF CROP PRODUCTIVITY

The assertion that productivity is constrained by development pattern, process physiology, management and environment leads to an hypothesis that actual productivity (as distinct from potential productivity) is controlled by interaction between these components within broadly defined limits. This view is essentially Waddington's notion of a "creode": a broad path of growth and development within which plants or animals travel through their life cycle (1962).

This hypothesis raises three points: (1) that a priori we should consider each constraint of equal importance, both as independent determinants of growth and in their interactions with other determinants; (2) that we should consider the strategies including "optimization" which plants adopt to adjust their productivity to changes in or as a response to management or environment; and (3) that to understand the "control of crop productivity" we have to consider interactions between components in a situation which is defined in terms of the components as well as being defined in the traditional way, in time.

A. Development Pattern

In annual agricultural crops, progression of an individual plant through its life cycle is an easily defined event which corresponds with the development of the crop. These progressions are described in general terms elsewhere (e.g. Evans, 1975a). In perennial systems the life cycle of the individual may coincide with that of the crop, as in orchards or planted forests, or the individual may not live as long as the crop, as in grasslands and most native communities. Regeneration and crop dynamics become integral features of the productivity of crops such as grasslands (Chapter 11) and native communities (Chapters 16, 17).

Generation of leaf area is arguably the most important aspect of crop development. Recognition of this led to detailed studies of leaf area development (see, for example, Milthorpe and Dale, 1983).

The quantitative importance of leaf area is given in equations such as

$$G = \varepsilon \, I[1 - \exp(- kL)] \qquad (1.1)$$

where G is net crop growth rate, ε is the efficiency with which radiation is used to produce dry matter, I is average daily PAR (photosynthetically active radiation of 400–700 nm wavelength) incident on the crop's uppermost surface, k is the canopy extinction coefficient, and L is leaf area index (Warren-Wilson, 1971). This type of analysis led to recognition that crop growth rate is, in a wide range of situations, proportional to intercepted PAR (e.g. Monteith, 1981). This is discussed further in Chapter 11 where D. R. Kemp shows that seasonal changes in pasture growth rate can be related to daylength inducing flowering which in turn affects leaf area expansion and canopy architecture.

For any given level of intercepted PAR, crop growth rate should increase with increasing photosynthetic rate per unit area (a higher ε) and decreasing extinction coefficient (a lower k). The latter led physiologists, accustomed to thinking in pigeon holes, to search for erect-leafed (low k) genotypes which might, but usually did not, give higher yields. The unimpressive performance of genotypes having low extinction coefficients is explained if we adopt a unitary view. Advantages of low k after canopy closure are, at least in maize, small (Chapter 10) and they are offset by the erect-leafed genotype intercepting less radiation per unit leaf growth early in its life cycle (e.g. Monteith, 1981).

Notwithstanding the importance of leaf area as a determinant of PAR interception, it is now apparent that ε, the "efficiency" of growth including dry matter distribution, contributes appreciably to determining productivity in particular situations. There is a large gap between calculated actual "efficiency" values (1.3–4.2 μg (dry weight) J^{-1} PAR) and the theoretical maximum (6.6 μg J^{-1}) (Charles-Edwards, 1982). Further, in some crops such as groundnut, improvement in yield associated with plant breeding has been largely as a result of increased ε through improved dry matter distribution ultimately to reproductive organs (Duncan *et al.*, 1977).

Patterns of dry matter distribution change as plants age and they change abruptly as a consequence of developmental events. These changes are of greater quantitative importance than the changes in distribution which may be brought about by non-catastrophic fluctuations in environment. Indeed, classical analyses of allometric ratios of parts established long ago the relative robustness of dry matter distribution among organs (D'Arcy Thompson, 1922, reproduced 1959). It has been suggested by Brouwer (1977 and references therein) and others that organ development and changes in dry weight partitioning which accompany such development are homeostatic; that is, they tend toward functional equilibria. Such homeostatis is apparent in Figure 1.2 and is intuitively desirable for the maintenance of the crop.

The concept of development homeostasis has been elaborated to produce an hypothesis of optimization of plant form for function. For example, roots may tend to optimize their shape for water uptake (Chapter 6). It is clear only that development and dry weight partitioning do adjust within limits (within a "creode"?) and the adjustments include marked homeostasis

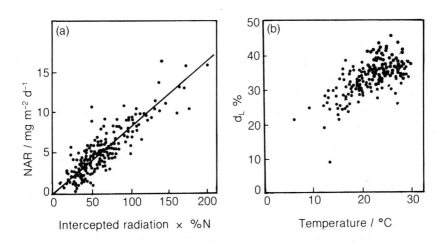

Fig. 1.2 Homeostasis of (a) net assimilation rate (NAR) and (b) partitioning of dry matter (to leaves relative to the change in whole plant dry matter, d_L) shown by growth analysis of rice in a range of environments in Japan. NAR is plotted against the product of intercepted radiation (calculated assuming k = 0.7) and nitrogen concentration of leaves (calculated from N = ae^{-bt} where N is concentration, t days from transplanting and a and b are empirical (site specific) constants ((a) from Murata, 1975 and (b) from Kumura, 1975).

through feedback such as seed abortion which occurs when fertility exceeds the post-anthesis supply of carbohydrates.

Developmental events, such as flowering and seed fertilization, are considered in chapters on wheat (representing a temperate cereal), common bean (representing a legume), maize and millet (representing tropical cereals) and temperate pastures. These chapters address the sequence which includes homeostatic feedbacks, in which environment and management affect developmental events which in turn affects both net dry weight gain and dry weight partitioning among plant parts. This approach is complementary to, but perhaps more instructive than, the traditional agronomic methodology of yield component analyses, which in a sense attempts to "look back" from final yield to the development events which gave rise to that productivity.

B. Process Physiology

This can be considered the second control of productivity through its instantaneous influence on ε and subsequent leaf area and light interception. Broadly, process physiology can be divided into dry weight acquisition (mainly carbon but also mineral) and its translocation and utilization. Many books and articles deal with these processes although F. L. Milthorpe and J. Moorby's *An Introduction to Crop Physiology* (1979) remains a standard text.

Our book develops the thesis that physiological machinery capable of controlling growth exists at many points (separated within the plant and in time) and that the importance of any particular control point changes depending on crop age, management or environment. Again, the robustness of the processes and the closeness with which they co-regulate, are noteworthy. To give an example, pearl millet retains a virtually identical relationship between relative dry weight gain and relative nitrogen uptake when grown over a temperature range from optimal to near lethal; this is despite sensitivity of rates of photosynthesis, translocation, leaf area expansion and nitrate assimilation to temperature and quantitative shifts between pathways of nitrogen flow at high and low temperature (Theodorides and Pearson, 1982, and references therein).

Close co-regulation, such as the example above with pearl millet, has led several plant biologists to suggest that crops may adopt strategies to optimize their physiological activity. More rigorous application of optimization theory confirms this possibility. Cowan (Chapter 2) demonstrates that stomata may affect a compromise between carbon gain and water loss, with growth emphasized when moisture is high and water loss being more important than growth when moisture is limiting.

Knowledge of such interactions and of the interactions between process physiology and management may lead to greater managerial efficiency and increased productivity.

C. Management and Environment

Interactions between development pattern and physiology and the external controls of productivity, namely management and environment, comprise Part II.

Growth analysis (Watson, 1952 and references therein) and yield component analysis of data collected at final harvest are appropriate tools to define problems, sharpen hypotheses and provide observations to test or validate models of crop productivity. These tools are applicable equally to row crops, orchards and native communities. Growth analysis has led to appreciation of the homeostasis within crops and of the overriding importance of some relationships, such as that between growth and PAR interception (Fig. 1.2).

We are, however, at a milestone in research into the response of crops to management and environment. We realize that growth and final yield analysis describe the net result of interactions between the factors that control productivity: they are tools which describe the net result of what *has* happened rather than demonstrate how crops respond, optimize or sacrifice in a *current* situation. Thus, measurement of dry weight and leaf area provide only a first step towards understanding productivity. They provide descriptions of net changes within particular systems, such as abound in classical agronomic research, but we must now build on these descriptions if we are to understand, rather than simply describe, "controls of crop productivity".

The building takes two forms; both lead from site specific observation to knowledge from which one can generalize. One form is the mathematical progression from growth analysis to generalized relationships, albeit net relationships, between interactive processes. This approach has been used elegantly by the Japanese International Biological Program (Murata, 1975), Hesketh and Jones (1979) and Charles-Edwards (1982). This currently seems to be the most appropriate way to develop simple simulation models of crop growth which can be used by extension agencies, farmers and wildlife managers to predict crop response to management and environment (Chapter 18).

A second approach is to look at a generalized net relationship such as Figure 1.2, and ask, how? How do crops control their development and physiological activity so that development and productivity, or at least life, are sustained in response to changes in management or environment?

"How" and "why" questions lead to recognition of the role that management may play in decreasing or increasing productivity. For example, we can define the effects of some common management practices such as the decapitation of maize during grain development that causes a decreased rate of grain filling in subsistence agriculture (Chapter 10). Likewise, integration of our knowledge of management as it affects environment, development and rates of physiological processes, have led to radical changes in orchard design which give earlier bearing and higher yields (Chapter 14) and to highly controlled environments for energy-efficient and cost-efficient production of flowers and vegetables (Chapter 15).

III. CONCLUSIONS

This book aims to present examples of how interactions between development, physiology, management and environment control crop productivity. In adopting a unitary view we have also chosen to discuss many crops, "cultivated" and "wild". The excitement of crop research comes in part from the realization that controls of productivity are interactive and robust, and lead to generalized relationships or outcomes such as in Figure 1.2. The challenge comes here too: how does the crop regulate such internal interaction? How can we manipulate it to our advantage?

REFERENCES

Brouwer, R. (1977). Root functioning. *In* "Environmental Effects on Crop Physiology" (J. J. Landsberg and C. V. Cutting, eds.), pp. 229–245. Academic Press, London.

Charles-Edwards, D. A. (1982). "Physiological Determinants of Crop Growth." Academic Press, Sydney.

Dale, J. E., and Milthorpe, F. L. (1983) "The Growth and Functioning of Leaves" Cambridge University Press, Cambridge.

D'Arcy Thompson, W. (1959). "On Growth and Form", 2nd edn. Cambridge University Press, Cambridge.

Duncan, W. G., McCloud, D. E., McGraw, R. L., and Boote, K. J. (1977). Physiological aspects of peanut yield improvement. *Crop Science* **18**, 1015–1020.

Evans, L. T. (1975a). "Crop Physiology, Some Case Histories." Cambridge University Press, Cambridge.

Evans, L. T. (1975b). Crops and world food supply, crop evolution, and the origins of crop physiology. *In* "Crop Physiology, Some Case Histories" (L. T. Evans, ed.), pp. 1–22. Cambridge University Press, Cambridge.

Hesketh, J. D., and Jones, J. W. (1979). Integrating traditional growth analysis techniques with recent modelling of carbon and nitrogen metabolism. *In* "Predicting Photosynthesis for Ecosystem Models Vol 1 (J. D. Hesketh, ed.), pp. 51–86. Chemical Rubber Company Press, Baton Rouge.

Kumura, A. (1975). Dry matter partition and climatic factors. *In* "Japanese International Biological Programme Synthesis: Crop Productivity and Solar Energy Utilization in Various Climates in Japan" (Y. Murata, ed.), pp. 49–59. University of Tokyo Press, Tokyo.

Milthorpe, F. L. (1956). "The Growth of Leaves". Butterworths, London.

Milthorpe, F. L., and Moorby, J. (1979). "An Introduction to Crop Physiology" 2nd edn. Cambridge University Press, Cambridge.

Monteith, J. L. (1981). Does light limit crop production? *In* "Physiological Processes Limiting Plant Productivity" (C. B. Johnson, ed.), pp. 23–38. Butterworths, London.

Murata, Y. (1975). The effect of climatic factors and aging on net assimilation rate of crop stands. *In* "Japanese International Biological Programme Synthesis: Crop Productivity and Solar Energy Utilization in Various Climates in Japan" (Y. Murata, ed.), pp. 172–186. University of Tokyo Press, Tokyo.

Norman, M. J. T., Pearson, C. J., and Searle, P. G. E. (1984). "The Ecology of Tropical Food Crops." Cambridge University Press, Cambridge.

Theodorides, T. N., and Pearson, C. J. (1982). Effect of temperature on nitrate uptake, translocation and metabolism in *Pennisetum americanum. Australian Journal of Plant Physiology* **9**, 309–320.

Waddington, C. H. (1962). "New Patterns in Genetics and Development". Colombia University Press, New York.

Watson, D. J. (1952). The physiological basis of variation in yield. *Advances in Agronomy* **4**, 101–146.

Warren-Wilson, J. (1971). Maximum yield potential. In *Transition from Extensive to Intensive Agriculture with Fertilizers,* Proceedings of 7th Colloqium of International Potash Institute pp. 34–56. International Potash Institute, Berne.

Part I

Processes Controlling Productivity

CHAPTER 2

Optimization of Productivity: Carbon and Water Economy in Higher Plants

I. COWAN

1. INTRODUCTION

We may visualize the physiological control of carbon metabolism operating at four points: (1) carbon dioxide exchange between the atmosphere and crop; (2) photosynthetic carbon fixation; (3) translocation; and (4) synthesis of a wide range of compounds including those which contain minerals. Physiological aspects of these four control points are considered in Chapters 3 to 6 respectively (although only carbon–nitrogen relations are described in Chapter 6).

Carbon metabolism and crop growth respond to aspects of the environment such as water deficits and management. Hanson and Hitz (1982) have

CONTROL OF CROP PRODUCTIVITY
ISBN 0 12 548280 9

identified two ways of viewing plant response to water deficits: as derangements that arise from stress-induced lesions at vulnerable sites in metabolism, or as potentially adaptive changes that reflect ordered operation of metabolic regulatory mechanisms and which favour the plant as a whole during or after stress. Crop physiologists, having a preoccupation with factors that limit growth and yield in artificial environments, often incline to the first view. Plant ecologists are more likely to be in sympathy with the second, recognizing that rapid growth may contribute to, but does not determine natural fitness, and may at times be incompatible with processes that enhance the probability of survival and propagation in plants. There are two categories of response to water deficits that enhance the probability that a plant will survive. There are those that increase the amount of water that the plant may extract from the soil before stress-induced lesions occur. Amongst them is osmotic adjustment (Hsiao et al., 1976), which promotes the maintenance of tissue turgor at reduced water potential. And there are those that moderate the rate of water loss from the plant as the soil becomes drier. Amongst these, and common to all higher terrestrial plants, are the movements of stomata. In this chapter, it will be shown how the responses of stomata to changes in the environment of the foliage and changes in the supply of water to the plant roots are adapted to achieve a compromise between carbon fixation and survival.

II. SOIL AND PLANT HYDRAULICS

A. A Model

The rate of evaporation of water in leaves, or transpiration, affects plant water relations in two ways. Firstly, it causes a "draw-down" of water potential in the leaf with respect to that in the soil, according to what is often called the "van den Honert hypothesis" of water transport (see also Chapter 11.4C). The effect is responsible for the transient loss of turgor pressure, in some species seen as wilting, that often takes place in the leaves of plants at the time of day when the rate of transpiration is potentially greatest. Secondly, transpiration causes a depletion of the water content of the soil and decrease in the potential of water available for absorption by the root system. Eventually if the soil water is not replenished this leads to permanent loss of turgor in the plant.

These two effects, together with an assumption about the control function of stomata, are the basis of the model illustrated in Figure 2.1. When the soil is wet, leaf water potential exceeds, throughout the day, the critical level at which turgor pressure is zero. Then the stomata remain fully open. Rate

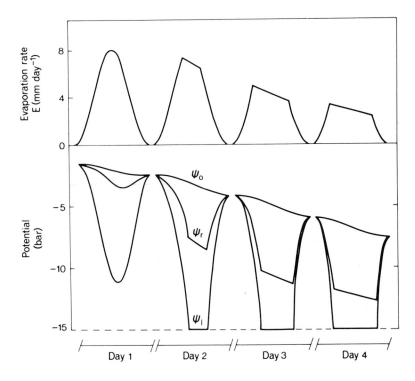

Fig. 2.1 Hypothetical time course of rate of transpiration, E, leaf water potential, Ψ_l, potential of water at the root surface, Ψ_r, and potential of water in the soil, Ψ_0, on successive days (adapted from Cowan, 1965; 1977).

of transpiration fluctuates with a maximum about midday because the climatic factors that influence it, principally radiation and temperature, vary in that way. Later, as the amount of water in the soil is depleted, leaf water potential falls to the critical level during periods that increase in length with each successive day. But the stomata respond so sensitively that they limit the rate of transpiration in these periods to prevent the potential of water in the leaf from decreasing any further. The rate of transpiration relative to the potential rate of transpiration (that which would occur if the leaves were turgid) is an indication of the degree of stomatal closure. One may think of the closure as a precaution against the shrinkage that occurs in tissue if leaf water potential is decreased below the point at which turgor pressure become zero. It has an immediate influence, in preventing transient dehydration of the leaves in the middle of the day. It has also a longer term influence: by decreasing the rate of water use, it delays the time at which turgor pressure in the leaves would be lost permanently if the soil were not rewatered. As shown in Figure 2.2a, the rate is reduced more severely the drier the soil and the greater the potential rate of transpiration.

B. Feedback Control of Leaf Turgor

This concept of plant hydraulics and the control function of stomata is supported by many observations. Stomata do often appear to act as high-gain feedback devices designed to prevent decrease of leaf water potential below the zero turgor pressure point (see e.g., Turner, 1974). Discoveries that abscisic acid (ABA) supplied to leaves causes stomatal closure (Little and Eidt, 1968; Mittelheuser and van Steveninck, 1969), and that ABA synthesis in leaves is enhanced by loss of turgor (Wright, 1969) pointed to a mechanism. Also, the concept of hydraulic control of stomatal aperture is consistent with the many observations, beginning with those of Loftfield (1921), of "midday stomatal closure". Figure 2.3 shows a sequence of diurnal variations in gas exchange in soybean, the variations in rate of transpiration resembling those idealized in Figure 2.1. Finally, the frequently cited data of Denmead and Shaw (1962) in Figure 2.2b exhibit trends similar to those indicated in Figure 2.2a.

However there is a significant discrepancy to be seen in comparison of Figures 2.2a and 2.2b. The three curves in Figure 2.2b, unlike those in Figure 2.2a, cross over before converging. This characteristic has not, as far as I am aware, received written comment. It is one of many indications that the influence of plant water status on stomatal aperture and rate of transpiration is sometimes tenuous and that midday depression of aperture can result from a different mechanism.

C. Feed Forward Control of Transpiration

There is evidence (see Hall *et al.*, 1976) that stomata tend to close with increase in the vapour pressure deficit of the external air with which they are in contact. The mechanism of the response is not clear. The most popular explanation is that it is due to effects of peristomatal transpiration; that is, water loss through the leaf external cuticle in the neighbourhood of the stomata. In any event the response is one which can cause midday depression of stomatal aperture, for it is generally near the middle of the day that the humidity deficit of the atmosphere is greatest. Variation of vapour pressure deficit in nature is predominantly associated with increase in air temperature; however, some important aspects of stomatal response to humidity

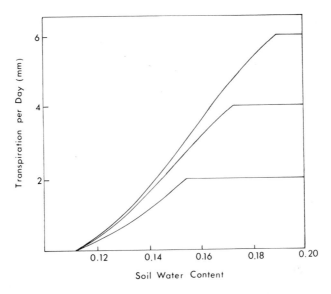

Fig. 2.2(a) Daily average rate of transpiration, Ē, as a function of soil water content, for three magnitudes of daily average potential rate of transpiration (computed from Fig. 2.1, see also Cowan, 1965).

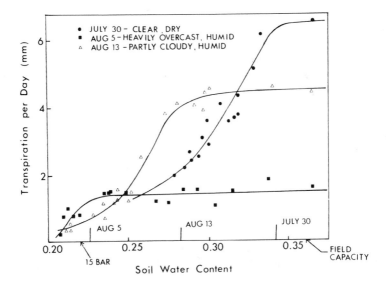

Fig. 2.2(b) Transpiration per day in corn plants in soils with differing water contents on three days of contrasting weather. Arrows indicate soil water contents below which wilting occurred (adapted from Denmead and Shaw, 1962).

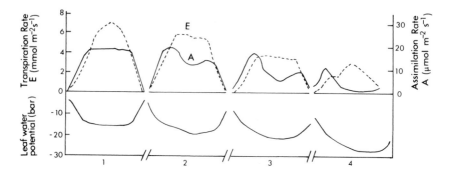

Fig. 2.3 Diurnal time courses of rate of transpiration, E, rate of assimilation, A, and water potential in leaves of soybean plants, 1, 2, 3, and 4 weeks after irrigation (adapted from Rawson and Constable (1980) with data from Rawson *et al.*, 1978, and Turner *et al.*, 1978).

deficit are best demonstrated by varying ambient vapour pressure while keeping temperature constant.

In *Prunus armeniaca* growing in the Negev Desert the resistance of stomata to vapour diffusion increased and the net rate of assimilation decreased as the difference in humidity between leaves and air was successively increased by decreased ambient humidity (Schulze *et al.*, 1972). In this respect the results are unremarkable. What is remarkable is that the increase in resistance on each occasion was so great that the rate of transpiration decreased despite the increase in humidity difference. Reduction in rate of transpiration was accompanied by an increase in the water content of the leaves. When, finally, the ambient humidity was increased, the rate of transpiration increased and the water content of the leaves declined. Similar results were obtained with the wild desert plants *Hammada scoparia* and *Zygophyllum dumosum*. Previously, Lange *et al.* (1971) had demonstrated the direct influence of change in ambient humidity on the aperture of stomata in epidermal strips of *Valeriana locusta* and *Polypodium vulgare*. Later investigations with intact plants, including crop species, also point to the existence of such an effect (Hall *et al.*, 1976; Schulze and Hall, 1982).

In so far as stomata respond to humidity deficit, they respond to external conditions which influence rate of transpiration from the leaf as distinct from leaf internal conditions that are influenced by rate of transpiration. This action can be described as feedforward control of water loss.

What is the adaptive role of feedforward control? Phenomenologically, it differs from a feedback response to leaf water potential in an important respect. A direct response can cause a decline in actual rate of evaporation when the potential rate of evaporation is increased. The feedback response cannot. This observation suggests that the short-term role of the changes in stomatal aperture is not so much preservation of turgor as economy of water.

However, conservation of water cannot be a dominant function of stomata—if it were so then the stomata should be continuously closed. The function of stomata in conserving water must be a compromise with their function in facilitating assimilation of carbon dioxide.

III. OPTIMAL CONTROL OF CARBON GAIN AND WATER LOSS

A. The Compromise

How much carbon fixation do plants forego in the interests of saving water? It has long been known that usually, in C_3 plants at least, rate of carbon dioxide (CO_2) assimilation is relatively less sensitive than rate of transpiration to change in stomatal aperture. There was enthusiasm once for the idea that water loss could be reduced without much affecting carbon uptake by using chemical inhibitors to partially close stomata (Zelitch and Waggoner, 1962). Whatever the agricultural relevance of this idea, it is unlikely that plants in the field use more water than they need in order to fix carbon. Taken over long periods of time, and many successive life cycles of each species, the net acquisition of carbon in a natural, stable community of plants tend to zero. Other things being equal, small differences in the capacities of individual plants and their progeny to acquire carbon must have been significant in competitive success or failure. However expensive, in terms of water, the last increment in rate of CO_2 fixation seems to be, it must be assumed that the cost and benefit would be found to balance if each could be expressed in terms of its influence on the fitness of the individual plant. This assumption leads to a proposition to do with optimal stomatal behaviour (Cowan, 1977; Cowan and Farquhar, 1977). In outlining the proposition here, I shall use language that is teleological: it is a premise in applying optimization to problems of biology that processes of natural selection have led to what has the appearance of intelligent design.

B. The Influence of Weather

Consider the time courses of rate of assimilation and rate of transpiration in a particular leaf during a particular day. They may be represented by a single trajectory across a surface such as that illustrated in Figure 2.4. The shape of the surface is determined partly by weather, and partly by properties of leaf structure and metabolism that are taken as fixed, for that day at least.

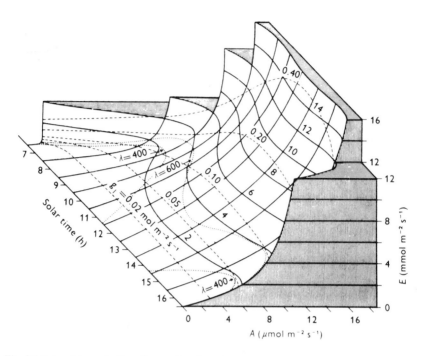

Fig. 2.4 Rate of transpiration, E, as a function of rate of assimilation, A, and time of day for certain assumed characteristics of leaf metabolism and environment. The magnitudes of E are indicated on the contours of the surface. The broken lines are trajectories on the surface representing the diurnal variation E and A for particular constant magnitudes of leaf conductance, g_l. The dotted lines are trajectories for which $\partial E/\partial A = \lambda$ is constant (reproduced from Cowan and Farquhar, 1977).

But it is assumed that the leaf has one degree of freedom. By varying the aperture of its stomata it may vary rate of assimilation and transpiration. In this way it can influence the form of the trajectory it takes. If it travels across the surface directly from east to west, as it were, then it maintains rate of assimilation, A, constant. If it follows one of the contours, it maintains rate of transpiration, E, constant. If it does not adjust to the terrain at all, then it will pass along one of the broken-line trajectories, representing a constant stomatal conductance to gas exchange. What it should do is to follow a trajectory, of the type indicated by the dotted lines, such that the slope of the surface, $\partial E/\partial A$ is maintained constant. These are optimal trajectories, in the following sense. If $\bar{A}$ and $\bar{E}$ are the average rates of assimilation and transpiration corresponding to a particular optimal trajectory, then the leaf could not possibly, by following a different trajectory, have increased $\bar{A}$ without increasing $\bar{E}$, or have decreased $\bar{E}$ without decreasing $\bar{A}$. Extension of the theory to the whole plant shows that not only should $\partial E/\partial A$ be maintained constant in each leaf, but the constant, λ say, should be the same for all leaves. The implication couched in terms of $\bar{A}$ and $\bar{E}$ then holds with $\bar{A}$ and $\bar{E}$ being the average rates of assimilation and transpiration per unit area of foliage in the whole plant.

Several optimal trajectories for A and E with time of day are shown in Figure 2.5. The variations in A closely resemble those in Figure 2.2. The comparison suggests that the decline in $\bar{A}$ and $\bar{E}$ with decrease in soil water content might be expressed in terms of a decline in λ. For each optimal trajectory defined by λ during a particular day, there are unique magnitudes of $\bar{A}$ and $\bar{E}$. Therefore there is a relationship of the kind shown in Figure 2.6. It will be supposed, for the time being, that the characteristics of the diurnal variation in plant environment are constant from day to day. Then the problem to be addressed is how should the plant move down the curve in Figure 2.1 as the amount of soil water available to it decreases. Before tackling this problem, however, it is necessary to qualify the rather simple picture that has been presented so far.

C. Some Remarks on λ and Its Significance

As seen in Figure 2.5, the optimal trajectories meet the $A = E = 0$ axis if λ is sufficiently small, implying that the stomata should be fully closed. This presents no difficulty; it is readily shown that, provided the surface is curved so that $\partial^2 E/\partial A^2 > 0$ as it is in Figure 2.4, then the optimal path is given by $\partial E/\partial A = \lambda$ wherever this equality can be met, and $E = A = 0$ wherever the magnitude of $\partial E/\partial A$ at the foot of the slope exceeds λ. One might visualize also an upper limit, associated with a maximum stomatal aperture. Undoubtedly there is a limit to aperture, determined by the physical properties of the stomatal apparatus. But there is no evidence that this limit is reached in intact plants out-of-doors, and it is preferable in terms of optimization to assume that there is no upper bound on A and E except that imposed by the economics of carbon gain and water loss. After all, it seems unlikely that there would have been an impediment to the evolution of an epidermis with more numerous, or larger stomata had such a development been of selective advantage.

If the surface is not everywhere curved as in Figure 2.4 then the optimal behaviour of the stomata becomes more complicated. Regions where $\partial^2 E/\partial A^2 > 0$ are most likely to occur in C_4 plants having broad leaves and at low windspeed and high temperature (Cowan, 1977). Optimization demands that such regions be avoided, the trajectories either surmounting the bump in the surface or passing along the foot of the slope (Cowan and Farquhar, 1977). There is the possibility that optimization will sometimes

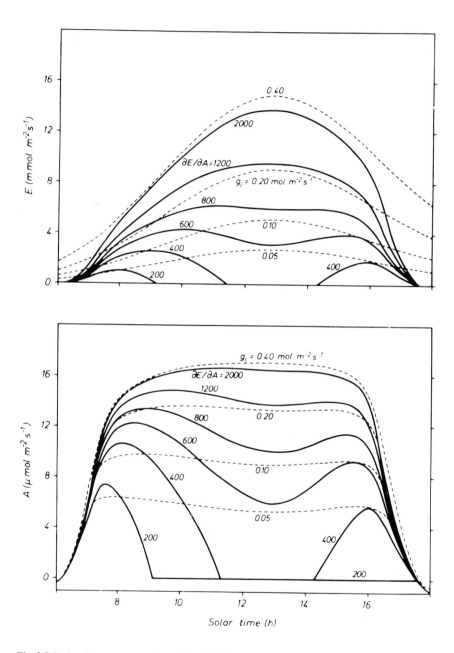

Fig. 2.5 Optimal time courses of rate of assimilation, A, and rate of transpiration, E, for various magnitudes of λ (solid lines). Also shown are A and E for various constant magnitudes of leaf conductance, g_l (reproduced from Cowan and Farquhar, 1977).

demand an instantaneous jump from one extreme to another. Of course, there are other circumstances, such as fluctuating irradiance associated with intermittent cloud, in which optimization would seem to demand rapid stomatal movements. Yet the most rapid movements have characteristic times of a few minutes, and movements in most species are much slower.

Why if speed of stomatal adjustment is potentially advantageous, are stomata relatively slow, taking minutes to change from one aperture to another? Perhaps the costs, in terms of energy, of stomatal movement are greater than currently supposed, and would be greater still if stomata were adapted to operate more quickly. Based on calculations of Raschke (1979), the glucose equivalent used in stomatal opening could be supplied by a few seconds, at the most, of photosynthesis in the leaf. But even this amount of carbon may not be negligible when taken in relation to fluctuations with periods of a few minutes.

There is a second reason why variations in stomatal aperture are unlikely to optimize gas exchange in precisely the way described. The expression $\partial E/\partial A = \lambda$, constant, should not be thought of as an equality. If stomata could sense $\partial E/\partial A$ directly, λ could be regarded as a reference "signal", and the difference between actual $\partial E/\partial A$ and λ as an "error signal". Stomatal movements could then be explored in terms of proportional, differential and integral control. One would not be surprised to find that the control system is imperfectly adapted to maintain the error signal small at all times, particularly as it has been shown (see Figure 2.6a) that the saving of water, or gain in carbon, associated with an optimal diurnal variation in stomatal aperture is usually small compared with the total amounts of water that would be lost and carbon that would be gained if stomatal aperture were constant. However, the interpretation of λ is more complicated because it is unlikely that stomata are able to sense $\partial E/\partial E$ directly. Insofar as they behave optimally, they do so by having a suitable set of "feedforward" responses to changes in the external variables, mainly irradiance, temperature, and humidity, that influence A and E (Cowan and Farquhar, 1977). Therefore λ is not an explicit reference signal: it is an abstract concept. The responses of stomata to changes in leaf environment, and the correlation of stomatal aperture with the capacity of the leaf mesophyll to fix carbon, tend to minimize $\bar{E}$ for any given $\bar{A}$, and maximize $\bar{A}$ for any given $\bar{E}$ (Cowan, 1982). This is equivalent to a tendency to maintain $\partial E/\partial A$ constant. It is useful to define a quantity λ as the equivalent reference signal.

D. The Influence of Soil Water Supply

The "penalty" incurred by plants when soil available water is exhausted is different in different species, and varies with the length of time the soil

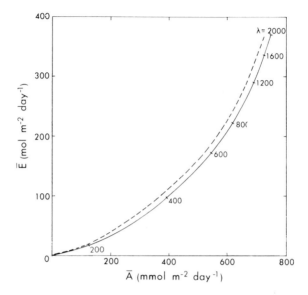

Fig. 2.6(a) Relationship between average rates of assimilation, Ā, and transpiration, Ē, computed for certain assumed characteristics of leaf metabolism and diurnal environment. The full curve is the optimal relationship, the slope of the curve at each point being λ, representing a constant magnitude of $\partial E/\partial A$. Each point on the broken curve corresponds to a constant magnitude of leaf conductance (reproduced from Cowan, 1982).

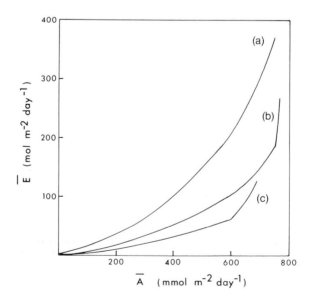

Fig. 2.6(b) Relationships between Ā and Ē computed for three days of contrasting weather: (a) as in Figure 2.6a, (b) and (c) with smaller humidity deficit and less light.

remains dry. At the very least they suffer a hiatus in growth. At the other extreme they die. The implications of this second extreme will be outlined briefly here; they have been explored in greater detail elsewhere (Cowan, 1982).

It is assumed that the times at which rain falls are completely unpredictable. The only statistic that can be defined is the average interval between successive falls, τ. For simplicity, it is taken that each fall of rain wets the soil completely. As some soil water is extracted by processes outside the control of the plant, such as cuticular transpiration, evaporation from the soil surface, drainage, and absorption by the root systems of competing plants, there would be a finite chance that the available water would be exhausted before the next rain fall, even if the plant did not itself take up water. Use of water by the plant decreases the time taken for the available water to be used up (t_1) and therefore increases the probability of death. Suppose that the magnitude of λ remained constant from day to day; that is $\bar{A}$ and $\bar{E}$ were constant. Then there would be a unique relationship between t_1 and $\bar{A}$. The faster the plant assimilated carbon, the more rapid the use of water, the shorter t_1, and the greater probability of death. But maintenance of constant λ is not the ideal strategy for the plant. The expectation of rain remains constant, irrespective of the time at which rain last fell, and therefore the drier the soil becomes, the greater becomes the risk of using more water. This suggests that λ should decline as the amount of available water decreases.

Optimal variations in λ, $\bar{A}$ and $\bar{E}$ with soil water deficit (Φ, the deficit being zero when the soil is wet and unity when the available water is exhausted) are shown in Figure 2.7. How these variations are computed has been described by Cowan (1982). Each variation corresponds to a particular critical period t_1, and a particular probable average rate of assimilation, $\hat{A}$, this being the probable average of $\bar{A}$ (taking into account the dependence of $\bar{A}$ on Φ and the statistics of rainfall) during any period in which a rain-free period greater than t_1 did not occur. Figure 2.8 shows $\hat{A}$ and t_1, as function of λ_0 the magnitude of λ when $\Phi = 0$. When λ_0 is small, $\hat{A}$ is small. However t_1 is large, implying a small risk of death. When λ_0 is large, $\hat{A}$ is large. Then t_1 is small and the risk of death is great. But whatever the particular balance between $\hat{A}$ and t_1 the plant could not, by varying λ with Φ differently have increased t_1 without decreasing $\hat{A}$ or have increased $\hat{A}$ without decreasing t_1. It is in this sense that the variations in Figure 2.7 are optimal.

The precise form of the variations in Figure 2.7 depend on several circumstances — the relationship between $\bar{A}$ and $\bar{E}$ which is determined by leaf metabolic characteristics and environment, the available water holding capacity of the soil in the plant root zone in relation to the frequency of rainfall and the area of plant foliage, and the rate at which water is extracted from the soil by processes outside the control of the plant. Because these circumstances differ greatly amongst different plant species in their natural

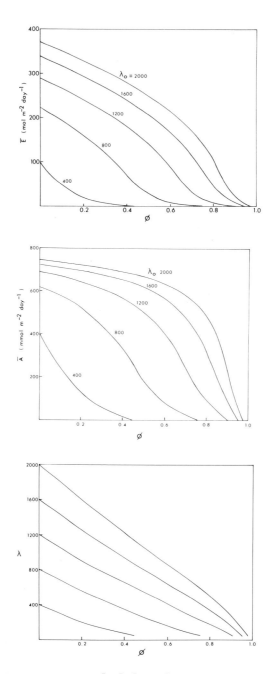

Fig. 2.7 Optimal variations in $\lambda = \partial\bar{E}/\partial\bar{A}$, $\bar{A}$ and $\bar{E}$ with soil water deficit (reproduced from Cowan, 1982).

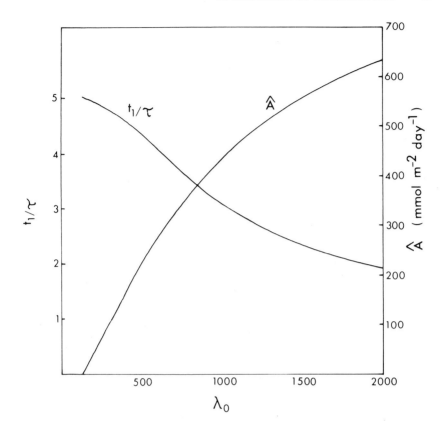

Fig. 2.8 Probable average rate of assimilation in undroughted plants, Â, and the critical period of drought relative to the average interval between rainfall, t_1/τ, as functions of λ_0, the magnitude of $\partial E/\partial A$ when $\Phi = 0$ (reproduced from Cowan, 1982).

habitats, the analysis may eventually contribute to an understanding of the diversity in the characteristics of stomatal control of gas exchange in higher plants. However, the mathematical precision of the analysis cannot be translated literally, because it is impossible precisely to quantify the environmental, edaphic and biotic conditions in which a particular species has become adapted. As with the theory relating to diurnal control of gas exchange, the analysis describes a limit to adaptation in idealized circumstances. In particular it demonstrates why stomata respond to changes in weather and soil water, and in so doing often restrict rate of assimilation even though the exigencies the responses are adapted to combat may not in fact occur in the life time of any particular individual plant.

E. Weather and Water Combined

An assumption made in discussing the response of the plant to change in soil water, that the environment of the foliage remains constant from day to day, will now be relaxed. If weather varies, the outcome (in terms of carbon fixed and water lost) of maintaining λ constant or varying it systematically with change in soil water content, is not determinate. Nevertheless there will be a probable outcome for a given plant species at a given stage of growth in a given environment—in much the same way as there is a probable outcome in terms of climatic averages of the various atmospheric processes that contribute to weather. The period t_1, previously a constant for any prescribed variation of λ with φ, becomes a probable critical period of drought, and there is an additional source of variation underlying the probable average rate of assimilation in undroughted plants, Â.

Three hypothetical relationships between Ā and Ē are shown in Figure 2.9. One of them is that used in previous computations; the other two relate to days with smaller humidity deficit and less light. We may associate these three relationships with the three days of contrasting weather on which the observations in Figure 2.2b were made. Small humidity deficit is manifest in a smaller slope, λ, at small magnitudes of Ā, and less light is manifest in a sharply increased slope at large magnitudes of Ā. Using the variation of λ corresponding to $\lambda_0 = 1200$ from Figure 2.7, I have calculated what Ē and Ā would be during these days of contrasting weather. The results are shown in Figure 2.9. The variations in Ē are the counterparts of those observed by Denmead and Shaw (1962). The essential feature of the experiment is reproduced. In contrast to the hydraulic model, the optimization hypothesis predicts that the curves will cross. If the available soil water is sufficiently reduced, Ē is greatest during humid days and least during the dry days.

Of course, as soon as the existence of the direct response of stomata to external conditions influencing transpiration had been established, the observations of Denmead and Shaw were no longer a puzzle in terms of plant mechanism. But what we have been concerned with here is the adaptive function of the mechanism. The optimization theory demonstrates that feedforward contributes to an effective compromise between carbon gain and water loss, the advantage of carbon gain having relatively greater weight when the soil is wet and the disadvantage of water loss having relatively greater weight when the soil is dry.

IV. DISCUSSION

There are other regulatory systems within plants that can be treated as compromises between carbon gain and water use. For example, the proportion of carbon allocated to the growth of roots is a compromise between the

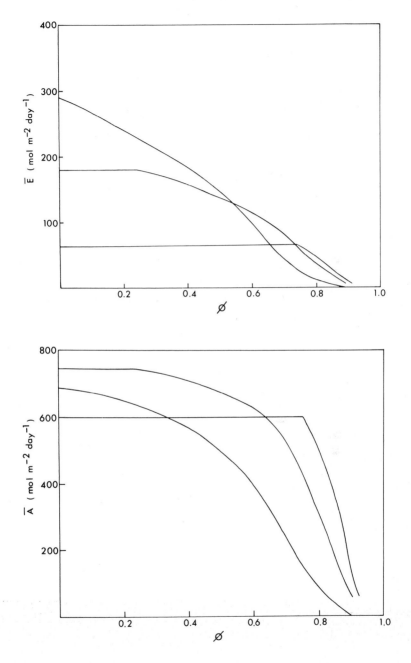

Fig. 2.9 $\bar{A}$ and $\bar{E}$ as optimal functions of soil water deficit, Φ, on three days of contrasting weather. Computed using the characteristics in Figure 2.6b and the variation in λ corresponding to $\lambda_0 = 1200$ in Figure 2.7.

cost, in terms of relative growth rate of the plant as a whole, inherent in the growth and maintenance of non-photosynthesizing tissues and the benefit, in terms of reduced probability of drought, of the water-holding capacity of the soil tapped by the root system. A second example relates to seasonality. Seasonal variation of conditions that influence the potential rates of CO_2 fixation and transpiration, and the supply of water to roots, must have influenced the evolution of life forms and life histories in plants in arid regions. Of particular interest to agriculture is the habit of annuals, which, in minimizing the risk of drought, not only forego the possibility of carbon fixation and growth altogether during a period of each year, but sacrifice 50% or more of the growth made during the previous vegetative phase of their existence. The length of the vegetative period is a reflection of the compromise between probable growth and seed yield on the one hand, and the risk of drought on the other. It should be possible to draw these different facets of the carbon and water economies of plants together within the framework of optimization theory. Such a development would not contribute to an understanding of the mechanisms of plant responses to water deficits, but it should help better to assess the individual contributions of various responses to the adaptation of particular species to water deficits.

Does the theory outlined in this chapter suggest how better to grow plants, or how to find better plants to grow in areas where yield is limited by an adverse balance of evaporation and precipitation? Clearly the compromise between probable growth rate and probable drought that emerges from natural selection is not necessarily that which is most favourable to agricultural production; nor are the circumstances in which crops are grown similar to those in which natural selection has taken place. At first sight then, it may seem to be a problem of engineering the characteristics of gas exchange control in crop plants to suit the climate in which they grow, and the economics of crop production. For example, we might try to identify imperfections in endogenous diurnal variations and short term responses of stomata — imperfections, that is, in the sense that $\partial E/\partial A$ is not held constant. Leaving aside, for the moment, the problem of using such information to select for better characteristics of control, there are three objections to this suggestion. Firstly, the determination of $\partial E/\partial A$ is exceedingly difficult. Secondly, constancy in $\partial E/\partial A$ is certainly no better than a first approximation to the ideal goal. Thirdly, the potential returns in terms of saving in water or increase in carbon fixation may be relatively small. In short, the technology and theory is not sufficiently advanced, and the potential benefits are not sufficiently great. Rather more attractive is the notion of altering the way in which the control of gas exchange responds to variation in soil water content. Where crops rely on sparse and irregular rainfall, an increased sensitivity of the stomata to water deficit may be desirable. Where crops are regularly irrigated, stomatal sensitivity to water deficit may be undesirable.

A difficulty in selecting for particular stomatal sensitivity is that the extent to which the guard cell mechanism operates autonomously or is integrated with function of other parts of the whole plant, is not known. It is probable that short-term responses of stomata to changes in humidity take place independently of the rest of the leaf. It may be, also, that short term responses to light and temperature receive little guidance from metabolic systems in the leaf mesophyll tissue and other parts of the plant that are also affected by these variables. However, it is very unlikely that long term responses to changes in light and temperature, and to changes in plant water status are not modulated by more pervasive systems of control within the plant as a whole.

To the extent that is so, it is impossible to select for patterns of stomatal behaviour without also unconsciously selecting other patterns of metabolic response. That, of course, is not necessarily disadvantageous. For example, suppose ABA acted as the effector for both stomatal closure and osmotic adjustment with decline in plant water content. In some circumstances both responses might be desirable. Then selection for stomatal sensitivity to water deficits would be efficacious, although selection for ABA synthesis under stress would be a more direct approach to the same end. On the other hand there may be circumstances in which it is desirable that stomata remain open while osmotic adjustment takes place. Selection for stomatal insensitivity to water deficits would not further this objective. It would be necessary to select for stomatal insensitivity to endogenous ABA. Without knowing the way in which the various responses to water deficits are "orchestrated" it is impossible to predict what selection based on physiological responses would achieve in terms of agricultural objectives. That is not a counsel of despair, however. A study of the comparative physiology of lines produced by selection of physiological characteristics is likely to provide the kind of information needed about the orchestration of the responses of plants to water deficits. Whether plant breeders can be prevailed upon to delay the primary objective of producing commercially viable cultivars in the interests of furthering our understanding of physiology is a matter of conjecture. In the long run, it might make the primary task much easier.

REFERENCES

Cowan, I. R. (1965). Transport of water in the soil–plant–atmosphere system. *Journal of Applied Ecology* **2**, 221–239.

Cowan, I. R. (1977). Stomatal behaviour and environment. *Advances in Botanical Research* **4**, 117–228.

Cowan, I. R. (1982). Regulation of water use in relation to carbon gain in higher plants. *In* "Physiological Plant Ecology II. Water Relations and Carbon Assimilation". (O. L. Lange, P. S. Nobel, C. B. Osmond and H. Ziegler eds.) Encyclopedia of Plant Physiology, New Series., Vol. 12B. pp. 589–613. Springer, Berlin.

Cowan, I. R., and Farquhar, G. D. (1977). Stomatal function in relation to leaf metabolism and environment. *In* "Integration of Activity in the Higher Plant," (D. H. Jennings, ed.), pp. 471–505. Cambridge University Press, Cambridge.

Denmead, O. T., and Shaw, R. H. (1962). Availability of soil water to plants as affected by soil moisture content and meteorological conditions. *Agronomy Journal* **54**, 385–390.

Hall, A. E., Schulze, E.-D. and Lange, O. L. (1976). Current perspectives of steady state stomatal responses to environment. *In* "Water and Plant Life, Problems and Modern Approaches." Ecol Stud. 19. (O. L. Lange, L. Kappen, and E.-D. Schulze, eds.), pp. 169–188. Springer, Berlin.

Hanson, A. D., and Hitz, W. D. (1982). Metabolic responses of mesophytes to plant water deficits. *Annual Review of Plant Physiology* **33**, 163–203.

Hsiao, T. C., Acevedo, E., Fereres, E. and Henderson, D. W. (1976) Stress metabolism, water stress, growth, and osmotic adjustment. *Philosophical Transactions of Royal Society, London* B **273**, 479–500.

Lange, O. L., Lösch, R., Schulze, E.-D., and Kappen, L. (1971). Responses of stomata to changes in humidity. *Planta* **100**, 76–86.

Little, C. H. A., and Eidt, D. C. (1968). Effect of abscisic acid on budbreak and transpiration in woody species. *Nature* **220**, 498–499.

Loftfield, J. V. G. (1921). The behaviour of stomata. *Carnegie Institution Publication* **314**, 1–104.

Mittelheuser, C. J., and Steveninck van, R. F. M. (1969). Stomatal closure and inhibition of transpiration induced by (rs)-abscisic acid. *Nature* **221**, 281–282.

Raschke, K. (1979). Movements of stomata. *In* "Physiology of Movements" Encyclopedia of Plant Physiology, New Series, Vol. 7. (W. Haupt and M. E. Feinleib, eds). pp. 383–441. Springer, Berlin.

Rawson, H. M., Turner, N. C., and Begg, J. E. (1978). Agronomic and physiological responses of soybean and sorghum to water deficits. IV. Photosynthesis, transpiration and water use efficiency of leaves. *Australian Journal of Plant Physiology,* **5**, 195–209.

Schulze, E.-D., Lange, O. L., Buschbom, U., Kappen, L., and Evenari, M. (1972). Stomatal responses to change in humidity in plants growing in the desert. *Planta* **108**, 259–270.

Schulze, E.-D., and Hall, A. E. (1982). Stomatal responses, water loss and CO_2 assimilation rates of plants. *In* "Physiological Plant Ecology II. Water Relations and Carbon Assimilation". Encyclopedia of Plant Physiology New Series, Vol. 12B, (O. L. Lange, P. S. Noble, C. B. Osmond, and H. Ziegler, eds.) pp. 181–230. Springer, Berlin.

Turner, N. C. (1974) Stomatal response to light and water under field conditions. *In* "Mechanisms of Regulation of Plant Growth" (R. L. Bieleski, A. R. Ferguson, and M. M. Creswell, eds). *Royal Society of New Zealand Bulletin* **12**, 423–432.

Turner, N. C., Begg, J. E., Rawson, H. M., English, S. D., and Hearn, A. B. (1978). Agronomic and physiological responses of soybean and sorghum to water deficits. III. Components of leaf water potential, leaf conductance, $^{14}CO_2$ photosynthesis and adaptation to water deficits. *Australian Journal of Plant Physiology* **5**, 179–194.

Wright, S. T. C. (1969). An increase in the "inhibitor-β" content of detached wheat leaves following a period of wilting. *Planta* **86**, 10–20.

Zelitch, I., and Waggoner, P. E. (1962). Effect of chemical control of stomata on transpiration and photosynthesis. *Proceedings National Academy of Science USA* **48**, 1101–1108.

Stomatal Physiology

N. THORPE

I. INTRODUCTION

The mechanism of stomatal movement has been studied for well over a century. Examples of the most accessible and readable early work are von Mohl (1856) and Lloyd (1908).

A myriad of hypotheses have arisen from this early work of which few were accepted even briefly. In this respect, this topic differs from the complementary topic of the role of stomata in regulating gas exchange, where very firm, usually incorrect, views have been widely held for long periods of time. Although early physiologists believed that stomata may play a significant role in gas exchange, many physiologists working in the first five decades of this century underestimated or dismissed the role of stomata. This underestimation was due in part to the neglect of the boundary layer by Brown and Escombe when interpreting their own experiments (see Penman and Schofield, 1951), the theory of "incipient drying" (Livingston and Brown, 1912) and the brilliantly conceived experiments of Knight (1917).

Later experiments by Pearse, Spencer and Milthorpe showed that the stomata inside the attached porometer cup used by Knight opened very widely and behaved differently from those on most of the leaf surface (Gregory *et al.,* 1950; Milthorpe and Spencer, 1957); moreover, there was

CONTROL OF CROP PRODUCTIVITY
ISBN 0 12 548280 9

a close relationship between transpiration and stomatal aperature, a finding that has since been well documented and accepted in a more balanced context (e.g. Chapter 2; Milthorpe and Penman, 1967; Monteith, 1973; Burrows and Milthorpe, 1976; Farquhar and Sharkey, 1982). The aberrant behaviour of stomata inside porometer cups was shown to result from a marked opening response of stomata at carbon dioxide concentrations below 300 μl l[-1] (Heath, 1950; Heath and Milthorpe, 1950). This response, although still unexplained, is central to current theories of stomatal physiology.

It is now accepted that changes in stomatal aperture follow changes in the difference in turgor between guard cells and adjacent cells. In turgid plants, these changes in turgor result from changes in osmotic potential due to influx and efflux of potassium ions (Outlaw and Lowey 1977; Raschke, 1979; MacRobbie, 1981). The main areas of interest are in the mechanisms which control the direction and magnitude of these fluxes.

II. CARBON DIOXIDE FIXATION AND ION BALANCE

Guard cells usually contain numerous well-developed mitochondria and starch-forming chloroplasts (e.g. Milthorpe, 1969; Pearson and Milthorpe, 1974). Guard cells have the ability to fix labelled carbon dioxide ($^{14}CO_2$) directly and also to accumulate photosynthetic products labelled after $^{14}CO_2$ fixation in the mesophyll. Rates of fixation in detached epidermis are about 0.4-1.3 nmol (CO_2) (g dry wt)[-1] s[-1] whereas rates of accumulation in attached epidermis are about 40–50 nmol (CO_2) (g dry wt)[-1] s[-1] in *Commelina cyanea* Thorpe and Milthorpe, 1977; Pearson and Milthorpe, 1974) and 80 nmol (CO_2) (g dry wt)[-1] s[-1] in broad bean (Pearson and Milthorpe, 1974).

Rates of accumulation of ^{14}C-fixation products by epidermis attached to leaves exposed to $^{14}CO_2$ are approximately linear with time during a ten minute feeding period; they are also linear with increasing the quantum flux density up to 1 nmol m[-2] s[-1] and with CO_2 concentration up to 300μl l[-1]. Rates of accumulation by attached epidermis in the light are about 100 times the rates in darkness (Pearson and Milthorpe, 1974; Thorpe and Milthorpe, 1977). However, rates of fixation by detached epidermis in the light are only twice those in the dark (Table 3.1) and are independent of quantum flux density and external CO_2 concentration greater than 100 μl l[-1] (Thorpe and Milthorpe, 1977). Furthermore, labelled intermediates recovered in soluble extracts of the attached epidermis reflect the labelled compounds of the underlying mesophyll, whereas malate and asparate are predominately the labelled compounds recovered from the detached epidermis (Willmer *et al.*, 1978; Thorpe, 1980). These observations suggest that there is a substantial short-term interchange of organic material between the mesophyll and the

Table 3.1

Rates of fixation of CO_2 by whole leaves, attached epidermis and detached epidermis in darkness and at the highest photon flux density used (1.08 mE m^{-2} s^{-1}) and with ambient CO_2 concentration of 355 ppm. (After Thorpe and Milthorpe, 1977.)

Units	Whole leaf		Attached epidermis		Detached epidermis	
	Dark	Light	Dark	Light	Dark	Light
ng CO_2 g(DW)$^{-1}$ s^{-1}	22.3	7980	20.0	1833	13.7	26.7
	±2.5	±183	±8.3	±67	±4.7	±6.7
μg CO_2 (g chlorophyll)$^{-1}$ s^{-1}	1.33	472	40	3667	27	53
μg CO_2 (g nitrogen)$^{-1}$ s^{-1}	0.29	104	0.82	75	0.56	1.09
ng CO_2 cm^{-2} s^{-1}	0.06	21.6	0.01	0.53	0.005	0.01

The values shown are means ± standard errors. Whole leaves and epidermis contain respectively 16.9 and 0.5 mg chlorophyll and 76.8 and 24.5 mg total nitrogen per gram dry weight and have weight : area ratio of 2.7 and 0.29 mg cm^{-2}.

epidermis, despite presumptive evidence that the exchange of potassium ions (K^+) is much less (Pearson, 1975).

A survey of enzymic activities in extracts of whole leaf and epidermis of *Commelina cyanea* showed that enzymes in the guard cells were specialized towards the metabolism of C_4 acids. The cells had a high ratio of phosphoenolpyruvate carboxylase to ribulose-bisphophate (Ru P_2) carboxylase–oxygenase activity and enhanced amounts of asparate aminotransferase, malate dehydrogenase (NADP$^+$), and NAD:malic and NADP:malic enzymes. Pyruvate, orthophosphate dikinase or phosphoenolpyruvate carboxykinase, enzymes possibly involved with C_4 acid metabolism, were not detected. Small but significant amounts of sugars, starch and various glycolysis and tricarboxylic acid (TCA) cycle compouds were, however, labelled in the isolated epidermis supplied with $^{14}CO_2$ or ^{14}C-alanine (Thorpe and Milthorpe, 1977; Willmer *et al.*, 1978).

If the reductive pentose phosphate pathway does not operate in guard cells (Raschke and Dittrich, 1977), then guard cells could function as an independent system only if the daily respiratory loss of carbon were replaced by the β-carboxylation reaction, and if a means exists for utilizing photophosphorylation to energize phosphoenolpyruvate formation. Confirmation of pyruvate and orthophosphate dikinase activity would thus seem to be key evidence concerning the ability of the guard cell to function independently. The high activity of malic enzyme in the epidermis is consistent with evidence for gluconeogenic conversion of malate into starch (Dittrich and Raschke, 1977; Thorpe and Milthorpe, 1977; Willmer and Rutter, 1977). Starch buildup and breakdown in guard cell chloroplasts has long been observed. Although the starch-to-sugar osmotic theory is now given little credence, it is possible that the capacity of this process (or for other polysaccharides to tie up to make available energy and 3-carbon skeletons), is fundamental to the operation of this type of guard cell.

Isolated epidermal tissue, induced to open and close its stomata by floating the tissue on high concentrations of potassium chloride and abscisic acid solutions respectively, shows a higher rate of $^{14}CO_2$ fixation than such tissue with either open or closed stomata (Thorpe et al., 1979). The proportions of the major fixation products, malate, aspartate, sucrose, sugar phosphates, glycine, serine and alanine, are similar during the opening, open and closed phases, but when closing, a higher proportion is diverted into sucrose and sugar phosphates. There is also a relatively greater leakage of labelled malate into the medium during the open, closed and closing phases than during opening (Thorpe et al., 1979).

When the tissues of leaves with open stomata were analysed for total contents, the concentrations (mol per unit of water in the tissues) of sucrose, malate, aspartate, glutamate, glycine, glutamine and asparagine in the mesophyll were about twice those in the epidermis; only glucose, serine and alanine had equivalent concentrations. This suggested that only about half of each of the substances in the first named group were readily available for diffusion into the epidermis. Over the time covered in these experiments, (2.5 h with stomata open), there were no detectable changes in concentration of substances in the epidermis, although studies of the change in the specific activity of the compounds indicated that sucrose, glucose and possibly both serine and glycine moved between the mesophyll and the epidermis. The feeding of $^{14}CO_2$ to intact leaves resulted in a rapid buildup of labelled materials in the mesophyll and a slower rise in the epidermis to less than the concentrations found in the mesophyll. After 2.5 hours, about 20% and 40% of the label was in ethanol-insoluble substances in the epidermis and mesophyll, respectively.

Epidermis peeled from the leaf, and also the residual mesophyll placed on 0.05 mM calcium sulphate solutions, showed efflux patterns that indicated that there were at least two compartments. One compartment consisted of rapidly diffusible substances and comprised about 10% of the total soluble compounds in the mesophyll and about 84% in the epidermis. Other efflux experiments, in which changes in sucrose, glucose, malate and amino acids were followed separately, indicated similar-sized, readily diffusible pools of each of these compounds in the epidermis with half-times of 3-5 minutes. These findings, together with the leakage of materials from isolated epidermis in other experiments, suggested that ordinary epidermal cells were extremely permeable to most substances.

Much interest centres around the roles of organic and inorganic anions, inorganic cations and phosphoenolpyruvate carboxylase. Malate is produced during stomatal opening whether the tissue is detached or attached (Pearson and Milthorpe, 1974; Thorpe, 1983). Isolated epidermal strips can be induced to open their stomata by incubation on solutions of sodium or K^+ (Willmer and Mansfield, 1969). Raschke and Schnabl (1978) showed the augmented use of the chloride ion decreased the malate concentration required as a counter-ion for the K^+. Guard cells can take up K^+ without accompanying anions, electroneutrality being maintained by excretion of protons (H^+) (Raschke and Humble, 1973). It is possible, using the non-absorbable anion iminodiacetate (IDA⁻), to ensure that the electroneutrality of the K^+ imported into guard cells is balanced predominantly by the endogenous production of malate (Raschke and Schnabl, 1978; Outlaw and Manchester, 1979).

The opening responses of stomata in isolated epidermal tissue of *Commelina cyanea* were saudied. The pieces of tissue were floated on buffered KC1, KIDA, and control buffer media to determine the changes in amounts of malate and citrate associated with stomatal opening. Tissues induced to open under these conditions were fed $^{14}CO_2$ and the rates of fixation and distribution of labelled compounds were analysed. The increases in the concentrations of malate and citrate under the two opening conditions were correlated with the augmented rates of $^{14}CO_2$ fixation by these tissues to determine the relationship between increased organic acid concentrations and the level of β-carboxylation. The findings are reported in Table 3.2. They indicate that phosphoenolpyruvate carboxylation is the major source of malate in the guard cells of the isolated *Commelina* epidermis. Some of the inconsistencies in the reported literature (Raschke and Dittrich, 1977; Travis and Mansfield, 1979) can be accounted for by the method of handling of the tissues, time of feeding and degree of isotope dilution.

Respiration rates for epidermis of *Commelina* with opening and closed stomata (30 nmol (O_2) g^{-1} s^{-1} and 25 nmol (O_2) g^{-1} s^{-1} respectively) are 10–12 times the observed rates of CO_2 fixation. The uptake of $^{14}CO_2$ is linear with time and the percentage distribution of labelled malate does not change over the time course (Willmer *et al.*, 1978); yet I did not detect a significant dilution of internal CO_2 by endogenous respiratory CO_2 (Thorpe, 1983).

In the attached epidermis, import of C is implicated. This is supported by higher, increased levels of malate associated with stomatal opening in the attached epidermis (opened 45 μmol g^{-1} and closed 24 μmol g^{-1}, respectively). It is probable that the detachment of the epidermis from the underlying mesophyll removes the guard cell from an internal control other than the supply of K^+, and thus accounts for many of the difficulties in trying to relate findings of stomatal responses in whole leaves to those in isolated epidermis.

The rate of accumulation of malate and proton exchange associated with stomatal opening in *Commelina* was analysed with micro-titration techniques. The analysis depended on the principle that protons were released in exchange for the imported K^+, and that when stomata in epidermal tissue were induced to open, the increased external proton concentration reflected

Table 3.2

Organic products which accumulate in isolated epidermal strips. Radioactivity in malate and citrate recovered after feeding $H^{14}CO_3^-$ – (7.4 MBq) for 10 min to strips of *Commelina cyanea* floating for 1 h in (i) control buffer (6.6 mM MES[a], 3.3 mM TRIS pH 6.5); (ii) buffer + 100 mM KIDA; and (iii) buffer + 100 mM KCl. (From Thorpe, 1983.)

	Control	+ KIDA	+ KCl
Rates of CO$_2$ fixation nmol g^{-1}s^{-1}			
ethanol soluble material	1.43 ± .42	2.96 ± .63	2.76 ± .68
ethanol insoluble material	0.06	0.18	0.16
recovered in medium	0.30	0.05	0.26
Synthesis by PEP carboxylation nmol g^{-1}s^{-1}			
tissue malate	0.87	1.78	1.57
citrate	0.02	0.03	0.03
medium malate	0.19	0.02	0.12
citrate	0.003	0.001	0.003
total malate	1.06	1.80	1.69
citrate	0.02	0.03	0.03
Total amount recovered µmol g^{-1} (cold companion series floated 1 h)			
tissue malate	18.4	31.3	26.0
citrate	9.06	8.96	8.96
medium malate	3.86	0.63	2.34
citrate	1.90	0.18	0.34
total malate	22.26	31.93	28.34
citrate	10.96	9.14	9.30
Malate nmol g^{-1} s^{-1}			
from PEP carboxylation		0.77	0.62
calculated from cold analyses		2.68	1.69
Minimum % malate from PEP carboxylase		29	37

[a]2(N-Morpholino)-ethane sulphonic acid

the extent of ionic exchange. A balance sheet can be prepared for the ions associated with stomatal opening in *Commelina* epidermis when tissue was floated on 100 mM KCl and 100 mM KIDA. The proton extrusion was 2.32 neq (H^+) g^{-1} s^{-1} µm^{-1}, when tissues were on KCl. This was balanced by 0.56 neq of malate ion and about 2 neq of Cl$^-$ (determined by [36]Cl$^-$ uptake studies). The proton extrusion of 2.32 neq associated with stomatal opening on 100 mM KIDA was nearly balanced by increased production of malate ion, 1.77 neq g^{-1} s^{-1} µm^{-1}. The findings would support the concept that unless Cl$^-$ is present in the detached epidermis, the osmotica necessary to balance the imported K^+ during periods of stomatal opening is provided by the production of malate. By contrast, studies of ^{14}C-alanine uptake indicate this molecule is used by epidermal tissues as a respiratory substrate. There was

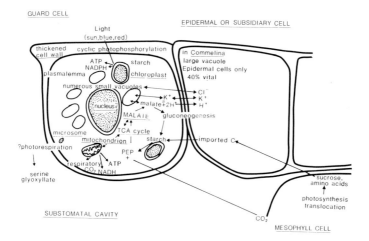

Fig. 3.1 Diagrammatic representation of the major metabolic pathways associated with stomatal movement that are operative in the guard cell and between the guard and adjacent cells of *Commelina*.

no evidence to suggest that it contributed directly to the accumulation of malate associated with stomatal opening.

III. CONCLUSIONS

The current concept of the physiology of stomatal movement is that K^+ is the major solute leading to guard cell movement and stomatal opening and that it is balanced, as required, by malate ions arising in the guard cell. The source of malate may be from both stored reserves and compounds imported from the mesophyll. Figure 3.1 shows a figurative representation of likely major metabolic pathways which operate in, and between, the guard and adjacent cells of *Commelina cyanea* to regulate stomatal movement. Metabolic pathways within the guard cells are presented in greater detail by Milthorpe *et al.* (1979).

Although phosphoenolpyruvate carboxylation is a major reaction in the guard cells and is responsible for the production of malate and aspartate, the uptake and loss of K^+ does not seem to be controlled through this reaction, as exemplified by the rates of CO_2 fixation being as high during closing as opening. However, leakage from the guard cells is much greater during closing, and it is suggested that malate concentrations are influenced to a large degree by changes in membrane permeability; these effects being secondary to the responses of K^+/H^+ pumps to CO_2 and light. The effect of low CO_2 concentration would seem to be on malate retention rather than formation. These findings, together with the demonstrated rapid exchange of metabolites between mesophyll and epidermis, suggest that control of stomatal movement in response to light and CO_2 may arise from changes in the relative concentrations of those metabolites formed by mesophyll reactions, these having direct effects on K^+/H^+ pumps. This suggestion is now being explored.

REFERENCES

Burrows, F. J., and Milthorpe, F. L. (1976). Stomatal conductance in the control of gas exchange. *In* "Water Deficits and Plant Growth" Vol. 4 (T. T. Kozlowski, ed.), pp. 103–152. Academic Press, New York.

Dittrich, P., and Raschke, K. (1977). Malate metabolism in isolated epidermis of *Commelina communis* L. in relation to stomatal functioning. *Planta* **134**, 77–82.

Farquhar, G. D., and Sharkey, T. D. (1982). Stomatal conductance and photosynthesis. *Annual Review of Plant Physiology* **33**, 317–345.

Gregory, F. G., Milthorpe, F. L., Pearse, H. L., and Spencer, H. J. (1950). Experimental studies of the factors controlling transpiration. I, II. *Journal of Experimental Botany* **1**, 1–14, 15–28.

Heath, O. V. S. (1950). Studies in stomatal behaviour V.1. *Journal of Experimental Botany* **1**, 29–62.

Heath, O. V. S., and Milthorpe, F. L. (1950). Studies in stomatal behaviour V.2. *Journal of Experimental Botany* **1**, 227–243.

Knight, R. C. (1917). The interrelations of stomatal aperture, leaf water content and transpiration rate. *Annals of Botany* **31**, 221–240.

Livingston, B. E., and Brown, W. H. (1912). Relation of the daily march of transpiration to variations in the water content of foliage leaves. *Botanical Gazette* **53**, 309–330.

Lloyd, R. E. (1908). The physiology of stomata. *Carnegie Institute of Washington Publication* **82**, pp. 142.

MacRobbie, E. A. C. (1981). Ion fluxes in isolated guard cells of *Commelina communis* L. *Journal of Experimental Botany* **32**, 545–562.

Milthorpe, F. L. (1969). The significance and mechanism of stomatal movement. *Australian Journal of Science* **32**, 31–35.

Milthorpe, F. L., and Penman, H. L. (1967). The diffusive conductivity of the stomata of wheat leaves. *Journal of Experimental Botany* **18**, 422–457.

Milthorpe, F. L., and Spencer, H. J. (1957). Experimental studies of the factors controlling transpiration III. *Journal of Experimental Botany* **8**, 413–437.

Milthorpe, F. L., Thorpe, N., and Willmer, C. M. (1979). Stomatal metabolism—a current assessment of its features in *Commelina*. *In* "Structure, Function and Ecology of Stomata" (D. N. Sen, ed.), pp. 121–142. Bishen Singh Mahendra Pal Singh, Dehra Dun.

Monteith, J. L. (1973). "Principles of Environmental Physics". Arnold, London.

Mohl, H. von (1856). Welche Ursachen bewirken die Erweiterung und Verengung der Spaltöffnungen? *Botanishches Zeitung* **14**, 697–704, 713–720.

Outlaw, W. H. Jr., and Lowry, O. H. (1977). Organic acid and potassium accumulation in guard cells during stomatal opening. *Proceedings of the National Academy of Sciences* **74**, 4434–4438.

Outlaw, W. H. Jr., and Manchester, J. (1979). Guard cell starch concentration quantitively related to stomatal aperture. *Plant Physiology* **64**, 79–82.

Pearson, C. J. (1975). Fluxes of potassium and changes in malate within epidermis of *Commelina cyanea* and their relationships with stomatal aperture. *Australian Journal of Plant Physiology* **2**, 85–89.

Pearson, C. J., and Milthorpe, F. L. (1974). Structure, carbon dioxide fixation and metabolism of stomata. *Australian Journal of Plant Physiology* **1**, 221–236.

Penman, H. L., and Schofield, R. K. (1951). Some physical aspects of assimilation and transpiration. *Symposia of the Society for Experimental Biology* **5**, 115–129.

Raschke, K. (1979). Movements of stomata. *In* "Encyclopedia of Plant Physiology" (New Series) Vol. 7 (W. Haupt and M. E. Feinleib, eds.), pp. 383–441. Springer Verlag, Berlin.

Raschke, K., and Dittrich, P. (1977). (^{14}C) carbon dioxide fixation by isolated leaf epidermes with stomata closed or open. *Planta* **134**, 69–76.

Raschke, K., and Humble, G. D. (1973). No uptake or anions required by opening stomata of *Vicia faba:* Guard cells release hydrogen ions. *Planta* **115**, 47–57.

Raschke, K., and Schnabl, H. (1978). Availability of chloride effects the balance between potassium chloride and potassium malate in guard cells of *Vicia faba* L. *Plant Physiology* **62**, 84–87.

Thorpe, N. (1980). Accumulation of carbon compounds in the epidermis of five species with either different photosynthetic systems or stomatal structure. *Plant, Cell and Environment* **3**, 451–460.

Thorpe, N. (1983). The role of PEP-carboxylase in the guard cell of *Commelina cyanea. Plant Science Letters* **30**, 331–338.

Thorpe, N., and Milthorpe, F. L. (1977). Stomatal metabolism: CO_2 fixation and respiration. *Australian Journal of Plant Physiology* **4**, 611–621.

Thorpe, N., Willmer, C. M., and Milthorpe, F. L. (1979). Stomatal metabolism: carbon dioxide fixation and labelling patterns during stomatal movement in *Commelina cyanea. Australian Journal of Plant Physiology* **6**, 409–416.

Travis, A. J. and Mansfield, T. A. (1979). Reversal of the CO_2-responses of stomata by fusicoccin. *New Phytologist* **83**, 607–614.

Willmer, C. M. and Mansfield, T. A. (1969). A critical examination of the use of detached epidermis in studies of stomatal physiology. *New Phytologist* **68**, 363–375.

Willmer, C. M. and Rutter, J. C. (1977). Guard cell malic acid metabolism during stomatal movements. *Nature* **269**, 327–328.

Willmer, C. M., Thorpe, N., Rutter, J. C., and Milthorpe, F. L. (1978). Stomatal metabolism: carbon dioxide fixation in attached and detached epidermis of *Commelina. Australian Journal of Plant Physiology* **5**, 767–778.

CHAPTER 4

Photochemical Capacity and Photosynthesis

N. TERRY and G. D. FARQUHAR

I. INTRODUCTION

Photosynthesis is composed of at least three processes: (1) the absorption and conversion of light energy into chemical energy; (2) the diffusion and transport of carbon dioxide to chloroplasts; and (3) the enzymatic fixation and reduction of carbon dioxide to carbohydrate. Over the last few years, considerable attention has been focused on processes (2) and (3) as potential rate-limiting components of photosynthesis under field conditions, but less attention has been given to light harvesting and electron transport capacity.

In this chapter we consider the involvement of photochemical capacity as a rate-controlling factor of photosynthesis. By photochemical capacity we refer to the capacity of the leaf to convert absorbed light into chemical energy. It is defined as the maximum rate per unit area at which a leaf can convert a nonlimiting supply of light energy to chemical product, for example, the maximum rate of adenosine triphosphate production and the reduced form of nicotinamide adenine dinucleotide phosphate production

CONTROL OF CROP PRODUCTIVITY
ISBN   0 12 548280 9

when their precursors (adenosine diphosphate, inorganic phosphate and nicotinamide adenine dinucleotide phosphate respectively) are present in saturating amounts. In Section II we present an experimental approach that has been used to determine the quantitative effects of photochemical capacity on the rate of photosynthesis. In Section III we present a mathematical approach to photochemical capacity whereas in Section IV we briefly summarize and compare the results of these two analyses.

II. AN EXPERIMENTAL APPROACH TO PHOTOCHEMICAL CAPACITY

In order to determine experimentally the effect of photochemical capacity on the rate of photosynthesis, a technique is needed that varies photochemical capacity with respect to other capacities of the photosynthetic apparatus. Recent studies have proposed that iron deficiency may be used for this purpose (Terry, 1979; 1980).

A. Effects of Iron Stress on the Thylakoid System

When young rapidly growing sugar beet plants are deprived of iron, they develop chloroplasts that have relatively few thylakoids (Fig. 4.1). Electron microscopy by Platt-Aloia *et al.* (1983) indicated that other parts of the leaf cell have normal organization; for example, the iron-containing mitochondria and microbodies are unaffected by iron stress.

The reduction in the thylakoid system as seen under the electron microscope is accompanied by a quantitive reduction in various components of the light harvesting and electron transport system. The light harvesting pigments, chlorophyll a, chlorophyll b, carotene and xanthophyll, decrease concomitantly under iron stress, as do the amounts of the photosystem (PS) I components, P_{700} and cytochrome f (Terry, 1980; Spiller and Terry, 1980). The primary acceptor of PS II, Q, also decreases during iron stress (Nishio and Terry, 1983). Other data show that iron stress reduced the amount of lamellar iron (Terry and Low, 1982), lamellar manganese (Young and Terry, in press), membrane lipids and proteins (S. E. Taylor, J. N. Nishio and N. Terry, unpublished) and chlorophyll–protein complexes (Machold, 1971). Thus, iron stress apparently results in a reduction of the total thylakoid system.

The influence of the decrease in thylakoid content on the absorption and photochemical conversion of light energy is illustrated in Figure 4.2. The absorption of white light by leaves is diminished to a small extent by the reduction in chlorophyll and carotenoid content. Leaves of control plants

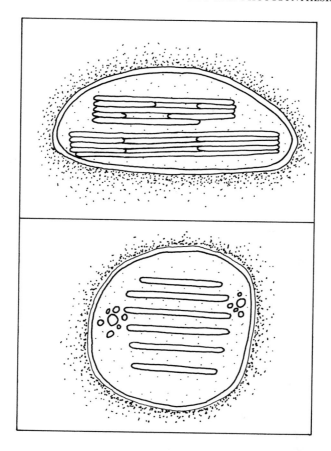

Fig. 4.1 Effect of iron stress on chloroplast structure. The upper drawing depicts a chloroplast from an iron-sufficient control plant. The lower drawing represents a chloroplast from a severely iron-stressed plant (after Spiller and Terry, 1980).

with chlorophyll contents greater than 0.56 mmol (chlorophyll) m^{-2} absorbed about 81% of the incoming light energy and when chlorophyll contents were reduced by 90%, absorption was still about 50%. This is because leaves are more efficient absorbers of light energy than are solutions of pigment of comparable concentration (Björkman, 1973; Terry, 1980).

Iron stress apparently had no effect on the photochemical conversion of absorbed light (quantum yield) provided it was measured under low light conditions (Fig. 4.2). However, since photosynthetic electron transport capacity decreased with iron stress (Fig. 4.2), iron-deficient leaves were less able to photochemically utilize absorbed quanta under high irradiances and light-saturated photosynthetic rates were diminished (Terry, 1980).

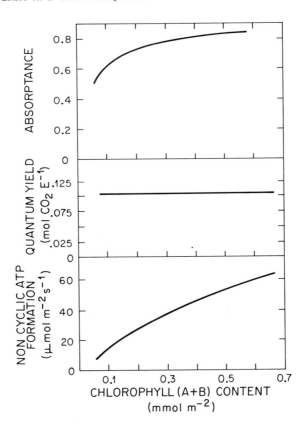

Fig. 4.2 The relationships of absorptance, quantum yield and photosynthetic electron transport with leaf chlorophyll content under conditions of iron stress (from Terry, 1980).

B. Specificity and Reversibility

So that iron deficiency is a valid technique to study photosynthetic limitation, it is essential that iron stress have little or no effect on photosynthetic apparatus other than the thylakoid system. If, for example, under iron stress the enzymatic capacity for carbon dioxide (CO_2) fixation and reduction decreased along with photochemical capacity, there would be no way of separating the effects on photochemical capacity from the effects on enzymes of the Calvin cycle. That iron stress had much less effect on Calvin cycle enzymes than on the thylakoid system is shown by iron stress reducing the thylakoid components of chloroplasts by 90%, while ribulose bisphosphate (RuP_2) carboxylase-oxygenase (Rubisco) activity was reduced only by about 30% (Terry, 1980). The extractable activities of other Calvin cycle enzymes;

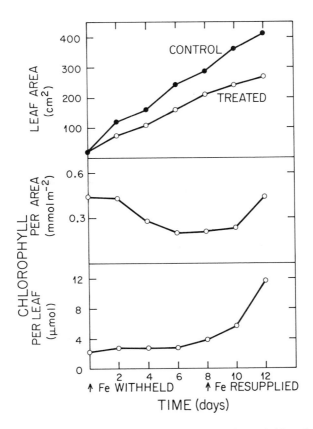

Fig. 4.3 Effects of withholding and resupplying iron on leaf expansion and chlorophyll content. Iron was withheld from the treated plant on day 0 and resupplied on day 8 (from Terry, 1979).

that is, fructose-bisphosphatase, glyceraldehyde-phosphate dehydrogenase, and phosphoribulokinase were not decreased by iron deficiency (Taylor *et al.*, 1982).

As mentioned earlier, electron microscopy revealed that iron deficiency reduced the extent of the thylakoid system while having no apparent effect on other parts of the cell. The view that the effects of iron stress are specific to chloroplasts is supported by other evidence (Terry, 1980). Iron deficiency had no effect on the number of chloroplasts per cell or per unit leaf area and chloroplasts apparently replicated at normal rates under iron stress. Other leaf growth attributes were unaffected by iron deficiency; the number of cells per unit leaf area and average leaf cell volume were unchanged. In fact, leaves grew about as fast under iron deficiency as leaves of control plants, and, at the same stage of leaf development, had the same thickness, fresh weight per area, and percentage intercellular space as normal leaves

(Terry, 1979). However, measurement of chloroplast volume and of chloroplast protein nitrogen indicated a 30% reduction in chloroplast size.

The effect of iron stress on the thylakoid system was not only specific but reversible. In an experiment in which iron was supplied to iron-deficient plants after eight days of iron deficiency, the chlorophyll content of the leaf returned to normal within 96 hours (Fig. 4.3). Leaf growth remained constant throughout the whole experiment, irrespective of whether iron was withheld or resupplied. Since the amount of chlorophyll per leaf did not change during the period of iron deficiency, it appears that the thylakoid material was conserved and diluted during leaf growth with chloroplast replication resulting in less thylakoid material per chloroplast. Electron transport components such as Q, P_{700}, cytochrome f, and lamellar iron and lamellar manganese also increased during iron resupply (Nishio and Terry, 1983; Young and Terry, 1983).

C. Influence of Photochemical Capacity on Photosynthetic Rate

By comparing the gas exchange of leaves from iron-sufficient and iron-deficient plants, it is possible to determine how changes in light harvesting and electron transport capacity influence photosynthetic rate in specific light and CO_2 environments. When the rate of photosynthesis was measured in an optimum environment [about 1000 μl (CO_2) l^{-1}], it was found to be almost linearly related to chlorophyll content (Fig. 4.4). This is also true at lower CO_2 concentrations of 600 and 300 μl (CO_2) l^{-1}. The data at 300 μl (CO_2) l^{-1} are particularly interesting because they are closest to photosynthesis under field conditions. The light-saturated rate of photosynthesis per area increased 36% with an increase in chlorophyll over the range of 0.45 to 0.73 mmol m^{-2}. These chlorophyll values were typical of iron-sufficient control plants, and there was no change in the extractable activities of Rubisco, chloroplast fructose-bisphosphatase, chloroplast glyceraldehyde-phosphate dehydrogenase, or phosphoribulokinase. Thus, the data at 300 μl (CO_2) l^{-1} suggest that the rate of photosynthesis under field conditions may be co-limited by light harvesting and electron transport capacity. At a lower CO_2 concentration, however, that is, 150 μl(CO_2) l^{-1}, there was an increase in photosynthesis with chlorophyll only over the lower part of the chlorophyll range (Fig. 4.4).

The increase in photosynthesis per unit area with increase in chlorophyll content at light saturation is attributed to an increase in the photochemical conversion of absorbed light rather than to an increase in light absorption (Terry, 1980). At irradiances that are limiting one might expect that an increase in chlorophyll content would enhance photosynthesis per unit area by increasing the quantity of light (photon flux density) absorbed. However,

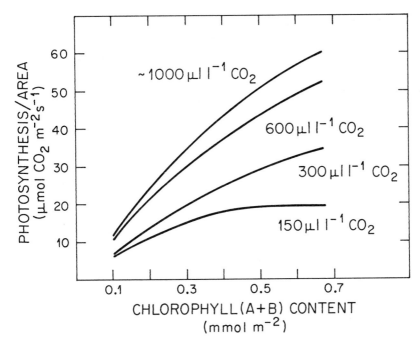

Fig. 4.4 The relationship of light-saturated photosynthesis with chlorophyll content at different ambient CO_2 concentrations (from Terry, 1983).

even at limiting irradiances the improvement in light absorption of photosynthetically active radiation (PAR) may be less important than the improvement in photochemical conversion. For example, photosynthesis per unit area at the limiting irradiance of 500 μE PAR $m^{-2} s^{-1}$ (Fig. 4.5) increased six-fold with increase in chlorophyll content from 0.056 to 0.67 mmol m^{-2}. Absorption of white light, however, increased much less, that is, from 51% to 81% over the same range of chlorophyll content (Fig. 4.2). Thus, although the increase in thylakoid components increased light absorption to some extent, it was clearly insufficient to produce the observed large increase in photosynthesis per unit area.

III. A THEORETICAL APPROACH TO PHOTOCHEMICAL CAPACITY

An alternative approach to the limitation of photochemical capacity is by mathematical modelling. This is not an easy undertaking because most of the work on integrated models of photosynthesis has centred on the diffusion of CO_2 and the kinetics of Rubisco. The role of light harvesting and electron

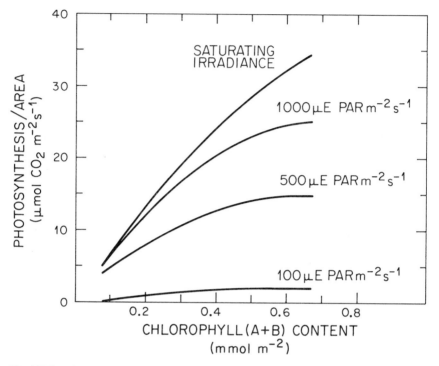

Fig. 4.5 The relationship of photosynthesis at an ambient CO_2 concentration of 300 μl (CO_2) l⁻¹ with chlorophyll content at different irradiances (from Terry, 1983).

transport capacity has received much less attention. In this section we provide an explanatory model that describes how leaf photosynthesis may be influenced by both the capacity for photosynthetic electron transport ("the light reactions") and the capacity of the Calvin cycle enzyme system for CO_2 fixation and reduction ("the dark reactions").

The net rate of photosynthetic CO_2 assimilation, A, is given by

$$A = V_c - F - R_d \qquad (4.1)$$

where V_c is the rate of carboxylation of RuP_2, F is the rate of photorespiratory CO_2 release, and R_d the rate of day respiration. In the photorespiratory carbon oxidation cycle, oxygenation of one mole of RuP_2 gives rise to the loss of 0.5 mole CO_2 by glycine decarboxylation (Lorimer *et al.*, 1978), and so

$$F = 0.5 V_o \qquad (4.2)$$

where V_o is the rate of RuP_2 oxygenation. Hence

$$A = V_c - 0.5V_o - R_d \tag{4.3}$$

Eqn 4.3 may be rewritten as

$$A = V_c(1 - 0.5\, V_o/V_c) - R_d \tag{4.4}$$

An expression for V_o/V_c was derived by Laing *et al.* (1974)

$$\frac{V_o}{V_c} = \frac{V_{omax}}{V_{cmax}} \frac{K_c}{K_o} \frac{O}{C} = \frac{2\Gamma_*}{C} = \frac{2\Gamma_* O}{C} \tag{4.5}$$

where O and C are the concentrations of oxygen and CO_2 at the sites of oxygenation and carboxylation respectively, K_o and K_c the respective Michaelis constants, and V_{omax} and V_{cmax} are the respective maximum rates of oxygenation and carboxylation of Rubisco. The CO_2 compensation concentration that would exist in the absence of day respiration is Γ_*, and, γ_*, its oxygen dependence (for elaboration of these terms see Farquhar and von Caemmerer, 1982). The above equation is independent of the degree of activation and of RuP_2 concentration. However, it does assume that RuP_2 binds before oxygen and substrate CO_2 (Farquhar, 1979).

$$V_{cmax} = k_c E_t \tag{4.6}$$

where k_c [mol(CO_2) per mol enzyme sites per second] is the catalytic turnover number of the carboxylase, and E_t is the density of enzyme sites (for example, if V_{cmax} is expressed as mol (of a particular component) per m^{-2}, then E_t would be expressed in mol m^{-2}).

The value of the derivation for V_o/V_c now becomes apparent. Using the estimate of V_o/V_c from eqn 4.5 we get

$$A = V_c(1 - \Gamma_*/C) - R_d \tag{4.7}$$

This expression permits us to calculate the rate of carboxylation of a leaf *in vivo*, that is,

$$V_c = \frac{A + R_d}{1 - \Gamma_*/C} \tag{4.8}$$

and is a measure of what has been referred to in the literature on gas exchange as "true photosynthesis" (Schnarrenberger and Fock, 1976). All the parameters on the right-hand side of eqn 4.8, A, Γ_*, C and R_d, can be measured. Γ_*, C and R_d can be determined approximately from measurements of CO_2 compensation point, intercellular CO_2 concentration, and dark respiration rate, respectively. In the case of Γ_*, an alternative method of estimation is to obtain the specificity factor enzymatically following the procedure of Jordan and Ogren (1981)

$$\text{Specificity factor} \quad = \quad \frac{V_c}{V_o} \quad \frac{O}{C} \quad = \quad \frac{1}{2\gamma_*} \qquad (4.9)$$

The specificity factor, like the compensation point, varies little between species (at any given temperature).

The rate of RuP_2 oxygenation, V_o, can be estimated from eqn 4.5 by substituting V_c from eqn 4.8.

Having derived estimates for V_c and V_o, we can calculate the rate of consumption of adenosine triphosphate (ATP) and the reduced form of nicotinamide adenine dinucleotide phosphate (NADPH) in the photosynthetic carbon reduction, photosynthetic carbon oxidation, and photorespiratory nitrogen cycles. For every RuP_2 consumed in carboxylation, 3 ATP and 2 NADPH are required, and in oxygenation, 3.5 ATP and 2 NADPH (Berry and Farquhar, 1978). Thus, the rate of ATP consumption

$$3V_c + 3.5V_o = (3 + 7\Gamma_*/C)V_c \qquad (4.10)$$

and the rate of NADPH consumption

$$2V_c + 2V_o = (2 + 4\Gamma_*/C)V_c \qquad (4.11)$$

Currently, it is thought that three protons are required to generate one ATP (Hangarter and Good, 1982). Thus, from eqn 4.10 the required rate of proton movement is

$$9V_c + 10.5V_o = (9 + 21\Gamma_*/C)V_c \qquad (4.12)$$

The movement of one electron through the whole electron transport chain results in the accumulation of two protons in the thylakoid spaces, one from the splitting of water in PS II and one from the shuttle of reduced plastoquinone across the membrane (Junge, 1977). This is equivalent to electron transport rate

$$4.5V_c + 5.25V_o = (4.5 + 10.5\Gamma_*/C)V_c \qquad (4.13)$$

On the basis of NADPH requirement alone, the electron transport requirement is smaller ($4V_c + 4V_o$). This point is considered further elsewhere (Farquhar and von Caemmerer, 1982).

In the discussion above, we showed how V_c may be estimated from gas exchange parameters (see eqn 4.8). Alternatively, V_c may be obtained from consideration of enzyme kinetics as

$$V_c = \frac{CM}{CM + K_{cm}} \frac{V_{cmax}C}{C + K_c(1 + O/K_o)} \frac{R}{R + K_r''} \qquad (4.14)$$

where M and R are the concentrations of free magnesium ions and RuP_2, respectively, in the stroma, and K_{cm} and K_r'' are kinetic parameters (Farquhar, 1979). The above formulation again assumes an ordered mechanism for both carboxylation and oxygenation. The Michaelis constant for RuP_2 effective when the enzyme is not fully activated is K_r'' and depends on C, O and M as

$$K_r'' = K_r' \frac{CM + (C + K_e)K_d}{CM + K_e K_r' K_d/K_f} \qquad (4.15)$$

where K_r' is the Michaelis constant effective for RuP_2 for the fully activated enzyme, and depends on C and O. K_d, K_e and K_f are kinetic parameters pertaining to inactivated sites. K_{cm} may be expressed more fully as $K_e K_r' K_d/K_f$ (Farquhar, 1979). Further complications arise when other phosphorylated compounds compete with RuP_2.

From the above estimate of V_c (eqn 4.14) we obtain A as

$$A = \frac{CM}{CM + K_{cm}} \frac{V_{cmax}(C - \Gamma_*)}{C + K_c(1 + O/K_o)} \frac{R}{R + K_r''} - R_d \qquad (4.16)$$

Under conditions of low C and adequate irradiance, Farquhar et al. (1980) suggested that R should be much greater than K_r'. This conclusion is supported by experiments of Collatz (1978) and Perchoriwicz et al. (1981). Farquhar et al. (1980) also assume that under these conditions Rubisco is fully activated which simplifies eqn 4.16 to

$$A = \frac{V_{cmax}(C - \Gamma_*)}{C + K_c(1 + O/K_o)} - R_d \qquad (4.17)$$

Comparisons of observed rates of CO_2 assimilation, A, with estimates of V_{cmax} obtained from in vitro measurements of Rubisco support this formulation (von Caemmerer and Farquhar, 1981; Seemann and Berry, 1982; Evans, in press).

Under conditions of high C and low irradiance, eqn 4.17 is unlikely to apply because the supply of ATP and NADPH is insufficient to regenerate RuP_2 at the rate that a fully activated and RuP_2-saturated Rubisco would use it. In principle this could occur in two ways. The concentration of Calvin cycle intermediates, including RuP_2, could decrease under low light, or, the level of activation of Rubisco may be decreased to allow the rates of carboxylation and oxygenation to match the rate of supply of RuP_2.

Photochemical energy supply can also be decreased by reducing photochemical capacity, that is, by decreasing the number of electron transport chains per area, as illustrated in Section II (or, by decreasing the photochemical efficiency per chain). The supply of photochemical energy is considered to be non-limiting in the derivation of eqn 4.17 so that A is determined solely by the kinetics of Rubisco. When the supply of photochemical energy becomes limiting (e.g. low irradiance, or diminished photochemical capacity), A is considered to be controlled by J, the potential electron transport rate where

$$A = \frac{J(C - \Gamma_*)}{4.5C + 10.5\Gamma_*} - R_d \qquad (4.18)$$

Thus, CO_2 assimilation rate is given by whichever eqn, 4.17 or 4.18, yields the lower value in any given instance (von Caemmerer and Farquhar, 1981). Electron transport capacity and the level of absorbed irradiance is determined by J. The assumption used to derive eqn 4.18 is that the actual rate of electron transport as given by eqn 4.13 is equal to J; that is,

$$(4.5 + 10.5 \frac{\Gamma_*}{C}) V_c = J \qquad (4.19)$$

Then

$$V_c = \frac{J}{4.5 + 10.5(\Gamma_* / C)} \qquad (4.20)$$

and eqn 4.18 follows by substituting this expression for carboxylation rate into the expression for net photosynthesis, eqn 4.7. This formulation has received support experimentally (von Caemmerer and Farquhar, 1981).

Perhaps the most uncertain aspect of all in modelling leaf photosynthesis is the derivation of an expression for J in eqn 4.18. One simple formulation (Kok, 1965) was similar to the Michaelis–Menten description of enzyme reactions, light being considered as substrate. Changing Kok's notation only slightly, the potential rate of electron transport, J, depends on absorbed photon flux, I, as follows

$$J = \frac{J_{max}I}{I + rJ_{max}} \tag{4.21}$$

where J_{max} is the maximum rate, and r is the quanta required per electron transported (Kok gives a value of 2.1). He noted that photosynthesis tended to be more linear in weak light and approached saturation more suddenly in strong light than is compatible with the rectangular hyperbola, eqn 4.21. Berry and Farquhar (1978) formulated ATP generation as increasing linearly with irradiance, and saturating abruptly. Electron transport was given an empirical dependence on irradiance using a quadratic function (Farquhar et al., 1980). This provides a useful description of the electron transport–irradiance relationship because it is intermediate between the above extremes.

A more mechanistic model was developed that included three limitations on capacity, each of which was related to electron fluxes through different parts of the chain (Farquhar and von Caemmerer, 1981). J_2 was approximately the rate of electron transport from water to plastoquinone, J_a was the maximum rate of electron transfer from plastoquinone to cytochrome f, and J_3 referred to the rate of electron transport from cytochrome f to $NADP^+$. Each of these fluxes was obtained as the mol (of a particular component) per m^2 of electron transport components involved divided by the time required for the electron to move through that part of the chain. The maximum rate of electron transport per unit leaf area, J_{max}, is related to the fluxes through the three components described above as follows:

$$1/J_{max} = 1/J_2 + 1/J_a + 1/J_3 \tag{4.22}$$

This implies that the delay associated with any specific carrier is not only dependent on the intrinsic delay associated with that carrier, but is also inversely dependent on the number of carriers involved. In this sense the kinetics of steady-state electron transport should differ from those of transient studies (e.g. flash experiments) from which time constants for specific carriers are derived.

IV. CONCLUSIONS

Perhaps the most important conclusion resulting from both the experimental and modelling approaches outlined above is that photochemical capacity can potentially limit photosynthesis under field conditions. The experimental approach (Section II) indicates that light-saturated photosynthesis of healthy sugar beets can be increased by as much as 36% with increase in photochemical capacity at the limiting ambient CO_2 concentration of 300 μl (CO_2) l^{-1}. This observation suggests that photosynthetic rate is colimited by several factors simultaneously so that an increase in any one factor independently of another results in an increase in photosynthetic rate.

The modelling approach (Section III) provides a basis for simultaneous colimitation. For example, eqn 4.18 indicates that an increase in J (electron transport rate), either by increasing irradiance or by increasing photochemical capacity, can result in an increase in A (photosynthetic rate) at values of C (intercellular CO_2 concentration) that may be limiting to photosynthesis. At very low values of C, the model predicts that A is determined by eqn 4.17 that is dominated by the kinetics of Rubisco. For the most part, however, we believe that values of A under natural conditions will lie close to the balance point between the Rubisco-dominated system (eqn 4.17) and the photochemical capacity-dominated system (eqn 4.18). Stomatal and mesophyll conductance to the diffusion of CO_2 will also colimit photosynthetic rate as they determine the CO_2 concentration at the site of carboxylation.

REFERENCES

Berry, J. A., and Farquhar, G. D. (1978). The CO_2 concentrating function of C_4 photosynthesis. A biochemical model. In "Proceedings of the 4th International Congress on Photosynthesis" Reading, England, 1977, (D. Hall, J. Coombs and T. Goodwin, eds.). The Biochemical Society, London.

Björkman, O. (1973). Comparative studies on photosynthesis in higher plants. In "Photophysiology: Current Topics in Photochemistry" Vol. 8 (A. C. Giese ed.), pp. 1–63. Academic Press, New York.

Caemmerer von, S., and Farquhar, G. D. (1981). Some relationships between the biochemistry of photosynthesis and the gas exchange of leaves. Planta 153, 376–387.

Collatz, G. J. (1978). The interaction between photosynthesis and ribulose-P_2 concentration effects of light, CO_2 and O_2. Carnegie Institute of Washington Yearbook 77, 248–251.

Evans, J. R. (in press). Nitrogen and phosphorus in wheat. Plant Physiology

Farquhar, G. D. (1979). Models describing the kinetics of ribulose bisphosphate carboxylase oxygenase. Archives of Biochemistry and Biophysics 193, 456–468.

Farquhar, G. D., and Caemmerer von S. (1981). Electron transport limitations on the CO_2 assimilation rate of leaves: A model and some observations in Phaseolus vulgaris L. In "Proceedings of the 5th International Congress on Photosynthesis". Vol. IV (G. Akoyunoglou, ed.), pp. 163–175. Balaban, Philadelphia.

Farquhar, G. D. and Caemmerer von, S. (1982). Modelling of photosynthetic response to environmental conditions. In "Physiological Plant Ecology II." Encyclopedia of Plant Physiology New Series Vol. 12B (O. L. Lange, P. S. Nobel and C. B. Osmond, eds.), pp. 550–587. Springer, Berlin.

Farquhar, G. D., Caemmerer von, S. and Berry, J. A. (1980). A biochemical model of photosynthetic CO_2 assimilation in leaves of C_3 species. Planta 148, 78–90.

Hangarter, R. P. and Good, N. E. (1982). Energy thresholds for ATP synthesis in chloroplasts. *Biochimica Biophysica Acta* **681**, 397–404.

Jordan, D. B. and Ogren, W. L. (1981). A sensitive assay procedure for simultaneous determinations of ribulose-1-5 bisphosphate carboxylase and oxygenase activities. *Plant Physiology* **67**, 237–245.

Junge, W. (1977). Physical aspects of light harvesting, electron transport and elctrochemical potential generation in photosynthesis of green plants. *In* "Photosynthesis I. Photosynthetic electron transport and photophosphorylation." Encyclopedia of Plant Physiology New Series Vol. 5 (A. Trebst and M. Avron, eds.), pp. 59–93. Springer, Berlin.

Kok, B. (1965). Photosynthesis: The path of energy. *In* "Plant Biochemistry" (J. Bonner and J. L. Varner, eds.)., 3rd edn., pp. 846–886. Academic Press, New York.

Laing, W. A., Ogren, W., and Hageman, R. (1974). Regulation of soybean net photosynthetic CO_2 fixation by the interaction of CO_2, O_2 and ribulose 1,5 diphosphate carboxylase. *Plant Physiology* **54**, 678–685.

Lorimer, G. H., Woo, K. C., Berry, J. A., and Osmond, C. B. (1978). The C_2 photorespiratory carbon oxidation cycle in leaves of higher plants: Paths and consequences. *In* "Photosynthesis 77" (D. O. Hall, J. Coombs and T. W. Goodwin, eds.), pp. 311–322. Biochemical Society, London.

Machold, O. (1971). Lamellar proteins of green and chlorotic chloroplasts as affected by iron deficiency and antibiotics. *Biochemica Biophysica Acta* **238**, 324–331.

Nishio, J. N., and Terry, N. (1983). Iron nutrition mediated chloroplast development. *Plant Physiology* **71**, 688–691.

Perchoriwicz, J. T., Raynes, D. A., and Jensen, R. G. (1981). Light limitation of photosynthesis and activation of ribulose bisphosphate carboxylase in wheat seedlings. Proceedings National Academy of Science (U.S.A.) **78**, 2985–2989.

Platt-Aloia, K. A., Thomson, W. W., and Terry, N. (1983). Changes in plastid ultrastructure during iron nutrition mediated chloroplast development. *Protoplasma* **114**, 85–92.

Schnarrenberger, C., and Fock, H. (1976). Interactions among organelles involved in photorespiration. *In* "Transport in Plants III. Encyclopedia of Plant Physiology New Series" Vol. 3 (C. R. Stocking and V. Heber, eds.), pp. 185–234. Springer, Berlin.

Seemann, J. R., and Berry, J. A. (1982). Interspecific differences in the kinetic properties of RuBP carboxylase protein. *Carnegie Institute of Washington Yearbook, 1981.* pp. 78–83.

Spiller, S., and Terry, N. (1980). Limiting factors in photosynthesis II. Iron stress diminishes photochemical capacity by reducing the number of photosynthetic units. *Plant Physiology* **65**, 121–125.

Taylor, S. E., Terry, N., and Huston, R. P. (1982). Limiting factors in photosynthesis III. Effects of iron nutrition on the activities of three regulatory enzymes of photosynthetic carbon metabolism. *Plant Physiology* **70**, 1541–1543.

Terry, N. (1979). The use of mineral nutrient stress in the study of limiting factors in photosynthesis. *In* "Photosynthesis and Plant Development", Proceedings of Symposium, Diepenbeek-Hasselt, Belgium, July 1978. (R. Marcelle, H. Clijsters and M. Van Poucke, eds.), pp. 151–160. Junk, The Hague.

Terry, N. (1980). Limiting factors in photosynthesis I. Use of iron stress to control photochemical capacity *in vivo*. *Plant Physiology* **65**, 114–120.

Terry, N. (1983). Limiting factors in photosynthesis IV. Iron stress mediated changes in light harvesting and electron transport capacity and its effects on photosynthesis *in vivo*. *Plant Physiology* **71**, 855–860.

Terry, N., and Low, G. (1982). Leaf chlorophyll content and its relation to the intracellular localization of iron. *Journal of Plant Nutrition* **5**, 301–310.

Young, T. F., and Terry, N. (1983). Kinetics of iron transport into the leaf symplast during recovery from iron stress. *Canadian Journal of Botany* **61**, 2496–2499.

CHAPTER 5

Carbon Partitioning

J. W. PATRICK

I. INTRODUCTION

The harvestable portion of most crops has limited photosynthetic capacity, and hence crop yield depends to a significant extent on carbon transported from other, more photosynthetically active, parts of the plant. Much of this movement occurs within the specialized sieve element–companion cell complex located in the phloem. Thus, partitioning of carbon to the harvestable portion of the plant includes the integrated operation of partitioning mechanisms in the source, path and sink organs. This chapter attempts to describe the control of these processes and to assess their significance as determinants of carbon transfer.

59

CONTROL OF CROP PRODUCTIVITY
ISBN 0 12 548280 9

II. CARBON PARTITIONING IN LEAVES

Following the reduction of carbon dioxide in the chloroplast stroma, the amounts of carbon (C) available for use by a sink will depend on how C is partitioned between storage and transport pools within the leaf (Moorby, 1977). Leaf processes that may regulate C partitioning include metabolic control of sucrose availability, sucrose exit into the leaf apoplast and active uptake of sucrose into the sieve element–companion cell (se–cc) complex (Geiger, 1979).

During the light period, the availability of C for transport depends upon net carbon dioxide fixation and is modified by the supply of C from mobilized reserves (mainly starch) in the leaf (Ho, 1979). Transfer of fixed C across the chloroplast envelope to the cytosol, largely in the form of dihydroxyacetone phosphate, is facilitated by the so-called phosphate translocator in exchange for inorganic phosphate (Heber and Heldt, 1981). The high transport capacity of the phosphate translocator is conceived as a key regulant of the partitioning of C between starch and sucrose synthesis (Heber and Heldt, 1981). That is, by permitting rapid exchange between inorganic phosphate and dihydroxyacetone phosphate, the phosphate translocator ensures that changes in the relative rates of carbon dioxide fixation and sucrose synthesis in the cytosol through the liberation of inorganic phosphate are immediately reflected in the relative pool sizes of inorganic phosphate, dihydroxyacetone phosphate and ultimately by feed back inhibition, 3-phosphoglycerate, in the chloroplast stroma. The ratio of 3-phosphoglycerate to inorganic phosphate in the stroma determines whether starch degradation (decreased ratio) or synthesis (increased ratio) is favoured (Preiss and Levi, 1979). Dislocation of the relationship between photosynthesis and translocation at higher photosynthetic rates (Ho, 1979) may result from limitations imposed on translocation sucrose synthesis being regulated by the activity of sucrose phosphate synthetase (Silvius et al., 1979) while further increases in photosynthetic products are partitioned into starch. At night, the levels of starch appear to be the chief factor governing sucrose level and export (Ho, 1979). Control of exchange of C between cytosol and vacuole remains to be investigated but, since transfer across the tonoplast is carrier mediated and energy dependent (Guy et al., 1979), the model for ion transport into the root (Pitman, 1977) may serve as a useful starting point.

The processes of sucrose efflux into the leaf apoplast and uptake into the se–cc complex are considered to be more responsive to conditions outside the source leaf than inside, such as sink demand (Geiger, 1979). Sucrose loading of the se–cc cell complex would appear to be mediated by coupled transport with protons moving down the electrochemical gradient generated by an outward-directed adenosine triphosphatase (ATPase) proton pump

(Giaquinta, 1980). Sucrose loading may regulate sucrose efflux into the apoplast through changing the apoplastic levels of potassium. Thus, low rates of sucrose loading would favour passive accumulation of potassium in the se–cc complex in order to dissipate the electrochemical gradient generated by the outward–directed ATPase proton pump. The accumulated potassium would serve to maintain phloem turgor under conditions of decreased sucrose loading and hence ensure continued export of C from the leaf (Smith and Milburn, 1980). The lowered potassium levels in the apoplast would decrease sucrose efflux (Doman and Geiger, 1979). Conversely, high rates of proton-coupled sucrose loading would dissipate the electrochemical gradient thereby releasing potassium from the se–cc complex. The attendant rise in apoplastic potassium levels would act to enhance sucrose efflux (Doman and Geiger, 1979).

The proposition that the effects of sink demand on phloem loading are mediated through sucrose levels in the se–cc complex (cf. Moorby, 1977) has been supported by the observation that rates of phloem loading are inversely related to sucrose levels in the se–cc complex of preloaded leaf discs (Giaquinta, 1980). However, an alternative proposition is that phloem loading is a turgor-sensitive process and turgor changes at the sink end of the pathway can be rapidly propagated to the loading sites (Smith and Milburn, 1980 and earlier papers).

III. CARBON PARTITIONING IN SINKS

Sink demand (i.e. lowered sucrose concentration or turgor potential in the se–cc complex) is considered to be regulated by the processes of phloem unloading and sink uptake (e.g. Moorby, 1977). The nature of these processes will, at least in part, be determined by the cellular pathway of C transfer between the phloem and sink tissues.

A. Symplastic Transfer

The movement of C in growing apices may follow a symplastic route from the protophloem termini of the vascular strands to the sink tissues (Gunning, 1976). Assuming that the cytoplasmic annuli of the plasmodesmata are open and of fixed diameter, then for a given sink tissue, the rate of C transfer would be governed by the sucrose concentration gradient between the se–cc complex and the sink cytosol. Metabolic consumption and vacuolar compartmentation of sucrose in the sink cells would contribute to the steepness of the gradient and hence the rate of phloem unloading. Sucrose pool size in the sink cytosol may be of less significance if the cytoplasmic annuli were occluded by neck constrictions, and desmotubules were continuous with the endoplasmic reticulum (Gunning, 1976); under these conditions, C transfer also may be governed by some form of facilitated transport across the endoplasmic membranes both in the phloem and sink cells.

B. Apoplastic Transfer

Transfer of C from maternal to embryonic tissues is universally apoplastic. This pathway also appears to apply for other dominant sinks such as sugar cane stems (Glasziou and Gayler, 1972), storage roots of sugar beets (Saftner and Wyse, 1980) and potato tubers (Mares and Marschner, 1980). Furthermore, radial movement of C may occur in the apoplast of both woody and herbaceous stems (Patrick and Turvey, 1981) and, in combination with a high capacity phloem loading system (Bieleski, 1966), may account for the impermeable nature of the phloem pathway to radial loss of C (Wardlaw, 1980).

Apoplastic unloading confers the potential for the control of C partitioning at the phloem membrane. Other than the studies of Wolswinkel (e.g. Wolswinkel, 1978), the mechanism of this membrane transfer has not yet been explored. Wolswinkel (1978) found that phloem unloading to the apoplast of broad bean stems parasitized by *Cuscuta* was sensitive to low temperatures and metabolic inhibitors, suggesting that phloem unloading was energy dependent and presumably carrier mediated. Whereas tightly controlled phloem unloading could be advantageous for a conservative sink (e.g. the transport axis of root and stem) such control, at least on superficial analysis, would be unexpected for terminal sinks committed to import (e.g. developing fruit). This proposition has been tested for the transfer of C from the seed coat to the cotyledons of developing ovules of bean. Since a gain in entire dry weight of the cotyledons involves secretions from the seed coat and 80% of this material is sucrose (Patrick and McDonald, 1980), measures of gain in cotyledon dry weight and the total membrane surface area of the seed coat allow estimates of sucrose fluxes. These were found to be in the order of 5 x 10^{-8} mol m^{-2} s^{-1} (Offler, C. E. and Patrick, J. W., unpublished data) and are in the range expected for facilitated membrane transfer (Gunning, 1976). The significance of facilitated phloem unloading in the bean ovule may be to maintain apoplastic sugar levels at saturation for the cotyledon uptake system (Patrick and McDonald, 1980; Patrick, 1981) and thus ensuring optimal growth of the cotyledons.

Transfer across the plasma membranes of sink cells may occur either as passive diffusion (e.g. sugar beet root, Saftner and Wyse, 1980) or by some form of facilitated transfer dependent on an energized membrane (Gifford and Evans, 1981). Diffusive entry would be driven by the sucrose concentration gradient between the apoplast and the sink cytosol. The control of gradient steepness by the sink would be determined by rates of metabolic interconversion or vacuolar compartmentalization of sucrose in the sink, the latter process depending on an energized tonoplast membrane (Saftner and Wyse, 1980). For facilitated transfer, energy coupling by cotransport with protons, that are pumped into the apoplast presumably by an ATPase, has been demonstrated for developing soybean cotyledons (Lichtner and Spanswick, 1981) and suspension cultures of storage parenchyma isolated from sugar cane stems (Komor et al., 1981). Interestingly, in the case of soybean cotyledons, a Michaelis–Menten constant (K_m) of 10 mM sucrose was found for cotransport (Lichtner and Spanswick, 1981). This indicates that the membrane carrier must be energized by some form of energy coupling other than proton cotransport between apoplastic sucrose concentrations of 20–100 mM (carrier saturation, Patrick, 1981). Since in vivo apoplastic concentrations of sucrose may in many circumstances exceed 20 mM sucrose (e.g. Patrick and McDonald, 1980), proton cotransport may contribute little to the total sugar uptake by sinks.

Irrespective of whether sink entry is by passive diffusion or facilitated transfer, apoplastic sugar concentrations would appear to be a major determinant of the rate of sugar uptake by sink tissues (Glasziou and Gayler, 1972; Gifford and Evans, 1981; Patrick, 1981). Sugar levels in the apoplast depend on the balance between phloem unloading and sink uptake (Patrick and McDonald, 1980; Patrick and Turvey, 1981) indicating that, in order to maintain apoplastic sugar levels, these two processes must be integrated. The stimulation of phloem unloading to the stem apoplast by Cuscuta (Wolswinkel, 1978) infers some form of sink control. Eschrich (1980) proposed that extracellular invertase may serve as the integrator. Thus, invertase activity would simultaneously regulate phloem unloading by determining the steepness of the sucrose concentration gradient and regulate sink uptake by determining the levels of hexoses available for accumulation. Sink control of enzyme activity, and hence phloem unloading, could be mediated by a outward-directed proton pumping ATPase that regulates apoplast pH (Eschrich, 1980). The ATPase also would create conditions favourable for proton cotransport of the hexoses (Komor et al., 1981). Without extracellular hydrolysis, the sink ATPase could provide the required force to drive a sucrose–proton antiport mechanism for phloem unloading (Wareing, 1977) and proton cotransport for sink uptake (Lichtner and Spanswick, 1981). For both models, the vectorial ATPase may be sensitive to hormones produced by the sink (Wareing, 1977), thus providing a link between the growth potential of a sink and the supply of C skeletons to it (Fig. 5.1).

The phenomenon of transinhibition of the membrane carrier by the pool size of sugars in the cytosol (e.g. Hampson et al., 1978; Komor et al., 1981) could form the basis of a feedback control, thereby linking sugar utilization with sugar uptake. Flexibility in sink utilization between metabolic conversion and temporary vacuolar storage of excess accumulated sugar permits

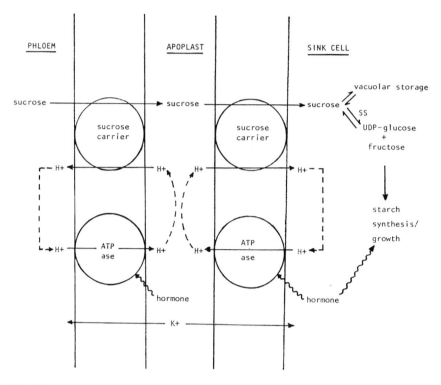

Fig. 5.1 Model of sucrose transfer across the phloem and the sink membranes and utilization of sucrose in the sink tissues. Note possible sites of action of sink-produced hormones which may coordinate membrane transfer of sucrose with growth and storage processes. Trans-inhibition (backward-directed arrows) by high levels of sucrose in the apoplast and sink symplast may serve to dampen sucrose transfer under conditions of decreased sink utilization. Xylem-delivered potassium may act to compensate for charge disequilibrium, Sections III and IV C. ss, source synthetase.

continued uptake when metabolism may otherwise limit accumulation (e.g. Hampson *et al.*, 1978; Mares and Marschner, 1980; Patrick, 1981). The activity of sucrose synthetase may be of significance in controlling sugar partitioning between metabolism and vacuolar storage in that, at cytoplasmic pH, the equilibrium of the catalyzed reaction is slightly in favour of synthesis (Pontis, 1977). Hence, complete cleavage of sucrose will occur only if the resulting uridine diphosphate (UDP) glucose or fructose were removed by being used in either synthetic or energy metabolism, thus leaving excess sucrose for vacuolar storage. Another factor, deserving of more research, is the contribution of sugar efflux to the determination of net accumulation by the sink tissues. In some circumstances, efflux may represent a significant proportion of sugar influx (Komor *et al.*, 1981; Patrick, 1981).

IV. CARBON PARTITIONING BETWEEN COMPETING SINKS

Carbon exported from a source leaf is distributed between the competing sink organs in reproducible patterns (Wardlaw, 1980). This observation implies that carbon partitioning between competing sinks is closely regulated at some point(s) along the source-sink path.

A. Source Control

The predominant role of the leaf would appear to be regulation of the amount of C available for sink import (Geiger, 1979). However, the finding that the light environment of the leaf may affect export patterns (Moorby, 1964), possibly mediated by the energy status of the leaf (Shiroya, 1968), deserves further investigation.

B. Path Control

For a fully differentiated vascular network, control of C partitioning could be imposed by the resistances to axial flow. These resistances could be in the vascular traces linking sources and sinks or in the vascuolar anastomoses interconnecting the vascular traces. While axial flow through the phloem pathway may be highly correlated with its cross-sectional area, causality of this relationship is rendered doubtful as partial restriction of the phloem was found not to impede C movement to the sinks (Wardlaw, 1980). This indicates that axial resistance of the pathway is relatively small. However, axial resistance may be significant at the sink end of the phloem path (Thornley et al., 1981) and this may account for the observed competitive advantage of source-sink proximity (e.g. Cook and Evans, 1978).

The distribution of C along phyllotaxic pathways (Wardlaw, 1980) is consistent with control being exerted by the relative resistances to lateral flow of the anastomoses between vascular traces. Whereas defoliation studies demonstrate that these resistances are minimal and are readily overcome (Wardlaw, 1980), their existence could serve to canalize carbon flow from a source to specified sinks and thus dampen inter-sink competition (Patrick, 1972).

From an analysis of growth characteristics of a number of sink types, Milthorpe and Moorby (1969) concluded that the phloem was more than adequate to meet the demands on it, except possibly, for the supply of sucrose to apical meristems and young primordia. In these tissues, C movement is through a presumably high resistance pathway of undifferentiated cells which may extend over distances of 300-400 μm (Wardlaw, 1980). Under these circumstances, C partitioning could depend on the sequence of vascular differentiation linking these primordial sinks with the established vascular system (Patrick, 1972). For sinks undergoing extension growth, the rate of sieve-tube destruction is exceeded by that of differentiation of new sieve tubes such that the cross-sectional area of these intercalary meristems does not appear to impose limitations on C transfer (Patrick, 1972).

C. Sink Control

Whereas it has been widely accepted that the predominant control of C partitioning resides in the competing sinks (Milthorpe and Moorby, 1969; Wareing and Patrick, 1975; Gifford and Evans, 1981), the properties of a sink that contribute to its ability to control C flow are ill defined and little understood. Sink effects on C partitioning may result from a sink stimulus that directly regulates a phloem transport process (Patrick and Wareing, 1980) or from sink depletion of C pools in the sink cytosol (Moorby, 1977). Phloem unloading and sink utilization, by affecting C removal from the se–cc complex, may contribute to the carbon concentration gradients found in the phloem path (Moorby, 1977). Although unsubstantiated, these gradients may generate sufficient differences in hydrostatic pressure to propel mass flow, or to act as directional signals to polarize some active transport mechanism. The nature of sink control on C transfer may, in part, be determined by the cellular pathway of C transfer between the se–cc complex and the site of utilization by the sink.

1. Symplastic Transfer

For symplastic transfer, effects of the sink on C partitioning may arise from sink depletion of sucrose pools in the cytosol which contributes to turgor-driven flow, or from a sink stimulus regulating some component of axial flow through the sieve tubes. In the case of growing apices, the combined hydraulic conductivity of the terminal phloem strands may be sufficiently low and the path through undifferentiated cells sufficiently long (Wardlaw, 1980) to impose significant resistances to turgor-driven flow. This high resistance pathway, together with small sink size (Cook and Evans, 1978), could disadvantage apical sinks within the competitive framework of the whole plant. These effects could be modified by direct sink regulation of phloem transport. Indeed, some evidence suggests that those auxins, synthesized in shoot apices and transported basipetally down the stem, may serve this function and account for the observed "priority demand" of shoot apices (Patrick and Wareing, 1980). Given that the cytoplasmic annuli are open, symplastic transfer in the apical tissues would permit the auxin-augmented flow to pass through the sink cells. In contrast, for apoplastic transfer, C flow could be limited significantly by the small area of phloem membrane available for exchange and an apoplast geometry (thin primary walls and path lengths of 300–400 μm, Wardlaw, 1980) of high resistance to C diffusion.

2. Apoplastic Transfer

Transfer through the apoplast would appear to be associated with large sinks (Section IIIB) which are served by a well-developed phloem pathway of presumed high hydraulic conductivity. These sinks and path properties would favour turgor-dependent flow such that any direct stimulus of axial phloem transport may contribute little to total C transfer. Thus, apoplastic transfer could be characterized by sink control of C partitioning being mediated by phloem unloading and sink utilization of C.

A striking example of a sink-mediated control of a phloem transport process is provided by the finding that enhanced flow of C to the parasite *Cuscuta* was associated with a stimulation of phloem unloading into the apoplast of the host's stem (Wolswinkel, 1978). The haustoria of *Cuscuta* were thus bathed in an elevated sugar concentration in the apoplast, and accordingly C gain by the parasite (sink equivalent) could be amplified. Similarly, the pattern of C accumulated by developing potato tubers was interpreted in terms of differential unloading from the phloem rather than by either uptake or metabolism by the tuber (Mares and Marschner, 1980). The demonstration that the synthetic cytokinin, kinetin, specifically stimulates C unloading to the apoplast (Hayes P. M. and Patrick, J. W., unpublished data) suggests that hormones may act as the sink stimulus that regulates the unloading process (Section IIIB and Fig. 5.1). Furthermore, sugar levels in the stem apoplast were elevated by kinetin, thus enhancing uptake by the stem storage cells (Patrick and Wareing, 1980). Based on these observations, it is proposed that sink-mediated regulation of phloem unloading would appear to serve as a mechanism to drive turgor-dependent carbon flow and simultaneously maintain apoplast C at levels for optimal sink uptake. This mechanism would be of particular significance in those situations where C levels may become limiting to sink processes if flow depended on the depletion of substrate pool sizes in the sink tissues.

Like phloem unloading, sink uptake from the apoplast, vacuolar compartmentation, and metabolism could ultimately lead to a reduction in carbon levels in the se–cc complex (Moorby, 1977). Sink uptake, vacuolar compartmentation, and metabolism together contribute to the strength of a sink. An operational definition of sink strength was proposed by Warren-Wilson (1972), and later refined by Wareing and Patrick (1975) as gain of an assimilate species by a sink under controlled conditions of assimilate supply. Sink strength was further defined as the product of two components; activity and size. Sink activity for C can be determind *in vitro* in terms of sugar uptake from the apoplast (Wareing and Patrick, 1975; Patrick, 1981) or for symplastic transfer by the activity of a key enzyme (e.g. sucrose synthetase) or a tonoplast carrier. Using this approach, it has been found that both temperature (Egli and Wardlaw, 1980) and light quality affects (Mor *et al.*, 1980) on C flow to the treated sink organ correlate well with commensurate changes in sink activity measured in terms of sugar uptake. Further progress in understanding the nature of these sink effects may profit from simultaneous measurements of the *in vivo* pool size of the sugar in the apoplast. Thus, under conditions where substrate supplies are limiting sink activity, lowering of the apoplast pool could be mediated by an increased affinity of the carrier system, perhaps by protonation. Furthermore, as the apoplast concentration fell, diffusion path length would become increasingly more important as a determinant of sink uptake, and if this varied between sinks, those with the shorter path length would be favoured. For this, and reasons advanced earlier (Section IVC) sink vasculature needs more thorough study. In contrast, at substrate levels saturating the carrier, the limiting influence of diffusion would be less significant, and sink activity would be governed largely by the maximal activity of the carrier. Variations in maximal activity between sink tissues appear to exist (Komor, 1977) and these could determine apoplast pool size. Reductions in sink activity would exert a negative effect on C flow as demonstrated by experimentally-induced decreases in sink metabolism (Wardlaw, 1980). Sink size also would contribute to the regulation of C partitioning by determining the relative size of the C pool in the se–cc complex. Indeed, manipulation of grain numbers (sink size) in two competing ears of wheat, demonstrated that C partitioning exhibited a pronounced bias in favour of the larger sink (Cook and Evans, 1978).

V. CARBON PARTITIONING AND CROP PRODUCTIVITY

Since the products of photosynthesis (carbon, hydrogen and oxygen) constitute some 90% of plant dry matter, it follows that whole-plant productivity must ultimately depend on whole-plant photosynthesis. Variations in whole-plant photosynthesis could arise from differences in photosynthetic rate per unit leaf area and differences in the rate at which new leaf surface is generated. Good and Bell (1980) proposed the following model to quantitatively describe the potential contribution of these two factors to plant dry matter production, in terms of photosynthetic apparatus (P) and storage material (S)

$$P + S = \frac{1}{d_p} P_0 e^{d_p A t} - \frac{1 - dp}{d_p} P_0 \qquad (5.1)$$

where

d_p = proportion of weight partitioned into photosynthetic apparatus
$1 - d_p$ = proportion of weight partitioned into storage material
P_0 = weight of the photosynthetic apparatus at t_0
A = rate of photosynthesis per unit P
t = time interval from t_0

This analysis established that the rate of photosynthesis, A, and the proportion of weight partitioned into photosynthetic apparatus, d_p, have the potential capacity to contribute equally to whole-plant growth (Good and Bell, 1980). However, the photosynthetic potential, A, will be expressed fully only under conditions of source-limited growth (Wareing and Patrick, 1975). Therefore, as verified experimentally, variations in whole-plant growth rates may, to a large extent, be accounted for by different rates of leaf expansion (Potter and Jones, 1977 and references therein). Amounts of C partitioned for whole-plant leaf production will reflect the number of active shoot apices, the rate of leaf initiation and the rate of leaf expansion and final leaf size. Until leaf emergence, the C utilized in leaf production is imported exclusively from the fully expanded foliage. Amounts of C transported to the shoot apices would appear to limit growth of the developing leaf primordia (Milthorpe and Moorby, 1969) possibly by some restriction imposed by the capacity of the transport system (Section IVC). Thus, future improvement in whole-plant growth rate may be obtained by modifications that enhance carbon flow during this phase of leaf development (Section IVC). Before the canopy achieves full interception of incident radiation, crop growth rates also would depend on C partitioning into leaf production.

After canopy closure, crop photosynthesis and hence crop growth rate would be expected to depend mainly on the rate of net carbon dioxide fixation per unit leaf area. However, there is no evidence that increases in yield during domestication and improvement of a wide variety of crops has resulted from indirect selection for increases in net photosynthesis per unit leaf area (Gifford and Evans, 1981). Rather, a number of studies have demonstrated that increases in yield potential may be attributed to an increased partitioning of C into the organs of agricultural importance. For instance, Duncan et al. (1977) have shown with a simulation model of peanut that significant increases in yield because of varietal changes, could largely be accounted for by greater partitioning of C to the developing fruit. The physiological basis of the improved bias of C partitioning into organs of agricultural importance remains to be elucidated. The finding that sink size is a potent determinant of C flow patterns (Cook and Evans, 1978) focuses attention on the initial phase of sink establishment when factors other than size must account for preferential partitioning of C to those sinks destined to be large and dominant. The significance of this consideration is reinforced by the finding that yield potential can be determined by C supply during an early phase of development of the storage organ (e.g. before anthesis in wheat, Gifford and Evans, 1981). At a later stage, expression of sink size potential and hence C partitioning may be determined by the supply of minerals, such as nitrogen (e.g. Kemp and Whingwiri, 1980) or inhibitors produced by more advanced sinks (e.g. Singh and Jenner, 1982). Similar shifts in factors regulating growth also have been found during the development of leaves (Milthorpe and Moorby, 1969). Therefore, both crop growth rate and yield may benefit from an improved understanding of carbon partitioning to small sinks so that the present bottlenecks (Section IVC2) in carbon skeleton supply may be removed to increase plant productivity.

REFERENCES

Bieleski, R. L. (1966). Accumulation of phosphate, sulphate and sucrose by excised phloem tissues. *Plant Physiology* **41**, 447–454.

Cook, M. G., and Evans, L. T. (1978). Effect of relative size and distance of competing sinks in the distribution of photosynthetic assimilates in wheat. *Australian Journal of Plant Physiology* **5**, 495–509.

Doman, D. C., and Geiger, D. R. (1979). Effect of exogenously supplied potassium on phloem loading in *Beta vulgaris* L. *Plant Physiology* **64**, 528–533.

Duncan, W. G., McCloud, D. E., McGraw, R. L., and Boote, K. J. (1977). Physiological aspects of peanut yield improvement. *Crop Science* **18**, 1015–1020.

Egli, D. B., and Wardlaw, I. F. (1980). Temperature response of seed growth characteristics of soybeans. *Agronomy Journal* **72**, 560–564.

Eschrich, W. (1980). Free space invertase, its possible role in phloem unloading. *Berichte der Deutschen Botanischen Gesellschaft* **93**, 363–378.

Geiger, D. R. (1979). Control of partitioning and export of carbon in leaves of higher plants. *Botanical Gazette* **140**, 241–248.

Giaquinta, R. (1980). Mechanism and control of phloem loading of sucrose. *Berichte der Deutschen Botanischen Gesellschaft* **93**, 187–201.

Gifford, R. M., and Evans, L. T. (1981). Photosynthesis, carbon partitioning, and yield. *Annual Review of Plant Physiology* **32**, 485–509.

Glasziou, K. T., and Gayler, K. R. (1972). Storage of sugars in stalks of sugar cane. *Botanical Review* **38**, 471–490.

Good, N. E., and Bell, D. H. (1980). Photosynthesis, plant productivity, and crop yield. *In* "The Biology of Crop Productivity" (P. S. Carlson, ed.), pp. 3–51. Academic Press, New York.

Gunning, B. E. S. (1976). The role of plasmodesmata in short distance transport to and from the phloem. *In* "Intercellular Communication in Plants: Studies on Plasmodesmata" (B. E. S. Gunning and A. W. Robards, eds.), pp. 203–227. Springer, Berlin.

Guy, M., Reinhold, L., and Michaeli, D. (1979). Direct evidence for a sugar transport mechanism in isolated vacuoles. *Plant Physiology* **64**, 61–64.

Hampson, S. E., Loomis, R. S., and Rains, D. W. (1978). Regulation of sugar uptake in hypocotyls of cotton. *Plant Physiology* **62**, 851–855.

Heber, U., and Heldt, H. W. (1981). The chloroplast envelope: Structure, function and role in leaf metabolism. *Annual Review of Plant Physiology* **32**, 139–168.

Ho, L. C. (1979). Partitioning of ^{14}C assimilate within individual tomato leaves in relation to the rate of export. *In* "Photosynthesis and Plant Development" (R. Marcelle, H. Clijsters and H. van Poucke, eds.), pp. 243–251. Junk, The Hague.

Kemp, D. R., and Whingwiri, E. E. (1980). Effect of tiller removal and shading on spikelet development and yield components of the main shoot of wheat and on the sugar concentration of the ear and flag leaf. *Australian Journal of Plant Physiology* **7**, 501–510.

Komor, E. (1977). Sucrose uptake by cotyledons of *Ricinus communis* L. : Characteristics, mechanism and regulation. *Planta* **137**, 119–131.

Komor, E., Thom, M., and Maretski, A. (1981). The mechanism of sugar uptake by sugarcane suspension cells. *Planta* **153**, 181–190.

Lichtner, F. T., and Spanswick, R. M. (1981). Electrogenic sucrose transport in developing soybean cotyledons. *Plant Physiology* **67**, 869–874.

Mares, D. J., and Marschner, H. (1980). Assimilate conversion in potato tubers in relation to starch deposition and cell growth. *Berichte der Deutschen Botanischen Gesellschaft* **93**, 299–313.

Milthorpe, F. L., and Moorby, J. (1969). Vascular transport and its significance in plant growth. *Annual Review of Plant Physiology* **20**, 117–138.

Moorby, J. (1964). The foliar uptake and translocation of caesium. *Journal of Experimental Botany* **15**, 457–469.

Moorby, J. (1977). Integration and regulation of translocation within the whole plant. *In* "Integration of Activity in the Higher Plant" (D. H. Jennings, ed.), pp. 425–454. Cambridge University Press, Cambridge.

Mor, Y., Halevy, A. H., and Porath, D. (1980). Characterization of the light reaction in promoting the mobilizing ability of rose shoot tips. *Plant Physiology* **66**, 996–1000.

Patrick, J. W. (1972). Distribution of assimilate during stem elongation in wheat. *Australian Journal of Biological Science* **25**, 455–467.

Patrick, J. W. (1981). An *in vitro* assay of sucrose uptake by developing bean cotyledons. *Australian Journal of Plant Physiology* **8**, 221–235.

Patrick, J. W., and McDonald, R. (1980). Pathway of carbon transport within developing ovules of *Phaseolus vulgaris* L. *Australian Journal of Plant Physiology* **7**, 671–684.

Patrick, J. W., and Turvey, P. M. (1981). The pathway of radial transfer of photosynthate in decapitated stems of *Phaseolus vulgaris* L. *Annals of Botany* **47**, 611–621.

Patrick, J., and Wareing, P. F. (1980). Hormonal control of assimilate movement and distribution. *In* "Aspects and Prospects of Plant Growth Regulators," (B. Jeffcoat, ed.), pp. 65–84. Joint British Crop Protection Council and British Plant Growth Regulator Group. Monograph 6.

Pitman, M. G. (1977). Ion transport into the xylem. *Annual Review of Plant Physiology* **28**, 71–88.

Pontis, H. G. (1977). Riddle of sucrose. *International Review of Biochemistry* **13**, 79–117.

Potter, J. R., and Jones, J. W. (1977). Leaf area partitioning as an important factor in growth. *Plant Physiology* **59**, 10–14.

Preiss, J., and Levi, C. (1979). Metabolism of starch in leaves. *In* "Photosynthetic Carbon Metabolism and Related Processes" Encyclopedia of Plant Physiology, New Series 6 (M. Gibbs and E. Latzko, eds.), pp. 282–312. Springer-Verlag, Berlin.

Saftner, R. A., and Wyse, R. E. (1980). Alkali cation/sucrose co-transport in the root sink of sugar beet. *Plant Physiology* **66**, 884–889.

Shiroya, M. (1968). Comparison of upward and downward translocation of ^{14}C from a single leaf of sunflower. *Plant Physiology* **43**, 1605–1610.

Silvius, J. E., Chatterton, N. Y., and Kremer, D. F. (1979). Photosynthate partitioning in soybean leaves at two irradiance levels. Comparative response of acclimated and unacclimated leaves. *Plant Physiology* **64**, 872–875.

Singh, B. K., and Jenner, C. F. (1982). Association between concentrations of organic nutrients in the grain, endosperm cell number and grain dry weight within the ear of wheat. *Australian Journal of Plant Physiology* **9**, 83–96.

Smith, J. A., and Milburn, J. A. (1980). Phloem turgor and the regulation of sucrose loading in *Ricinus communis* L. *Planta* **148**, 42–48.

Thornley, J. H. M., Gifford, R. M., and Bremner, P. M. (1981). The wheat spikelet—growth response to light and temperature: Experiment and hypothesis. *Annals of Botany* **47**, 713–725.

Wardlaw, I. F. (1980). Translocation and source–sink relationships. *In* "The Biology of Crop Productivity" (P. S. Carlson, ed.), pp. 297–339. Academic Press, New York.

Wareing, P. F. (1977). Hormonal regulation of assimilate movement. *In* "Opportunities for Chemical Plant Growth Regulation" pp. 481–487. Joint British Council and British Plant Growth Regulator Group. Monograph 4.

Wareing, P. F., and Patrick, J. (1975). Source–sink relations and the partition of assimilates in the plant. *In* "Photosynthesis and Productivity in Different Environments" (J. P. Cooper, ed.), pp. 481–499, Cambridge University Press, Cambridge.

Warren-Wilson, J. (1972). Control of crop processes. *In* "Crop Processes in Controlled Environments" (A. R. Rees, K. E. Cockshull, D. W. Hand and R. G. Hurd, eds.), pp. 7–30. Academic Press, London.

Wolswinkel, P. (1978). Phloem unloading in stem parts parasitized by *Cascuta:* The release of ^{14}C and K^+ to the free space at 0°C and 25°C. *Physiologia Plantarum* **42**, 167–172.

CHAPTER 6

Nitrogen Uptake

C. J. PEARSON and W. A. MUIRHEAD

I. INTRODUCTION

Mineral imbalance or deficiency constrains carbon productivity. For example, nitrogen amounts to 0.5% to 5% of crop dry weight and its deficiency has been estimated to reduce potential annual productivity by 60% in an Australian barley grass–subterranean clover pasture (Cocks, 1980). In this chapter we consider briefly nitrogen uptake and distribution, emphasizing the interactions between nitrogen and carbon, and we discuss the management of nitrogen availability in order to optimize crop productivity.

II. NITROGEN UPTAKE

Most non-legumes, except those that grow in reduced environments (e.g. rice), use nitrate as their principal source of nitrogen (N). In this section we will confine our discussion to nitrate.

CONTROL OF CROP PRODUCTIVITY
ISBN 0 12 548280 9

The rate of nitrate uptake, the time it stays in the root and the site of its reduction vary with species, time of day, environment and plant age. There is not space here to discuss these variations; we have suggested elsewhere that they are adaptations to, or at least act together with, changes in the whole-plant carbon (C) economy (Theodorides and Pearson, 1982). It is tempting to link this view—that the plant modifies its N metabolism to optimize its C economy—with other recent proposals of optimization strategies in plant growth. Some of these have been discussed by Cowan in Chapter 2 whereas Fowkes and Landsberg (1981) have developed mathematical models of root-growth strategies designed to optimize water uptake. The Fowkes and Landsberg analysis is pertinent because models of water uptake approximate those of N flux where there is a large flow of nitrate in the soil–plant-transpiration stream. Their analysis suggests that, initially (when each root acts as an isolated absorber) for the same amount of C being used for root growth, a root system consisting of many fine roots will have higher flow conductance than a system of few long roots; later, when there is interference between roots for uptake of water (or N), the low order (few long roots) system is preferable (Fig. 6.1). Considered in these terms, the pattern of root growth of most plants appears intuitively to optimize the allocation of C for maximum water uptake and mass flow of N because main roots and laterals together combine the time and space advantages of each system (Fig. 6.1).

The rate of nitrate uptake, U, may be defined as the product of efficiency of uptake, $R\alpha$, root length, L, and ion concentration at the root–environment interface, C_r (e.g. Milthorpe and Moorby, 1979)

$$U = 2\pi\, R\alpha\, L\, C_r \qquad (6.1)$$

where R is the mean radius of the roots and α a transfer coefficient. Values of $R\alpha$ calculated from nitrate uptake and concentrations in soil, lie within those found for plants growing in solution culture (Bhat et al., 1979). This is significant because current models of nitrate uptake rely heavily on experiments with plants growing in solution culture at nitrate concentrations which are usually much higher than those found in the field.

A. Influx and Efflux

One qualitative view or schema of nitrate assimilation in non-legumes is given by Jackson (1978), based on Hodges (1973), Smith (1973) and others. Within this schema nitrate uptake occurs as a result of active influx and concurrent efflux back to the external medium. Nitrate is either stored within the root or loaded as nitrate into the xylem, or reduced in the root and stored, or reduced in the root and loaded as reduced N into the phloem and xylem

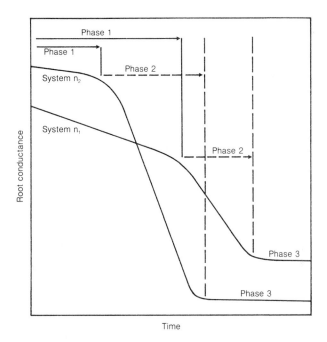

Fig. 6.1 Root conductance changes in two hypothetical root systems derived from the same amount of carbohydrate. The system with many roots (n_2) initially has higher conductance but its conductance declines rapidly when there is interference between roots so that ultimately a system with few long roots (n_1) is advantageous (after Fowkes and Landsberg, 1981).

for long-distance transport to other parts of plant. Influx is driven by a pH gradient across the plasmalemma. This gradient is probably maintained by (1) production of free H^+ ions by an adenosinetriphosphatase (ATPase), (2) decarboxylation of organic acids and (3) N reduction within the cytoplasm (Jackson, 1978, and references therein). Thus, rate of nitrate influx is probably determined by three concurrent processes, each intimately linked to overall C metabolism. In contrast, nitrate efflux is thought to be a leakage from the cytoplasmic nitrate pool (Morgan *et al.*, 1973).

Influx and efflux, and hence net uptake, vary dramatically with time of day (Fig. 6.2) (Pearson and Steer, 1979; Theodorides and Pearson, 1982). Some of this variation may be due to feedback from C metabolism: the availability of C determines not only the provision of energy to facilitate uptake and reduction but also the provision of skeletons for amino acid synthesis. Decapitation or girdling to stop C supply to the root restricts nitrate uptake within a few hours (Koster, 1963; Morgan *et al.*, 1972; Bowling *et al.*, 1978; Lee, 1978; Breteler *et al.*, 1979); that is, within the time taken

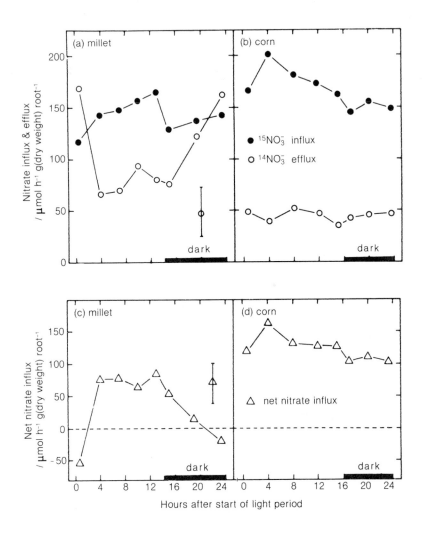

Fig. 6.2 Hourly rates of ^{15}N-nitrate influx and ^{14}N-nitrate efflux in pearl millet (a) and corn (b) and net nitrate uptake (c,d) during a 24 h cycle (from Pearson *et al.*, 1981).

to detect major changes in root–cell transmembrane potentials (Graham and Bowling, 1977). Decapitation causes efflux to fall only after about 12 h (Minotti and Jackson, 1970). Likewise, where C metabolism is curtailed by CO_2-free air, influx is severely retarded, but efflux is less affected (Aslim *et al.*, 1979). Darkness in a normal diurnal cycle reduces influx, and it may or may not stimulate efflux (as in pearl millet but not in maize; Fig. 6.2). Prolonged darkness causes influx to be halved (Table 6.1).

Table 6.1

$^{15}NO_3$ influx and $^{14}NO_3$ efflux into pearl millet seedlings in normal 14/10 h light/dark cycle and after 36 h darkness

	Normal	36 h dark	SE
$^{15}NO_3$ influx μmol g(DW) root^{-1}h^{-1}	258.7	107.6	7.3
% of total ^{15}N in shoot[a]	46.5	41.1	5.1
$^{15}NO_3$ concentration μmol g(DW) organ^{-1}[a]			
shoot	19.8	8.6	1.2
root	142.5	63.3	2.3
$^{14}NO_3$ efflux			
μmol g(DW) root h^{-1}	110.7	83.3	12.0
net NO$_3$ uptake			
μmol g(DW) root h^{-1}	148	24.3	

[a] ^{15}N distribution and concentrations refer to 3 h after seedlings were exposed to $^{15}NO_3$.

The second physiological control operating on rates of influx and efflux is residence time of nitrate and perhaps its concentration in the cytoplasm. Influx is positively correlated with translocation out of the root (Pearson *et al.*, 1981); decreased transpiration increases the residence time for nitrate within the root, and this may stimulate its reduction (Rufty *et al.*, 1981). However, where energy, reductant or carbon skeletons are in short supply, as in darkness, reduction cannot siphon off the excess nitrate. In this situation, nitrate concentrations within the root will increase, and unless the nitrate can be compartmentalized, we might expect increased concentrations to stimulate efflux (as in pearl millet; in Fig. 6.2).

A third potential physiological control over nitrate uptake is the enzyme mediated reduction of nitrate. The modulation of nitrate reductase is complex (Beevers and Hageman, 1980). It is not understood *in vivo*, although it is partially controlled by phytochrome (Jones and Sheard, 1972; Duke and Duke, 1978) and a plethora of effectors (e.g. ATP, NADH) (Hewitt *et al.*, 1979). In leaves, both nitrate and nitrite reductase are spatially linked with the C_3 photosynthetic pathway, as these enzymes are found in mesophyll but not in bundle sheath cells (Moore and Black, 1979, and references therein). Jessup and Fowler (1977) argued that the pentose phosphate pathway provided most of the energy for nitrate reduction, whereas, according to Woo *et al.* (1980) and Beevers and Hageman (1980), reducing equivalents and energy come from shuttles involving the TCA cycle. In roots, nitrate and nitrite reductase operate together throughout 24 h day/night cycles (e.g. Yoneyama *et al.*, 1980; Pearson *et al.*, 1981) and there is no specific linkage between nitrate reduction and C metabolism (e.g. Hanisch ten Cate and Breteler, 1981). Likewise, at the whole-plant or crop level of organization, it is clear that nitrate reductase alone does not control nitrate uptake. To cite one example, nitrate reductase may remain high when uptake declines after anthesis in dryland wheat (Simmons and Moss, 1978).

B. Distribution

The pathway of recently acquired N is not unidirectional from roots to tops but involves cycling of a variable proportion of total amino acids through the phloem to the roots (e.g. Theodorides and Pearson, 1982; Simpson et al., 1983).

Cycling of recently acquired N is overlaid by mobilization and translocation of previously sequestered N. This translocation, shown in Figure 6.3, is analogous to that of carbon redistribution (Chapter 5) and is particularly important during grain development. Translocation and N metabolism during grain development determine yield and grain protein, because from anthesis onwards the parameters of nitrate uptake (eqn 6.1) are declining or zero. By anthesis, wheat contains about 80% of the N which is present in its above-ground parts at maturity (Austin et al., 1977). There is considerable genetic variation within wheat in N uptake and translocation but only a weak correlation (r = 0.31) between the uptake of N during grain filling and the duration of the grain filling period (Austin et al., 1977). Furthermore, N uptake and translocation during grain filling are not closely linked with changes in dry weight (or at least, the linkage is not the same in all 47 genotypes examined by Austin et al., 1977). The relationship between post-anthesis N uptake and C metabolism is one which may be related to genetically influenced crop longevity, senescence and the speed of senescence, and changes in dry weight. For example, N uptake during grain filling was greatest in wheat genotypes that lost the least dry weight from stems, leaves and sheaths.

The lack of close linkage between N uptake and translocation and dry weight changes—and thus, the complexity of their underlying co-regulation—applies equally to the whole plant or to its developing grain. This is demonstrated elegantly in Figure 6.4 which shows N, C and water fluxes during the growth of fruit of white lupin. Here, 1 unit by weight of N, 12 of C and 600 of water flow into the fruit over its growth cycle. In Figure 6.4 these net inflows into the fruit are each expressed as 100 arbitrary units. Importation via the phloem contributes differing percentages of each: 89% of N inflow, virtually all carbohydrates but only 40% of water. As the seed ripens the pod senesces but again, according to the lupin data, there is no simple relationship between N and C fluxes during late grain development: most (80% or 16/20 arbitrary units) of N resident in the pod is retranslocated to the seed while a much higher proportion of C remains in the pod until death. Thirteen-fourteenths of the pod's water is lost as it senesces (Fig. 6.4).

Despite such complexity, the importance of N translocation late in the life of the crop and the concomitant decline in N acquisition and in leaf N concentration, led Sinclair and de Wit (1975, 1976) to propose that, in some species, senescence is associated with, and its rate is controlled by,

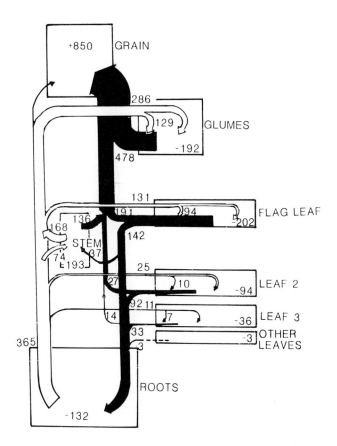

Fig. 6.3 A model of translocation and utilization of N in wheat plants for one day during the linear phase of grain filling (15 d after anthesis). Open arrows represent translocation of N in the xylem; black arrows represent translocation of N in the phloem. The widths of arrows are proportional to the amount of N translocated. Amounts of N are expressed in μg d^{-1} plant^{-1} close to the translocation channel to which they apply. Values within boxes representing plant parts that are not associated with a translocation channel are increments or decrements of N in that plant part on day 15 (from Simpson *et al.*, 1983).

remobilization of N from leaves to grain. This "self-destruct hypothesis" is currently popular. For example, Hageman *et al.* (1981) varied rates of senescence in maize by removing ears and concluded that senescence was related to the rate of decline in leaf nitrate reductase and chlorophyll. There are undoubtedly correlations between senescence and leaf N acquisition and translocation, but as with N uptake (Section IIA), it seems simplistic to believe that these correlations provide evidence of control of crop ontogony by a single enzyme within the N pathway.

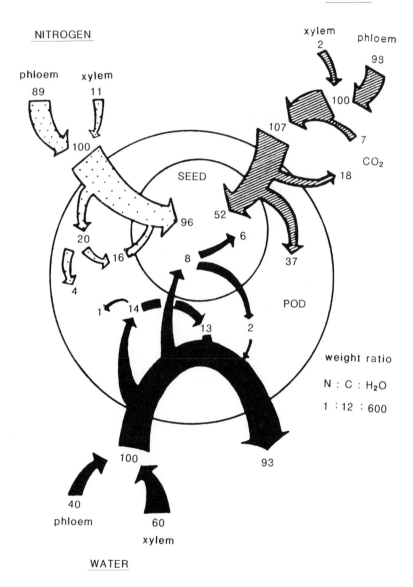

Fig. 6.4 Economy of nitrogen, carbon and water over the growth cycle of fruit of white lupin. Xylem and phloem deliveries are expressed relative to a net inflow of 100 units of a specific commodity. Net photosynthesis, respiration loss, transpiration, dehydration losses in ripening, and nitrogen mobilization from pod to seed are shown. The ratios of weight change of C, N and water were 12:1:600 (from Pate *et al.*, 1977).

III. NITROGEN-PRODUCTIVITY RELATIONS

A. Nitrogen and Dry Weight

The pattern of crop N and dry weight accumulation follows a sigmoid curve when environmental conditions are not limiting (Fig. 6.5a). Although much above-ground N may be taken up early in the life of the crop (e.g. as much as 70% of final N can be taken up before floral initiation in sunflowers; B. T. Steer, pers. comm., 1982), generally 80% of the increase in above-ground N and C occurs in the period from midway between emergence and anthesis to midway between anthesis and physiological maturity. The start of this period corresponds to tiller elongation following floral initiation in grasses and to floral initiation in sunflower (Fig. 6.5).

At this point $R\alpha$ is declining and root length is increasing, usually to reach a "gross" maximum at about anthesis. However, the "effective" or absorbing length probably reaches its maximum sometime earlier. External N concentrations decline throughout growth, unless replenished by abnormal mineralization or through fertilization (Fig. 6.5b). Nitrogen concentrations decline with crop age (Fig. 6.4a) because $R\alpha$ declines with age and N remobilization within the plant increases. The decline in concentration in leaf blades of rice was described by Murata (1975) as

$$\% \, (N) = \% \, (N_0) + e^{-Kt} \tag{6.2}$$

where % N is the percentage of N (on a dry weight basis) early in the season and K is the slope of $\ln[\%(N)]$ versus time.

Per cent of leaf N multiplied by intercepted radiation accounted for over 90% of seasonal and site variation in net assimilation rates of rice, maize and soybean in the Japanese International Biological Program (Murata, 1975).

Such correlations make it tempting, and useful predictively, to relate gain in dry weight to an N factor. For example (Seligman *et al.*, 1975; Seligman and van Keulen, 1981)

$$G = E \times WU \times D \tag{6.3}$$

where G is daily dry matter production (kg ha^{-1} d^{-1}), E is evaporation (mm d^{-1}), WU is water use efficiency (kg ha^{-1} mm^{-1}) and D is a dimensionless reduction factor for N limitation. The reduction factor relates to leaf N concentration, as in the Japanese work, except that Seligman and his colleagues introduced the idea of % N varying between a biologically permissible maximum M and minimum T concentrations (T being defined as the threshold below which there is no growth). Thus:

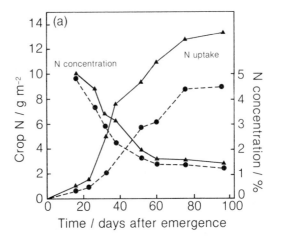

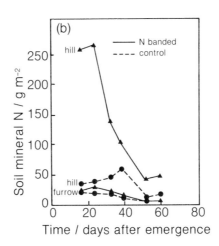

Fig. 6.5 (a) Nitrogen uptake by sunflowers growing at Griffith, NSW, Australia, under conditions of native mineral nitrogen and 100 kg (N) ha⁻¹ placed as an anhydrous ammonia band 20 cm below the top of hills spaced 90 cm apart; (b) changes in mineral nitrogen to 30 cm depth below the hill in the vicinity of the band (10 cm² sampling circle) and the furrow.

$$D = 1 - \left\{ 1 - [(N - T)/(M - T)]^2 \right\}^{0.5} \qquad (6.4)$$

In eqn 6.4 N, T and M are expressed as kg (N) per kg dry matter; representative values for M for temperate cereals are 0.08 to 0.01 as the crop progresses from emergence to maturity, and for T, 0.02 to 0.005.

This simple analysis of the correlation between dry weight gain and N concentration appears to have general applicability. For example, when the differing growth responses of a C_3 and C_4 grass to N starvation are expressed as fractions of a maximum net assimilation rate, the responses are essentially the same, the rates are halved at slightly above 1% leaf N (Wilson, 1975). Moreover, by using different values for N, T and M, the analysis may readily be extended to responses of physiological processes (e.g. leaf expansion) to N.

B. Influence of Fertilizer Management on Crop Productivity

When N is between T and M (eqn 6.4) there is a linear relationship between growth rate of a crop and the relative nitrogen uptake (e.g. Theodorides and Pearson, 1981). Nitrogen fertilization will affect the rates of both these processes until further addition of N will increase the external (soil) concentration of N but reduce crop growth (Ingestad and Lund, 1979; Ingestad, 1983). A constant ratio of growth rate to relative N uptake does not imply that N concentrations remain constant in the plant. These decline with crop age (as mentioned in Section III.A). The initial concentration and the relative rate of decline are higher under high fertilization than under low fertility (e.g. Saiette and Lemaire, 1981).

Ultimately, however, the response of crop grain yield to fertilizer N depends as much, or more, on N redistribution within the plant as on uptake during the "grand period" of sigmoid growth. A comparison of effects of fertilizer from limiting to luxury N levels on three irrigated crops is shown in Figure 6.6. Wheat, with a stable harvest index (HI) of around 0.5, partitioned most dry matter to grain under low N, presumably because semi-dwarf wheats have been selected for high HI. In contrast, sunflower cultivars vary in both their HI and their response to N fertilizer (Muirhead et al., 1982). The relatively low HI of sunflower in Figure 6.6b was associated with the highest N concentrations in above-ground parts at physiological maturity (Fig. 6.6c) and the lowest translocation of N to grain (Fig. 6.6d). The apparent recovery of fertilizer N, defined as (crop uptake−control crop uptake)/amount applied, decreased from 70% for wheat grown with 50 kg (N) ha^{-1} to 35% for all crops receiving 150 kg (N) ha^{-1} and decreased further to 25% for maize receiving 300 kg (N) ha^{-1}. As only part of the crop N is in the grain, the apparent recovery in the harvested portion of each crop would be lower than these values.

Management, through varying the type, amount, position and time of fertilizer application, can alter the responses shown in Figure 6.5. Fertilizer is most commonly applied as urea or anhydrous ammonia banded below the seed at, or just before, sowing. Synchronizing nitrification of the fertilizer

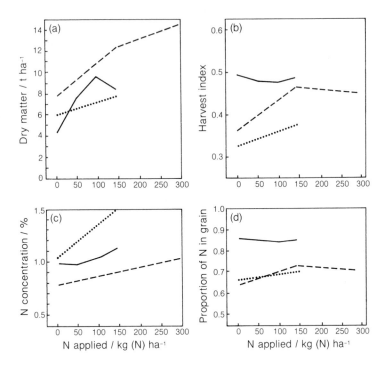

Fig. 6.6 Effect of increasing rates of nitrogen fertilizer on (a) dry matter production (b) harvest index (c) nitrogen concentration in above-ground parts and (d) proportion of N in the grain, for wheat (–), maize (– – –) and sunflower (. . .) crops at various sites at Griffith, NSW, Australia.

with the physiological requirement of the crop, and minimizing of N losses (through reducing ammonia volatilization, immobilization, leaching and denitrification), should improve crop N utilization. Management techniques aimed at this have met with some success: band spacing and granule size can control the rate of nitrate formation from fertilizers producing ammonium ions (e.g. Wetselaar *et al.*, 1972). Other means are by coating granules with a diffusion barrier (e.g. sulphur-coated urea), using urea-aldehydes that are sparingly soluble (Oertli, 1980), or by using inhibitors to delay the enzymatic hydrolysis of urea to ammonium and the microbial conversion of ammonium to nitrate. All methods, except fertilizer banding and the use of large granules, significantly increase the cost of the fertilizer. These methods will be more attractive as the cost of fertilizer increases, and the techniques for programming N supply to match the physiological needs of the crop improve.

Split application of fertilizer N during crop growth improves the efficiency of N use under irrigation. Split applications on rice generally increase the HI and reduce excessive vegetative growth and lodging (de Datta, 1981). Response to split N applications is less clear-cut in upland crops; generally no yield advantage occurs where the crops are quick-maturing, or on sites with high N fertility, or where the yield achieved with N fertilizer is relatively low (e.g. Christensen and Killorn, 1981). If there is a benefit, it may be associated with less lodging and reduced incidence of leaf diseases (Ellen and Spiertz, 1980).

The application of N fertilizer during crop growth presents some difficulties. Surface application of ammonium-forming fertilizers is prone to loss through volatilization, and application below the soil surface can damage the root system. Applying the fertilizer in the irrigation water provides a carrier which can effectively place the fertilizer in the root zone. Urea, which is readily soluble in water, is an ideal fertilizer for this purpose as it is non-polar and is carried into the root zone with the water before undergoing hydrolysis to the ammonium ion. Irrigating with urea has produced maize yields up to 27% higher than those achieved with anhydrous ammonium banded into the soil before sowing (Muirhead et al., 1980). Although banded fertilizer produced higher N concentrations in the crop during early growth, the rate of N uptake in the two weeks before tasselling was higher for the irrigation with urea treatment than for the banded treatment (Muirhead, W. A., unpublished). This may result in a higher HI and increased apparent recovery of N.

IV. CONCLUSIONS

Biochemists and physiologists can qualitatively describe nitrate uptake and distribution within the plant. We lack knowledge, however, of both the qualitative and quantitative control of N–C interactions, of how nitrate is compartmentalized, and how its flow is regulated in situ. For example, it is exciting to realise that, despite a great deal of experimental data, we don't understand the physiological basis of why N fertilization commonly has little effect on N concentrations within the plant but increases the absolute amounts of both crop N and dry matter. Hopefully more widespread use of ^{15}N will provide answers to problems such as these, and provide a physiological basis for fertilizer management to optimize crop productivity.

REFERENCES

Aslim, M., Huffaker, R. C., Rains, D. W., and Rao, K. P. (1979). Influence of light and ambient carbon dioxide concentration on nitrate assimilation by intact barley seedlings. *Plant Physiology* **63**, 1205–1209.

Austin, R. B., Forde, M. A., Edrich, J. A., and Blackwell, R. D. (1977). The nitrogen economy of winter wheat. *Journal of Agricultural Science* **88**, 159–167.

Beevers, L., and Hageman, R. H. (1980). Nitrate and nitrite reduction. *In* "Biochemistry of Plants Vol. 5 Amino Acids and Derivations" (B. J. Miflin, ed.), pp. 116–168. Academic Press, New York.

Bhat, K. K. S., Brereton, A. J., and Nye, P. H. (1979). The possibility of predicting solute uptake and plant growth response from independently measured soil and plant characteristics VIII. *Plant and Soil* **53**, 193–201.

Bowling, D. J. F., Graham, R. D., and Dunlop, J. (1978). The relationship between cell electrical potential difference and salt uptake in the roots of *Helianthus annuus*. *Journal of Experimental Botany* **29**, 135–140.

Breteler, H., Hanische Ten Cate, C., and Nissen, P. (1979). Time course of nitrate uptake and nitrate reductase activity in nitrogen-depleted dwarf bean. *Physiologia Plantarum* **47**, 49–55.

Christensen, N. W., and Killorn, R. J. (1981). Wheat and barley growth and N fertilizer utilization under sprinkler irrigation. *Agronomy Journal* **73**, 307–312.

Cocks, P. S. (1980). Limitations imposed by nitrogen deficiency on the productivity of subterranean-clover based annual pasture in southern Australia. *Australian Journal of Agricultural Research* **31**, 95–107.

de Datta, S. K. (1981). "Principles and Practices of Rice Production". Wiley, New York.

Duke, S. H., and Duke, S. O. (1978). *In vitro* nitrate reductase activity and *in vivo* phytochrome measurements of maize seedlings as affected by various light treatments. *Plant and Cell Physiology* **19**, 481–489.

Ellen, J., and Spiertz, J. H. T. (1980). Effects of rate and timing of nitrogen dressings on grain yield formation of winter wheat (*T. aestivum* L.) *Fertiliser Research* **1**, 177–190.

Fowkes, N. D., and Landsberg, J. J. (1981). Optimal root systems in terms of water uptake and movement. *In* "Mathematics and Plant Physiology" (D. A. Rose and D. A. Charles-Edwards, eds.), pp. 109–128. Academic Press, London.

Graham, R. D., and Bowling, D. J. F. (1977). Effect of the shoot on the transmembrane potentials of root cortical cells of sunflower. *Journal of Experimental Botany* **28**, 886–893.

Hageman, R. H., Below, F. E., and Christensen, L. E. (1981). Nitrate, nitrate reductase and reduced nitrogen in leaves and stalks of crop plants during seed development. XIII International Botanical Congress: Sydney. Abstracts p. 71.

Hanisch ten Cate, C. H., and Breteler, H. (1981). Role of sugars in nitrate utilization by roots of dwarf bean. *Physiologia Plantarum* **52**, 129–135.

Hewett, E. J., Huckleaby, D. P., Plann, A. F., Notton, B. A., and Rucklidge, G. J. (1979). Regulation of nitrate assimilation in plants. *In* "Nitrogen Assimilation of Plants" (E. J. Hewett and C. V. Cutting, eds.), pp. 255–287. Academic Press, London.

Hodges, T. K. (1973). Ion absorption by plant roots. *Advances in Agronomy* **25**, 163–207.

Ingestad, T. (In press). Nutrient addition rate and external concentration; driving variables used in plant nutrition research. *Plant, Cell and Environment*.

Ingestad, T. and Lund, A. B. (1979). Nitrogen stress in birch seedlings 1. Growth technique and growth. *Physiologia Plantarum* **45**, 137–148.

Jackson, W. A. (1978). Nitrate acquisition and assimilation by higher plants: Processes in root system. *In* "Nitrogen in the Environment" Vol. 2 (D. R. Nielsen and J. G. MacDonald, eds.), pp. 45–86. Academic Press, New York.

Jessup, W., and Fowler, M. W. (1977). Interrelations between carbohydrate metabolism and nitrogen assimilation in cultured plant cells. *Planta* **137**, 71–76.

Jones, R. W., and Sheard, R. W. (1972). Nitrate reductase activity: Phytochrome mediation of induction in etiolated peas. *Nature New Biology* **238**, 221–222.

Koster, A. L. (1963). Changes in metabolism of isolated root systems of soybean. *Nature* **198**, 709–710.

Lee, R. B. (1978). Inorganic nitrogen metabolism in barley roots under poorly aerated conditions. *Journal of Experimental Botany* **29**, 693–708.

Milthorpe, F. L., and Moorby, J. (1979). "An Introduction to Crop Physiology" 2nd. edn. Cambridge University Press, Cambridge.

Minotti, P. L., and Jackson, W. A. (1970). Nitrate reduction in the roots and shoots of wheat and shoots of wheat seedlings. *Planta* **95**, 36–44.

Moore, R., and Black, C. C. (1979). Nitrogen assimilation pathways in leaf mesophyll and bundle sheath cells of C_4 photosynthesis plants formulated from comparative studies with *Digitaria sanguinalis* (L) Scop. *Plant Physiology* **64**, 309–313.

Morgan, M. A., Volk, R. J., and Jackson, W. A. (1972). Nitrate absorption and assimilation in ryegrass as influenced by calcium and magnesium. *Plant Physiology* **50**, 485–490.

Morgan, M. A., Volk, R. J., and Jackson, W. A. (1973). Simultaneous influx and efflux of nitrate during uptake by perennial ryegrass. *Plant Physiology* **51**, 267–272.

Muirhead, W. A., Low, A., and White, R. J. G. (1982). The response of irrigated sunflower cultivars to nitrogen fertilizer. "Proceedings 10th International Sunflower Conference, Surfers Paradise, Australia, March 14–18, 1982" pp. 82–85.

Muirhead, W. A., White, R. J. G., and Lockhart, J. T. (1980). Management of nitrogen fertilizers in maize and sunflower. CSIRO Division of Irrigation Research. Research Report 1978–80. pp. 71–72.

Murata, Y. (1975). "Japanese International Biological Programme Synthesis: Crop Productivity and Solar Energy Utilization in Various Climates in Japan." Vol. 2. University of Tokyo Press, Tokyo.

Oertli, J. J. (1980). Controlled-release fertilizers. *Fertiliser Research* **1**, 103–123.

Pate, J. S., Sharkey, P. J., and Atkins, C. A. (1977). Nutrition of a developing legume fruit. Functional economy in terms of carbon, nitrogen, water. *Plant Physiology* **59**, 506–510.

Pearson, C. J., and Steer, B. T. (1979). Daily changes in nitrate uptake and metabolism in *Capsicum annuum. Planta* **137**, 102–123.

Pearson, C. J., Volk, R. J., and Jackson, W. A. (1981). Daily changes in nitrate influx, efflux and metabolism in maize and pearl millet. *Planta* **152**, 139–142.

Rufty, T. W., Jackson, W. A., and Raper, C. D. (1981). Nitrate reduction in roots as affected by the presence of potassium and by flux of nitrate through the roots. *Plant Physiology* **68**, 605–609.

Salette, J., and Lemaire, G. (1981). Sur la variation de la teneur en azote des Gramin'ees fourragères pendant leur croissance: formulation d'ure loi de dilution. *C. R. Academé de Science, Paris, Série III* **292**, 875–887.

Seligman. N. G., and van Keulen, H. (1981). Prapan: a simulation model of annual pasture production limited by rainfall and nitrogen. *In* "Simulation of Nitrogen Behaviour in Soil-Plant Systems" (M. J. Frissel and J. A. van Veen, eds.), pp. 192–222. PUDOC, Wageningen.

Seligman, N. G., van Keulen, H., and Gourdriaan, J. (1975). An elementary model of nitrogen uptake and redistribution by annual plant species. *Oecologia* **21**, 343–361.

Simpson, R. J., Lambers, H., and Dalling, M. J. (1983). Nitrogen redistribution during grain growth in wheat (*Triticum aestivum L.*) IV. Development of a quantitative model of the translocation of nitrogen to the grain. *Plant Physiology* **71**, 7–14.

Simmons, S. R., and Moss, D. N. (1978). Nitrate reductase as a factor affecting N assimilation during the grain filling period in spring wheat. *Crop Science* **18**, 584-586.

Sinclair, T. R., and de Wit, C. T. (1975). Photosynthate and nitrogen requirements for seed production by various crops. *Science* **189**, 565-567.

Sinclair, T. R., and de Wit, C. T. (1976). Analysis of the carbon and nitrogen limitations of soybean yield. *Agronomy Journal* **68**, 319-324.

Smith, F. A. (1973). The internal control of nitrate uptake into barley roots with differing salt content. *New Phytologist* **72**, 769-782.

Theodorides, T. N., and Pearson, C. J. (1981). Effect of temperature on total nitrogen distribution in *Pennisetum americanum*. *Australian Journal of Plant Physiology* **8**, 201-210.

Theodorides, T. N., and Pearson, C. J. (1982). Effect of temperature on nitrate uptake, translocation and metabolism in *Pennisetum americanum*. *Australian Journal of Plant Physiology* **9**, 309-320.

Wetselaar, R., Passioura, J. B., and Singh, B. R. (1972). Consequences of banding nitrogen fertilizers in the soil. I. Effects on nitrification. *Plant and Soil* **36**, 159-175.

Wilson, J. R. (1975). Comparative response to nitrogen deficiency of a tropical and temperate grass in the interrelation between photosynthesis, growth and the accumulation of nonstructural carbohydrate. *Netherlands Journal of Agricultural Science* **23**, 104-112.

Woo, K. C., Jokinen, M., and Carvin, D. T. (1980). Reduction of nitrate via dicarboxylate shuttle in a reconstituted system of supernatant and mitochondria from spinach leaves. *Plant Physiology* **65**, 433-436.

Yoneyama, T., Iwata, E., and Yazaki, T. (1980). Nitrate utilization in the roots of higher plants. *Soil Science and Plant Nutrition* **26**, 9-23.

Part II

Control of Productivity in Specific Crops

CHAPTER 7

Water Deficit and Wheat

D. ASPINALL

I. INTRODUCTION

In many climates water is the major constraint on productivity of temperate cereals. Economic responses to irrigation can frequently be obtained in an apparently wet environment such as Britain (French and Legg, 1979). Most of the wheat produced in the Indian sub-continent depends upon irrigation. In dryland farming areas in Australia and the USA, extreme fluctuations in crop yield can occur as the annual rainfall varies.

It is possible to measure variation in plant growth, development and, ultimately, yield in response to changes in the environment or to other factors through analysis of the gain in carbon through photosynthesis and its distribution. This analysis is the basis of most contemporary interpretations of data on yield production in cereals and is the foundation of most crop models (see Chapter 18).

Such analyses are descriptions of crop response and not necessarily an exposition of the underlying physiological mechanisms. Apart from the primary processes concerned in carbon acquisition a variety of plant responses may be involved in the complex reaction to decreased water supply. Each response may have consequences that can be described in terms

CONTROL OF CROP PRODUCTIVITY
ISBN 0 12 548280 9

of assimilation. For instance, a response that results in an inhibition of shoot apex growth, will limit the size of a potential sink for assimilate, alter carbon distribution and possibly decrease carbon acquisition.

Whether or not any of the responses of cereals to water deficit that result in reduced yield are caused by an inhibition of carbon assimilation is not yet resolved. Evidence is difficult to obtain and most experiments are open to alternative interpretations. For instance, Fischer and Turner (1978) described an experiment in which plant water deficit for three days, commencing two days before ear emergence, reduced wheat grain number in proportion to a simultaneous reduction in inflorescence dry weight. Such evidence can be used to support an argument that grain set is one manifestation of inflorescence growth that is limited by lack of assimilate. It can be interpreted equally validly as an effect of inflorescence water potential, which reduces both grain number and inflorescence dry weight despite an abundant supply of assimilate. If it could be established that growth is limited by assimilate supply in the absence of water deficit, a primary role for assimilate limitation of organ growth during water stress would be more easily accepted. The view that grain growth is limited by assimilates (a "source" limitation) has been strongly held, but evidence does not support this idea for either early inflorescence growth (Mohapatra et al., 1982) or grain filling (Jenner, 1982b). Moreover, the evidence suggests that the progressive increase in harvest index (grain weight as a proportion of total shoot weight) which accounts for the major part of yield improvement in modern times is incompatible with the concept of an assimilate limited system.

II. CROP DEVELOPMENT

Environment affects growth during each phase of development, but how the response is controlled may well differ between phases. Descriptions of the effects of water stress at different stages of development on yield production in wheat are scattered through the literature and conform, in general, to the responses of barley detailed by Aspinall et al. (1964). Such studies do not establish differences in sensitivity to water deficit between different developmental stages in the strict sense of differences in response to the same water potential, as it is virtually impossible to compare plants of different ages and sizes at an equivalent water potential. Such experiments can, however, identify periods of water stress that have specific effects on yield and those that may be particularly damaging to production. Most experiments have been carried out with plants grown in containers or in otherwise artificial conditions and it can be argued that it is unwise to extrapolate from such data to responses in the field.

The following sections will (1) outline the information available on the response of the wheat plant to water deficit at specific growth phases, (2) delineate the physiological mechanisms that control these responses, (3) examine the likely effect of stress during a growth phase on the final yield, and (4) attempt to compare the response in controlled experiments with responses in the field.

A. Early Vegetative Growth to Terminal Floret Formation

Wheat sown into a dry soil, or subjected to drought shortly after, will face difficulty in establishment. This may have an effect on yield if the stand that eventuates is too thin to allow effective compensation by increased tillering. Milthorpe (1950) recorded the increase in drought sensitivity, measured in terms of tissue survival, that rapidly ensued upon germination and root and shoot elongation. In some instances, cultural practices are modified to ensure that seeds germinate at a favourable water status, for instance, by deep sowing beyond the range of surface soil drying. Failure of establishment in a dry season may lead to re-sowing later after rain but usually results in a decreased yield because of the shorter growing season.

In the stages leading to terminal floret initiation, two processes occur that have direct bearing on final grain yield: tiller establishment and spikelet initiation. Cereals, including wheat, initiate more tiller buds in the axils of the leaves than become established as emerged tillers, even in favourable circumstances. Environmental modification of tiller number occurs mainly by affecting tiller bud elongation, not through affecting initiation. Tiller establishment is generally determined by inter-plant competition, particularly for light and nitrogen (Puckridge and Donald, 1967; Puckridge, 1968). Water stress during this phase of growth will limit tiller establishment, but in most circumstances this is of lesser importance than effects on tiller survival later in development when the evaporative demand of the established crop is much greater.

The well-watered crop will establish more tillers than survive to form a productive ear. Water deficit during ear formation, if followed by irrigation or rain during the succeeding phase of floret initiation and internode elongation can lead to the emergence of new tillers late in development. Clearly, unexpanded tiller buds can withstand a period of water deficit even though their elongation is inhibited. A cardinal stage in ear formation is double-ridge initiation when the floral organs are first initiated. Water deficit inhibits floral initiation in plants that respond to a single cycle of inductive photoperiod; for example, *Lolium temultentum* and *Pharbitis nil* (Aspinall and Husain, 1970; King and Evans, 1977). Although a similar inhibition may occur in a quantitative long-day plant such as wheat, it is difficult to demonstrate and the few reports of effects on water deficit on the time to ear emergence are best explained as effects on responses subsequent to floral initiation (Angus and Moncur, 1977).

After floral initiation, spikelets are initiated sequentially from the base of the dome of the apical meristem until the process is completed with the initiation of a terminal spikelet by the transformation of the apical dome. The apical meristem is remarkably resilient to water stress during this period, surviving at water potentials as low as -6 MPa which result in the death of the expanded leaves (Barlow *et al.*, 1977). The water potential of the apex falls in parallel with that of the leaves but the turgor of the apex is maintained by the accumulation of solutes such as sucrose and free amino acids (Munns *et al.*, 1979). This accumulation of solutes represents a real gain in mass; it is not derived from insoluble polymers already in the apex. Evidently, the translocation of solutes to the apex is maintained, even under conditions of severe water deficit where photosynthesis is presumably strongly inhibited (Kriedemann and Downton, 1981). Apex growth in these circumstances is unlikely to be substrate limited and it is more likely that the growth inhibition in the apical cells at low water potential results from disruption of protein synthesis (Barlow *et al.*, 1977).

Whether or not water stress during this period has a lasting effect on the number of spikelets produced, as has been suggested for barley (Husain and Aspinall, 1970) is less certain. Barley has an indeterminate axis with determinate branches on the apical meristem. Stress causes the premature termination of primordium initiation at the dome of the apex and its subsequent senescence. In contrast, the wheat shoot apex is determinate with indeterminate lateral branches. It would seem less likely that stress would induce the premature formation of the terminal spikelet of the wheat ear. More likely, the initiation of florets in the spikelets may be inhibited.

A further general consequence of water deficit during early development is that shoot growth and expansion is inhibited more than root growth (e.g. Connor, 1975). The shift in root–shoot dry weight ratio in favour of root growth has been interpreted as advantageous, with the change in the relationship between root and leaf surface area, or disadvantageous because of the diversion of a supposedly limited assimilate pool to "unproductive" root growth. Such arguments are more philosophical than scientific, however, and lead to little advance in our understanding of crop response.

A simple mechanistic reason for this apparent difference in sensitivity to water deficit of roots and shoots might be that when plants growing in soil are subjected to stress, the water potential of the plant tissues will decline from the roots to the leaves as a consequence of transpiration and the subsequent water flow from the roots along a water potential gradient. If we suppose that root and shoot growth are equally sensitive to tissue water potential, the roots will be at an advantage for the continuation of growth as stress develops. This may result in the apparent diversion of assimilate to support this continued root growth, as determined by the ability of the competing sinks to continue growth (Gifford and Evans, 1981). There is little evidence that such a diversion will *per se* affect the yield of the plant.

In general, it may be concluded that wheat is resistant to effects of water deficit during early development, and has the capacity to compensate later if the environment improves.

B. Terminal Floret Formation to Anthesis

This is a period of rapid growth of the developing ear, involving important morphogenetic events in the growing florets, elongation of the upper internodes on the shoot and is accompanied by the senescence of a proportion of tillers. This proportion is increased by water stress and may be a major determinant of yield. In a glasshouse experiment, where care was taken to simulate field conditions, the difference in yield between well-watered plants and those subjected to a water stress from terminal floret formation to the end of grain filling, was predominantly determined by the number of tillers that survived to form a fertile ear (well-watered 84% of maximum tiller number, stressed 69%; Parameswaran, *et al.*, 1981). In the field, Connor (1975) found that tiller density at maturity was the determinant of yield in Sherpa wheat grown in successive seasons, one with abundant water supply to earing, the other with a continuous water deficit throughout this period. In this case, however, no attempt was made to distinguish between tiller initiation and tiller survival. Fischer and Turner (1978) attribute the failure of tillers to survive, both in well-watered and stressed crops, to competition within the plant for a limited supply of assimilate. This is an attractive hypothesis as all shoot apices are growing rapidly at this stage as are the elongating internodes and these together presumably compete increasingly for assimilate. Moreover, tillers which fail exhibit death of the apical meristem before senescence of the expanded leaves (unpublished observations). Definitive evidence is not available, however, and the experiments of Kemp and Whingwiri (1980) suggest that the competition between shoots which occurs is not for a limited supply of assimilate.

A feature of wheat crops that have been subjected to severe drought during this period and grain filling is that the ears are generally small, with dead, bleached spikelets towards the extremities (Fig. 7.1). As the ear appears tapered towards both the tip and the base, with spikelets apparently arrested in development progressively earlier the further from the centre, it seems that much of this response occurs before anthesis and the start of grain filling.

CONDOR

Fig. 7.1 Ears of wheat cv. Condor obtained from a commercial crop grown at Port Wakefield, South Australia, in a year of severe drought that commenced before anthesis. Note the inhibited spikelet development towards the extremities of the ears and sterile, but apparently fully developed, spikelets closer to the central region.

Spikelet death can occur immediately after ear emergence if the stress is particularly severe (Morgan, 1971). The response is measurable at less severe levels of water stress, together with a reduction in the number of grains per fertile spikelet, but is not as important as ear number in determining yield (Table 7.1). Fischer and Turner (1978) attribute these responses to a reduced assimilate supply and competition between developing organs but, as with tiller survival, little critical evidence is available and the response merits further investigation.

Water deficit between terminal floret formation and anthesis may also lead to a loss in yield without any obvious effects on ear morphology. Udol'skaja (1936) was the first to report a severe reduction in wheat yield associated with water stress during meiosis in the sex organs. Her report was followed by a series of papers by Russian workers (summarized by Skazkin, 1961) who found that cereals were highly susceptible to drought from the appearance of stamens to fertilization. The observations of a reduction in grain set in wheat associated with water deficit in this period have been supported by workers elsewhere (Dubetz and Bole, 1973; Fischer, 1973; Morgan, 1980; Saini and Aspinall, 1981) and attributed largely to induced male sterility. For instance, Bingham (1966) reported that water deficit occurring

Table 7.1

Yield and components of grain yield of wheat (cv. Gabo) subjected to water stress. (From Parameswaran *et al.*, 1981)

	Water stress	No water stress	
Number ears/plant	2.2	2.7	[a]
Number fertile spikelets/ear	14.6	15.0	[a]
Number grains/fertile spikelet	2.1	2.4	[a]
Number grains/ear			
1. main culm	34.9	40.9	[a]
2. tillers	22.4	24.5	[a]
Wt/grain (mg)			
1. main culm	45	40	[a]
2. tillers	42	40	n.s.
Grain wt/plant (g)	2.7	3.3	[a]

[a]Means significantly different (P = 0.05).
n.s. Means not significantly different.

either during or a few days before meiosis caused a reduction in grain set which could be eliminated by hand-pollination with pollen from well-watered plants.

The effect of water deficit on male fertility has been examined in detail and more is now known about this response than about most of the reactions of the wheat ear to stress. When wheat plants (cv. Gabo) were subjected to single short episodes of stress, effectively lasting for five days, at consecutive intervals extending from immediately before meiosis in the developing pollen mother cells to just after anthesis, grain set in self-pollinated ears was decreased only by treatments that covered the period of pollen mother-cell meiosis and subsequent microspore release (Fig. 7.2). Neither stress before meiosis nor at anthesis caused any effect on grain set, in contrast to other reports (Bingham, 1966; Wardlaw, 1971; Brocklehurst *et al.*, 1978). These differences in response may have been as a result of differences in the level of water deficit between experiments, difficulties in synchronizing plant development or to differences between cultivars. For Gabo wheat, however, the response seems clearly confined to this relatively short period and may be of a more specific nature as differences in time of meiosis between spikelets would have widened the apparent period of susceptibility.

Bingham (1966) showed the effect is a result of pollen sterility only; stressed plants are as fertile as well-watered plants when hand pollinated with viable pollen (see also Chapter 8.2). The anthers on stressed plants are of two types: in some florets they are apparently normal whereas in other florets the anthers are considerably smaller, contain no viable pollen and do not dehisce (Fig. 7.3). Assessed by staining with tetrazolium salts, pollen

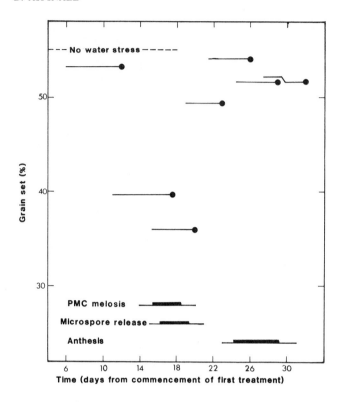

Fig. 7.2 Grain set of wheat ears (cv. Gabo) subjected to short periods of water deficit at different times during pollen development and anthesis (from Saini and Aspinall, 1981). Horizontal lines indicate when lower leaves were wilted, terminating at the time indicated (●) when the plants were re-watered. Plants remained well-watered thereafter. Key indicates when spikelets were at specific stages of development, the thicker lines indicating when most ears were at the specified stage. Note that the only two treatments to significantly reduce grain set were subjected to water deficit during pollen mother cell (PMC) meiosis and microspore release.

viability, even in apparently normal anthers on stressed plants is considerably reduced (Table 7.2). The combination of the two effects would have reduced the viable pollen count on stressed ears in this experiment to about a third of that on well-watered ears.

This difference in the response of individual anthers to water stress during the sensitive period was also evident when the development of anthers was followed microscopically (Saini, 1982). The anthers developed apparently normally up to the stage of the first pollen grain mitosis (Bennett *et al.*, 1973). In those most severely affected, the developing microspores lost contact with the tapetum, microspore nuclei stained unusually heavily and the cells of the anther filament began to stain heavily and show signs of degeneration (Fig.

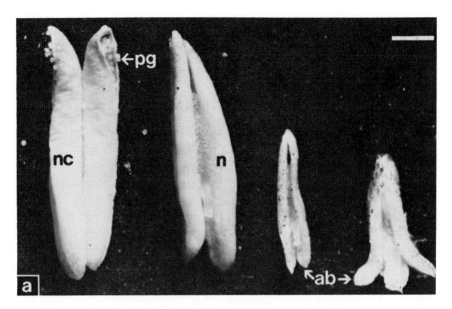

Fig. 7.3 Anthers from wheat (cv. Gabo) plants subjected to water deficit in the period of pollen mother cell meiosis and microspore release (from Saini and Aspinall, 1981). **nc**, normal anther from a well-watered plant; **n**, apparently normal anther from a stressed plant; **ab**, abnormal anther reduced in size, non-dehiscent and shrivelled; **pg**, pollen grain. Bar represents 0.5 mm.

7.4a). Although the tapetum subsequently degenerated in the normal manner, the pollen grains in these anthers remained shrivelled and accumulated little or no starch. The filament and the vascular bundle of the anther eventually degenerated and the mature anthers were small, containing only shrivelled pollen grains with little cytoplasm (Fig. 7.4b). Other anthers that were not as badly affected showed similar responses except that filament degeneration did not occur and at least some pollen grains accumulated starch. These anthers contained a mixture of normal and abnormal pollen grains at anthesis (Fig. 7.4c).

These responses in the anthers do not seem to have been evoked by a decline in the water status of the spike during water stress. Even though water potential of the flag leaf declined to –2.4 MPa, spike water potential remained unchanged at –0.3 MPa (Saini and Aspinall, 1981). Similarly, spikelet turgor was unaffected while leaf turgor fell (Morgan, 1980). This suggests that the source of the response must be sought outside the spike, although changes in anther water status may have been undetected.

One possibility is that pollen development was inhibited by the accumulation of an excessive concentration of abscisic acid in the spike during water stress; the abscisic acid synthesized in the stressed leaves and translocated

Table 7.2

The effect of water deficit on anther morphology and pollen viability in wheat (cv. Gabo). (From Saini and Aspinall, 1981)

	No water stress	Water stress
Florets containing abnormal anthers (%)	3.8	41.0[a]
Pollen viability in normal anthers (%)	91.1	57.8[a]

[a]Significantly different from values without water stress (P = 0.01)

to the spike. Certainly, abscisic acid applied to plants at meiosis causes pollen sterility (Morgan, 1980). The period of sensitivity to abscisic acid is similar to that of water stress except that abscisic acid applied three days before meiosis also causes sterility, an effect perhaps attributable to the persistence of abscisic acid in treated plants (Saini and Aspinall, 1982a). Moreover, the endogenous abscisic acid content of the spikes of water-stressed plants is comparable to the amount in the spikes of plants supplied with sufficient abscisic acid to cause a similar degree of pollen sterility. Abscisic acid has no effect on female fertility. In one apparent discrepancy from the effects of stress, the application of abscisic acid to anthers results in complete pollen sterility accompanied by anther abnormality or has no effect at all (Saini and Aspinall, 1982a). Some differences between the effects of stress and applied abscisic acid also appear in microscopic studies of the developing anthers (Saini, 1982). These apparent differences may be caused by differences in the distribution of abscisic acid within the plant and the effective concentrations reaching the pollen mother cells, but they do raise the possibility that abscisic acid is not the prime cause of sterility in stressed plants.

The significance of this potential cause of yield loss in field crops has not been carefully evaluated, but floret sterility is a common feature of droughted wheat crops (Fig. 7.1). The degree to which partial pollen sterility alone can cause failure of fertilization, when there is normally a gross excess of pollen produced, needs assessment. The effects of water deficit at this stage may be significantly increased if drought is accompanied by high temperature, a frequent coincidence in most environments in which wheat is grown. High temperature causes pollen sterility and failure of grain set independently of the effects of water deficit (Fig. 7.5). Moreover, high temperature also causes sterility in a proportion of the developing embryo sacs (Saini and Aspinall, 1982b; Saini, 1982). This latter response is potentially far more damaging than pollen sterility but, although it occurs at temperatures as low as 30°C in plants grown in a controlled environment, its significance in the field remains to be assessed.

In summary, water deficit, particularly severe deficit, between terminal floret initiation and anthesis has a number of serious effects on wheat yield,

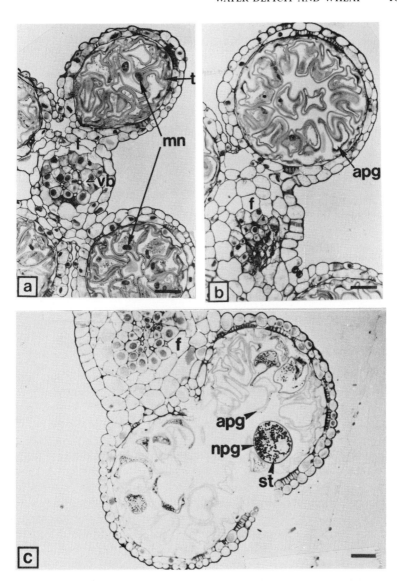

Fig. 7.4 The development of the anthers on wheat (cv. Gabo) plants subjected to water deficit during pollen mother cell meiosis and microspore release (from Saini, 1982); (a) early stage of degeneration in anthers that will be non-dehiscent and completely sterile. Note heavy staining in the microspore nuclei and cells of the vascular bundle; (b) late stage of degeneration in a similar anther. Note complete degeneration of vascular bundle and abnormal pollen at maturity; (c) transverse section of an apparently normal anther from a stressed plant. Note normal filament development but partial pollen sterility at maturity. **mn**, microspore nucleus; **t**, tapetum; **vb**, vascular bundle; **f**, filament; **apg**, abnormal pollen grain; **npg**, normal pollen grain; **st**, starch. Bar represents 50 μm.

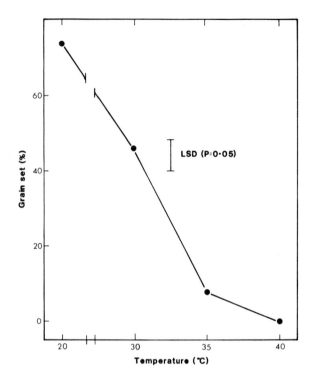

Fig. 7.5 The response of grain set in wheat (cv. Gabo) to 24 h exposure to high temperature during pollen mother cell meiosis and microspore release (from Saini and Aspinall, 1982a).

reducing ear numbers, spikelet numbers and fertility of surviving spikelets. As development proceeds the plant progressively loses the ability to compensate for these effects. The crop does not compensate for earlier tiller loss by renewed tillering at anthesis if rain falls, for example, as the few new tillers that are produced are generally small and produce an insignificant yield. Spikelet and floret loss also cannot be compensated for at this late stage and yield may be further reduced by floret sterility resulting from the effects on male gametogenesis. Bingham (1966) claims that this reduction may be partially compensated by grain set in more distal florets, but Saini and Aspinall (1981) found no such compensation.

Fischer and Turner (1978) suggested that the sensitivity to water deficit of potential grain production during this period has evolutionary significance. It is envisaged that the sensitivity evolved as a means to limit the numbers of grains set so as to maintain grain size despite the continuation of water deficit during the grain-filling period. Adequate grain size is viewed as important in ensuring establishment of the following generation, and is likely to have been selected by early agriculturalists. This is an attractive speculation, impossible to either deny or verify experimentally, but rests on the assumption that grains filling during water deficit will vary in size according to the number of grains set. This remains to be ascertained.

C. Anthesis to Maturity

There is some evidence that grain number may be reduced by water deficit occurring at or immediately after anthesis (Wardlaw, 1971; Brocklehurst et al., 1978; Morgan, 1971) although fertilization and grain set may not be as susceptible to stress as the earlier process of male gametogenesis.

This response apart, water deficit at this period will affect yield through the components of grain growth; that is, cell division in the endosperm, grain filling which principally concerns starch accumulation in the endosperm and the termination of grain growth. Variation in any of these as a result of water deficit may result in a reduction in grain size and a consequent reduction in grain yield.

Although so-called "pinched grain" and an increase in the screening percentage are features of commercial crops grown under drought, most experimental evidence suggests that variation in grain size is not as significant a determinant of grain yield during drought as is variation in grain number (e.g. Connor, 1975). Under some circumstances an increase in mean size may reflect shifts in the population sampled rather than actual changes in grain size.

The extent of cell division in the endosperm has been suggested as a major determinant of grain size, particularly in comparisons between genotypes (Brocklehurst, 1977; Gleadow et al., 1982). The number of endosperm cells increases during the first 15–20 days of grain growth (Briarty et al., 1979). During at least the first 10 days of this process, grain water status is responsive to plant water deficit (personal observation). The main component of the grain at this time is the succulent pericarp and the major part of any fluctuation in grain water may be in this tissue. However, changes in plant water status may potentially affect endosperm cell division directly.

During the linear phase of grain growth, expansion of endosperm cells is accompanied by deposition of starch and proteins within those cells (Jenner, 1982b). Although a careful comparison of the rates of grain growth at different levels of water deficit has not been carried out, it would appear that, relative to other plant processes, grain filling is resistant to stress. Indeed, in a recent experiment, the rate of grain dry matter increase in the first 29 days after anthesis was unaffected by water deficit despite a fall in flag leaf water potential to -2.2 MPa 17 days after anthesis and the subsequent premature senescence of those leaves (Brooks et al., 1982). This lack of response was also apparent in the accumulation of starch and the number of granules per endosperm being unaffected until the onset of the maturation phase.

The comparative resistance of grain growth to the effects of water stress can be construed as a useful evolutionary adaptation to habitats prone to a progressive development of drought during plant growth. In the wheat plant in the absence of water stress, most of the carbon accumulated in the enlarging grains is derived from current assimilation, much of the carbon fixation taking place in the leaves. This source of carbon is greatly reduced during stress, apart possibly from assimilation in the floral organs, particularly the awns (Evans *et al.*, 1972). Grain growth is maintained, however, with mobilization of previously fixed carbon from the stems and other plant organs becoming of greater significance (Passioura, 1976).

The ability of grains to continue growth during episodes of water deficit which severely affect other functions suggests either that the grains escape the changes in water status which occur in the rest of the plant or that the metabolism of the grains is less affected by a change in water status than the metabolism of the leaves. Morgan (1977) found that the water potential of wheat spikelets was considerably higher than that of the flag leaf, particularly at low levels of water potential. The work of Barlow *et al.* (1980) and of Brooks *et al.* (1982) suggests that, in addition to this difference between the flag leaf and spikelets, a major difference in water potential occurs between the floral organs and the grains. Differences as great as 1.0 MPa between the grains and their enclosing bracts are found during water stress. In an experiment where wheat plants growing in pots were allowed to lose water from 20 days after anthesis onwards, the water potential of the rachis and peduncle was linearly related to that of the flag leaf throughout, although generally 1.0 MPa higher (Fig. 7.6). The water potentials of both the floral organs and the grains were virtually independent of flag leaf water potential until flag leaf potential fell to –3.5 MPa. There was then a precipitate fall in bract water potential to –2.5 MPa and a less dramatic fall (of 0.35 MPa) in grain water potential. These results taken together, suggest that the grain's synthetic function exists in a protected environment during water stress. Only when stress is extreme does the water potential of the grain fall, and the grain loses water rapidly once maturation ensues.

The insulation of grain water status from extreme changes in the plant's water potential could only be achieved by restriction of water loss from the grain surface coupled with a large resistance to water flow back into the plant along the steep water potential gradients which develop. The limitation of water loss from the grain to the atmosphere is presumably as a result of the enclosure of the grains within bracts together with the paucity of stomata in the pericarp (Gifford and Bremner, 1981). There is no direct vascular connection between the maternal tissues of the parent plant and the endosperm. The vascular bundle servicing the grain runs through the furrow in the

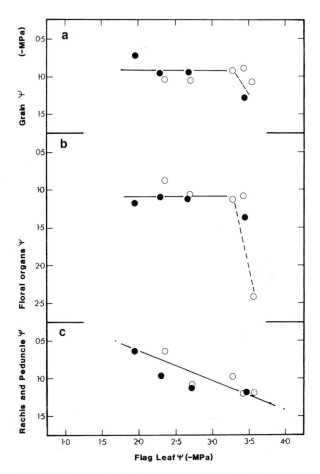

Fig. 7.6 The relationship between flag leaf water potential (ψ) and the water potential of the grain (a), floral organs (glumes, palea and lemmae—b) and rachis and peduncle (c) in wheat (cv. Gabo) during grain filling and maturation.

pericarp. The phloem is continuous with the phloem of the floral axis, but the continuity of the xylem is interrupted in the "neck" region of the grain by a core of thick-walled cells (Zee and O'Brien, 1970). Additionally, Barlow *et al.* (1980) reported the deposition of lipid-like materials in the vascular bundle in this region during water deficit. These observations suggest that there is a considerable restriction to the movement of water through this region, presumably in both directions, and raises the possibility that there is only slow turnover of water within the grain during development. Certainly, the grain water content remains substantially unchanged throughout most of the period of grain filling. If growth can continue without substantial water exchange, it follows that sucrose transport into the grain must also be substantially independent of water movement. Jenner (1982a), using grains attached to the rachis but enclosed within artificially manipulated atmospheres, demonstrated that the movement of assimilate was indeed independent of the direction and scale of water movement between the grain and the rachis. This system would allow continued growth despite virtual isolation of the grain water pool from that of the rest of the plant.

Despite the resistance to water stress of the process of grain filling, grain weight is restricted if water deficit is sufficiently severe during grain growth. The restriction appears mainly as a result of premature maturation (Brooks et al., 1982). Termination of grain growth is marked by cessation of starch synthesis and a fall in grain water content in both well-watered and water-stressed plants. These events may be coincident (Jennings and Morton, 1963; Brooks et al., 1982). The cessation of starch synthesis at maturity in well-watered plants is not a result of a limitation in assimilate supply consequent upon leaf senescence (Jenner and Rathjen, 1975). There is ample sucrose available within the endosperm at this time but the endosperm loses the capacity to convert sucrose to starch. It also appears that the premature maturation of water-stressed grains is not a consequence of lack of assimilate, despite the evident leaf senescence, as the sucrose concentration in the grains of water-stressed wheat rose sharply as dry matter accmulation ceased (Brooks et al., 1982). Similarly, grain growth is not terminated as a result of a shortage of amino acids (Brooks et al., 1982). It has been suggested that termination of grain growth is controlled by the concentration of abscisic acid within the grains (King, 1976; Radley, 1976). There is an increase in grain abscisic acid content at this time, the increase occurring earlier in stressed plants. Whatever the mechanism which induces the premature maturation of water-stressed grains, it is evident from the "pinched" or shrivelled appearance of the harvested grains that starch synthesis ceases before the space available for accumulation has been completely filled (Brooks et al., 1982).

III. CONCLUSIONS

This account has concentrated on a consideration of the responses to water deficit of the morphogenetic and physiological processes that are specifically concerned with generation of grain yield. Little attention has been paid to such processes as leaf expansion. This approach directs attention to the cause of the event in the developing plant that influences yield. Armed with such information it should be possible, at least theoretically, to work back to the ultimate reason for the response. This approach has the advantage over research conducted in the reverse direction of inviting no initial preconceptions concerning the mechanism. It has the disadvantage, however, that it adopts an essentially short-term view and focuses attention on the immediate causes of the response measured. Carry-over effects and complex interactions, such as the effects of previous stress on root and leaf growth, are difficult to elucidate in this manner.

Our knowledge of the causes of responses by wheat to water deficit is fragmentary. It is sufficient, however, to suggest strongly that different mechanisms are involved in the various growth and developmental responses: no single mechanism accounts for all observed responses.

The wheat plant seems to be adapted so that effects of water deficit during early growth can be compensated for by growth later in development if conditions ameliorate. The effects of water deficit during grain filling seem to be strongly resisted. Crops grown even in favourable environments encounter water deficit during grain filling and there is some evidence which suggests that stomatal behaviour alters at this time so that stomata do not close as far after anthesis as before at a given leaf water potential (Morgan, 1977). This response would exacerbate the stress faced by the plants during grain filling. The most vulnerable growth phase, in so far as comparisons are possible, appears to be that commencing with the initiation of internode expansion and terminating at, or just after, anthesis. At this stage, the plant has passed beyond the period where growth compensation can occur but has not developed the resistance to the effects of water deficit which characterize the ear during grain-filling.

These conclusions must be viewed against the environment in which the crop is grown and the progressive increase in evapotranspiration with development. Wheat crops in southern Australia, for instance, grow in an environment where the risk of drought increases rapidly later in crop development. The wheat plant is well adapted to such an environment as long as the growth cycle of the plant is carefully fitted to the average seasonal cycle. Problems then occur only in seasons that depart appreciably from the average such that drought occurs during the earlier, less tolerant phase of growth.

REFERENCES

Angus, J. F., and Moncur, M. W. (1977). Water Stress and phenology in wheat. *Australian Journal of Agricultural Research* **28**, 177–181.

Aspinall, D., and Husain, I. (1970). The inhibition of flowering by water stress. *Australian Journal of Biological Science* **23**, 925–936.

Aspinall, D., Nicholls, P. B., and May, L. H. (1964). The effects of soil moisture stress on the growth of barley. I. Vegetative development and grain yield. *Australian Journal of Agricultural Research* **15**, 729–745.

Barlow, E. W. R., Munns, R., Scott, N. S., and Reisner, A. H. (1977). Water potential, growth and polyribosome content of the stressed wheat apex. *Journal of Experimental Botany* **28**, 909–916.

Barlow, E. W. R., Lee, J. W., Munns, R., and Smart, M. G. (1980). Water relations of the developing wheat grain. *Australian Journal Plant Physiology* **7**, 519–525.

Bennett, M. D., Rao, M. K., Smith, J. B., and Bayliss, M. M. (1973). Cell development in the anther, the ovule and the young seed of *Triticum sativum* L. var. Chinese Spring. *Philosophical Transactions of the Royal Society London. B* **266**, 39–81.

Bingham, J. (1966). Varietal response in wheat to water supply in the field, and male sterility caused by a period of drought in a glasshouse experiment. *Annals of Applied Biology* **57**, 365–377.

Briarty, L. G., Hughes, C. E., and Evers, A. D. (1979). The developing endosperm of wheat–a stereological analysis. *Annals of Botany* **44**, 641–668.

Brocklehurst, P. A. (1977). Factors controlling grain weight in wheat. *Nature* (London) **266**, 348–349.

Brocklehurst, P. A., Moss, J. P., and Williams, W. (1978). Effect of irradiance and water supply on grain development in wheat. *Annals of Applied Biology* **90**, 265–276.

Brooks, A., Jenner, C. F., and Aspinall, D. (1982). Effects of water deficit on endosperm starch granules and on grain physiology of wheat and barley. *Australian Journal of Plant Physiology* **9**, 423–436.

Connor, D. J. (1975). Growth, water relations and yield of wheat. *Australian Journal of Plant Physiology* **2**, 353–366.

Dubetz, S., and Bole, J. B. (1973). Effect of moisture stress at early heading and of nitrogen fertilizer on three spring wheat cultivars. *Canadian Journal of Plant Science* **53**, 1–5.

Evans, L. T., Bingham, J., and Roskams, M. A. (1972). The patterns of grain set within ears of wheat. *Australian Journal of Biological Science* **25**, 1–8.

Fischer, R. A. (1973). The effect of water stress at various stages of development on yield processes in wheat. *In* "Plant Responses to Climatic Factors" (R. O. Slatyer, ed.), pp. 233–241. UNESCO, Paris.

Fischer, R. A., and Turner, N. C. (1978). Plant productivity in the arid and semiarid zones. *Annual Review of Plant Physiology* **29**, 277–317.

French, B. K., and Legg, B. J. (1979). Rothamsted irrigation 1964–76. *Journal of Agricultural Science* **92**, 15–37.

Gifford, R. M., and Bremner, P. M., (1981). Accumulation and conversion of sugars by developing wheat grains. II. Light requirement for kernels cultured *in vitro*. *Australian Journal of Plant Physiology* **8**, 631–640.

Gifford, R. M., and Evans, L. T. (1981). Photosynthesis, carbon partitioning and yield. *Annual Review of Plant Physiology* **32**, 485–509.

Gleadow, Roslyn M., Dalling, M. J., and Halloran, G. M. (1982). Variation in endosperm characteristics and nitrogen content in six wheat lines. *Australian Journal of Plant Physiology* **9**, 539–551.

Husain, I., and Aspinall, D. (1970). Water stress and apical morphogenesis in barley. *Annals of Botany* **34**, 393–407.

Jenner, C. F. (1982a). Movement of water and mass transfer into developing grains of wheat. *Australian Journal of Plant Physiology* **9**, 69–82.

Jenner, C. F. (1982b). Storage of Starch. *In* "Plant Carbohydrates 1" Encyclopedia of Plant Physiology, New Series Vol. 13A (F. A. Loewus and W. Tanner, eds.), pp. 700–747. Berlin, Springer-Verlag.

Jenner, C. F., and Rathjen, A. J. (1975). Factors regulating the accumulation of starch in ripening wheat grain. *Australian Journal of Plant Physiology* **3**, 311–322.

Jennings, A. C., and Morton, R. K. (1963). Changes in carbohydrate, protein and non-protein nitrogenous compounds of developing wheat grain. *Australian Journal of Biological Science* **16**, 318–331.

Kemp, D. R., and Whingwiri, E. E. (1980). Effect of tiller removal and shading on spikelet development and yield components of the main shoot of wheat and on the sugar concentration of the ear and flag leaf. *Australian Journal of Plant Physiology* **7**, 501–510.

King, R. W. (1976). Abscisic acid in developing wheat grains and its relation to grain growth and maturity. *Planta* **132**, 43–51.

King, R. W., and Evans, L. T. (1977). Inhibition of flowering in *Lolium temulentum* L. by water stress: A role for abscisic acid. *Australian Journal of Plant Physiology* **4**, 225–233.

Kriedemann, P. E., and Downton, W. J. S. (1981). Photosynthesis. *In* "The Physiology and Biochemistry of Drought Resistance in Plants" (L. G. Paleg and D. Aspinall, eds.), pp. 283–314 Academic Press, Sydney.

Milthorpe, F. L. (1950). Changes in the drought resistance of wheat seedlings during germination. *Annals of Botany, London, N.S.* **14**, 79–89.

Mohapatra, P. K., Aspinall, D., and Jenner, C. F. (1982). The growth and development of the wheat apex: The effects of photoperiod on spikelet production and sucrose concentration in the apex. *Annals of Botany* **49**, 619–626.

Morgan, J. M. (1971). The death of spikelets in wheat due to water deficit. *Australian Journal of Experimental Agriculture and Animal Husbandry* **11**, 349–351.

Morgan, J. M. (1977). Changes in diffusive conductance and water potential of wheat plants before and after anthesis. *Australian Journal of Plant Physiology* **4**, 75–86.

Morgan, J. M. (1980). Possible role of abscisic acid in reducing seed set in water-stressed wheat plants. *Nature* **285**, 655–657.

Munns, Rana, Brady, C. J., and Barlow, E. W. R. (1979). Solute accumulation in the apex and leaves of wheat during water stress. *Australian Journal of Plant Physiology* **6**, 379–389.

Parameswaran, K. V. M., Graham, R. D., and Aspinall, D. (1981). Studies on the nitrogen and water relations of wheat. I. Growth and water use in relation to time and method of nitrogen application. *Irrigation Science* **3**, 29–44.

Passioura, J. B. (1976). Physiology of grain yield in wheat growing on stored water. *Australian Journal of Plant Physiology* **3**, 559–565.

Puckridge, D. W. (1968). Competition for light and its effect on leaf and spikelet development of wheat plants. *Australian Journal of Agricultural Research* **19**, 191–201.

Puckridge, D. W., and Donald, C. M. (1967). Competition among wheat plants sown at a wide range of densities. *Australian Journal of Agricultural Research* **18**, 193–211.

Radley, M. (1976). The development of the wheat grain in relation to endogenous growth substances. *Journal of Experimental Botany* **27**, 1009–1021.

Saini, H. S. (1982). Physiological studies on sterility induced in wheat by heat and water deficit. Ph.D. thesis, University of Adelaide, Adelaide.

Saini, H. S., and Aspinall, D. (1981). Effect of water deficit on sporogenesis in wheat (*Triticum aestivum* L.). *Annals of Botany* **48**, 623–633.

Saini, H. S., and Aspinall, D. (1982a). Sterility in wheat (*Triticum aestivum* L.) induced by water deficit or high temperature: Possible mediation by abscisic acid. *Australian Journal of Plant Physiology* **9**, 529–537.

Saini, H. S., and Aspinall, D. (1982b). Abnormal sporogenesis in wheat (*Triticum aestivum* L.) induced by short periods of high temperature. *Annals of Botany* **49**, 835–846.

Skazkin, F. D. (1961). The critical period in plants as regards insufficient water supply. *Timiryazevskie Chteniya Akad. Nauk. SSSR* **21**, 1–51.

Udol'skaja, N. L. (1936). Drought resistance of spring wheat varieties (in Russian). Omgiz:Omsk. (cited by Salter and Goode, 1967. Crop responses to water at different stages of growth. Commonwealth Bureau of Horticulture and Plant Crops, Research Review No. 2, pp. 15–48).

Wardlaw, I. F. (1971). Early stages of grain development in wheat: Response to water stress in a single variety. *Australian Journal of Biological Sciences 24,* 1047–1055.

Zee, S. Y., and O'Brien, T. P. (1970). A special type of tracheary element associated with "xylem discontinuity" in the floral axis of wheat. *Australian Journal of Biological Science* **23,** 783–791.

CHAPTER 8

Flowering in Wheat

J. L. DAVIDSON and K. R. CHRISTIAN

I. INTRODUCTION

Flowering is a crucial event in the life of the wheat plant, which must compromise between the conflicting requirements of restricting the life cycle to a short period compatible with plant survival and of allowing the maximum time for reproductive development. The time of flowering determines not only the potential number of grains, but also the conditions and time that remain for their growth and maturation.

As wheat has been cultivated in many parts of the world for centuries, a wide range of genetic material inevitably evolved with adaptation to widely contrasting climates. In selecting genotypes for specific environments particular attention has been devoted to time of heading, which largely determines the suitability of a genotype for specific regions, and to the number of spikelets produced, which has frequently been associated with potential yield.

CONTROL OF CROP PRODUCTIVITY
ISBN   0 12 548280 9

II. DEVELOPMENTAL STAGES

The period from germination to flowering (anthesis) may be divided into three physiological phases: (1) vegetative, ending in the initiation of the first spikelet; (2) spikelet initiation, ending in the formation of the terminal spikelet; and (3) ear development, including stem elongation, ear emergence, floret initiation and development, culminating in anthesis and fertilization.

During the vegetative stage, the apex is ringed by raised primordia from which leaves expand in order of appearance. Under uniform conditions, primordia are initiated at a constant rate. The start of the spikelet initiation phase is marked by a three-fold increase in the rate of primordia initiation (Kirby, 1974). Since this transition can be identified only retrospectively, either the elongation of the apical dome (Stern and Kirby, 1979) or the double ridge stage, at which spikelet primordia first become obvious, is usually taken as the indicator of the change-over from vegetative to floral development.

The number of spikelets initiated is, in general, related to the length of the vegetative stage (Rawson, 1970; Halloran, 1977) and to the number of leaves that have been formed previously. The spike length at floral initiation bears a strong relation to the final spikelet number (Lucas, 1972; Rahman and Wilson, 1977), indicating that the latter is largely determined at this time. However, spike length may be modified up to the time of terminal spikelet initiation by changes in the environment (Rahman and Wilson, 1977; Mohapatra et al., 1982).

The length of time from sowing to heading depends on the environment as well as on genotype. By 1857 it was known that temperatures just above freezing applied to germinated seed would enable winter wheat sown in spring to mature early and produce a good crop (McKinney and Sando, 1935). The technique was used extensively in the USSR (McKinney and Sando, 1933), from where the term "vernalization" was derived. The effects of daylength were not linked with the control of flowering until much later. In 1920 Garner and Allard established that flowering in plants is controlled by the relative length of day and night. Two years later Wanser (1922) claimed there was a critical daylength below which newly sown winter wheat would not joint.

Traditionally, wheats are described as winter or spring types according to whether or not they must be sown before winter in order to flower early enough to produce a harvest in the following summer. Although winter and spring types are identified by their performance in the field, they have become regarded simply as late or early varieties, respectively (Aitken, 1966), or more frequently as varieties with or without substantial vernalization requirements, respectively (Murfet, 1977; Wiegand et al., 1981). Many of the early varieties have an appreciable response to vernalization whereas other varieties fail to flower in low or medium latitudes because their daylength requirements are not satisfied. We will describe wheats broadly as either daylength-responsive, or vernalization-responsive. Almost all wheats use the physiological mechanisms of daylength and vernalization responses to delay flowering, but to varying extents depending on variety. This results in a large spread of flowering dates under natural conditions. Even the latest varieties will probably flower within 40–60 days after vernalization and under continuous light.

Many varieties show a strong interaction between daylength and vernalization responses. For example, an Australian wheat, Isis, flowered about 10 days earlier when plants were either vernalized or subjected to 16 h daylengths, but when given both treatments it flowered another 69 days earlier (J. L. Davidson, D. B. Jones and K. R. Christian, unpublished).

Whereas time to flowering is also influenced by general conditions affecting plant growth, particularly water relations and mineral nutrient supplies, these effects are usually comparatively minor.

III. ENVIRONMENTAL EFFECTS ON FLOWERING

A. Daylength

Growth of the whole wheat plant is proportional to the total light energy received each day (MacDowall, 1977). However, the rate of floral development increases with daylength from 8 h to 24 h with the greatest change found usually between 10 h and 16 h.

Wheat is generally looked upon as a quantitative long-day plant, and even genotypes regarded as daylength-insensitive display small responses to long days (Rahman, 1980). However, the quantitative nature of daylength sensitivity has not been established for many genotypes. Rawson (1970) found in the variety Triple Dirk that one long day had no effect on time to floral initiation, whereas two long days reduced the period by 15 days.

The perception of what constitutes long or short days differs markedly between varieties, and there is no sharp distinction between them (Forster *et al.*, 1932). Aitken (1966) has pointed out that contrary to a common assumption floral initiation can occur at low to medium daylengths. Using time to floral initiation as the criterion, Rawson (1971) found no further response to daylength above 10 h in Triple Dirk and Sonora 64 whereas Chinese Spring, Jufy 2 and Witspitskop responded to at least 16 h. The varieties Park and Pitic require a daylength of 17 h to reach ear emergence in the minimum time (Major, 1980).

The duration of all three physiological phases is reduced in sensitive varieties by long days. The shorter period of vegetative growth results in fewer leaves (Allison and Daynard, 1976; Rahman and Wilson, 1977; Williams and Williams, 1968); for example, Allison and Daynard (1976) found that the mean leaf number was about 6–8 during long days in all varieties studied, whereas at daylengths below 10 h the difference between varieties was much greater, ranging from 9 to 16.

Similarly, the duration of the spikelet initiation phase may be greatly reduced under long days, tending to a lower limit of about 6–9 days in continuous light (Rahman and Wilson, 1977). Although the rate of initiation is higher, the net result is a reduction in spikelet number with increasing daylength. Apices of plants grown under short-day regimes are more elongated, with many more foliar ridges (Williams and Williams, 1968). In short daylengths, the spikelet phase is longer, the rate of initiation slower, and the number of spikelets rises asymptotically to a maximum value of about 30 (Allison and Daynard, 1976). The rate of development appears to depend on the current daylength, though with a lag of several days.

The time from terminal spikelet formation to ear emergence is also longer under short days, which may prevent flowering in many sensitive varieties and result in abnormalities such as elongation of the internodes of the rachis and transformation of basal spikelets to foliar structures (Rahman and Wilson, 1977). Combined with high temperatures, even 12–14 h daylengths may lead to death of the reproductive shoot apices after spikelet initiation (Aitken, 1966). The transfer of plants from long days to short days after floral initiation may result in abnormalities such as ovules forming anthers and ovaries forming inflorescences (Fisher, 1972).

The development of florets is accelerated by continuous light, though the total number formed is unaffected (Langer and Hanif, 1973).

Overall, daylength has been found to be the primary factor controlling heading in spring wheats (Levy and Peterson, 1972). As Martinic (1975) observed, varieties that are very responsive to long days are usually unsuited to low latitudes because they generally flower late. However, low sensitivity is not synonymous with earliness, and slightly sensitive varieties may head later than highly responsive types.

B. Light Intensity

Morphological development is not retarded by low light intensities to the same extent as total plant growth (McKinney and Sando, 1935), but the duration of all three phases is lengthened in much the same way as by short days (Rahman et al., 1977). Leaf primordia formation is slower, and the final leaf number is slightly reduced (Friend et al., 1963). Rates of spikelet initiation and apex extention are slower under low light, resulting in small plants with short shoots and small ears with fewer spikelets (Friend, 1965).

Considerable variation in tolerance of reduced light levels has been found among varieties and may be important in determining the effects of shading brought about by plant competition (Rahman *et al.*, 1977). Although shading during the time up to ear initiation may have little effect on spikelet number, reducing light between ear initiation and anthesis by artificial shading or increased plant density can sharply reduce the proportion of fertile spikelets per ear and fertile florets per spikelet (Puckridge, 1968; Fischer, 1975).

C. Vernalization

Vernalization-responsive wheats are those that flower earlier if the seed or seedling is exposed to temperatures just above freezing. The main effect of low temperatures on seed before sowing is to hasten flowering among responsive wheats by reducing the duration of the vegetative phase (Gott, 1961; Halse and Weir, 1970; Rawson, 1970). Vernalization also increases the rate of spikelet initiation, but this is more than outweighed by the shorter time period (Rawson, 1970) so that fewer spikelets are produced (Halloran, 1977; Rahman, 1980).

Partial vernalization may take place during seed maturation in cool weather (Gotoh, 1976; Pugsley, 1963). Riddell and Gries (1958b) found that variable flowering dates from seed harvested in different years could be eliminated by cold treatment. Immature wheat embryos can be vernalized by chilling spikes harvested 8–12 days after anthesis (Weibel, 1958).

By demonstrating that the excised apex grown on an agar medium can be vernalized, Ishihara (1961) identified the apical bud as the operative site for vernalization effects. Although McKinney and Sando (1935) reported that light does not influence vernalization, other work has suggested that the process may be promoted by light (Chujo, 1969; Gotoh, 1976) or the addition of sucrose (Chujo, 1969; Krekule, 1961).

Gott (1961) found that plants up to the 6–7 leaf stage could be vernalized, with older plants requiring shorter periods of treatment. Gotoh (1976) noted that differences among genotypes was greatest when plants were vernalized at the one-leaf stage, whereas Ahrens and Loomis (1963) and Chujo (1966a) confirmed that the youngest seedlings showed the greatest acceleration; after 10 days increasing age led to decreasing responses (Chujo, 1966a).

In controlled environment studies, it is convenient to use germinated seeds, which can be treated in large numbers on Petri dishes in a refrigerator for 4–8 weeks. To prevent excessive elongation of seedlings, the temperature must be kept around 1°C–2°C, which may not be optimal for vernalization.

Observations to determine optimum vernalization temperatures are rare, and experimental temperatures mentioned in the literature range between

0° and 11°C. Ahrens and Loomis (1963) found 1°C to be effective for vernalizing wheat, whereas -2°C had no effect. Junges (1959) reported that the optimum temperature was 3°C for a winter wheat, but 11°C for a spring wheat. Trione and Metzger (1970) found that 7°C was the most effective temperature for an intermediate wheat, with a rapid decrease in effect above 9°C or below 3°C. Chujo (1966a) showed that some classes of winter wheats responded most to 8°C and 11°C, whereas a class of winter wheat responded most at 4°C or 8°C. The effect of cold treatment may be completely removed by subjecting the plants to temperatures about 30°C for several days, a process known as devernalization (Dévay et al., 1976).

Requirements for vernalization in the field have not been intensively studied, but Ishihara (1960) reported that under mean temperatures of 7°C–15°C, vernalization took place at a similar rate to that in germinated seeds treated at 5°C. Ishihara (1963) also concluded that it is the low temperature at night that causes vernalization. Working with the variety Norin 64 he found vernalization proceeded even if day temperatures were above 30°C as long as night temperatures were below 10°C. With fluctuating temperatures, vernalization was completed within four weeks at mean temperatures of 19°C. This suggests that devernalization would be unlikely under field conditions. Again varietal differences may be important. Chujo (1966b) found with Norin 4 that fluctuating and constant temperatures were equally effective at the same mean temperature. With Norin 27, which has a greater vernalization response, constant temperatures were the more effective treatment, and if 16 h of high temperature alternated with 8 h of low temperature, the vernalizing effect was lost.

The quantitative requirement for low temperature and the sensitivity to temperature for promotion of flowering differ widely between varieties. In a typical winter wheat, 2–3 weeks of chilling has little or no effect on time to flowering, whereas each day of the succeeding 4–6 weeks steeply reduces the time until the minimum value is reached. In varieties with less response the initial stage may be absent, whereas in spring wheats, vernalization may be complete within 2–3 weeks. In general, even winter wheat will eventually flower without experiencing vernalizing temperatures (Gott, 1957; Ahrens and Loomis, 1963; Rahman, 1980). Although vernalization is often stated to be essential for flowering in winter wheats, it merely modifies the speed of flowering, enabling field crops to mature at a suitable time.

Numerous biochemical changes have been observed during vernalization, although many of these may be merely a reflection of low-temperature growth or general non-specific metabolic activity. Vernalization comprises extensive modifications that prepare the plant for the differentiation of reproductive tissue, but the nature of the processes involved is still largely unknown. It is a matter of conjecture whether or not effects on time to flowering produced in certain varieties by such treatments as high temperatures, short days or prolonged chilling (e.g. Friend and Purvis, 1963) are merely reactions invoked by unaccustomed conditions, since their relevance to development under natural conditions has not been demonstrated.

Ahrens and Loomis (1963) regarded vernalization as a low temperature requirement for the initial reactions of floral initiation in winter wheats. It has been suggested that vernalization is a slow physiological response to low temperatures that renders the plant susceptible to the daylength effect (Trione and Metzger, 1970; Evans and Wardlaw, 1976), but little evidence supports this hypothesis. Halloran (1975) found that in seven out of eight varieties examined, prior vernalization did not enhance the daylength response in terms of rate of spikelet formation. The various mechanisms so far proposed (e.g. Friend and Purvis, 1963) have neither been confirmed nor refuted experimentally, and hence they have contributed little to our understanding.

Dévay et al., (1976) concluded from experiments involving devernalization and re-vernalization that several thermolabile phases alternate with intermediate stages which are relatively insensitive to temperature. Tan et al. (1981) further suggested on the basis of inhibition experiments that at least three consecutive processes are involved: (1) carbohydrate oxidation and phosphorylation; (2) nucleic acid metabolism; and (3) mainly protein metabolism. A multi-phase mechanism is needed to account for the diversity of the phenomena observed.

D. Temperature

Once vernalization requirements have been satisfied, increasing temperature generally reduces the length of all developmental phases until these reach values which may represent the lower limits for particular genotypes (Friend et al., 1963; Halse and Weir, 1974; Rahman, 1980). Reports of slower development at higher temperatures (Riddell and Gries, 1958a; Aitken, 1966; Rawson, 1970; Midmore et al., 1982) have apparently resulted from a lack of vernalizing conditions for responsive varieties.

During the vegetative stage, when the shoot apex is below ground level, the rate of formation of leaf primordia is most closely associated with soil temperature (Hay and Wilson, 1982) and has been found to be a linear function of temperature accumulated above 0°C in the winter wheat, Maris Huntsman (Baker et al., 1980).

Rates of spikelet initiation increase as temperature rises from low to moderate values. This tends to counter the decrease in duration of initiation, and there may be little change in spikelet number over the range from 10°C to 20°C (Halse and Weir, 1974; Wall and Cartwright, 1974; Rahman and Wilson, 1978; Pirasteh and Welsh, 1980). There is often a marked decline in spikelet numbers above 20°C, but with substantial differences among varieties (Halse and Weir, 1974; Kolderup, 1979).

The change in spikelet number with temperature may also depend on vernalization treatment. Vernalized plants of Mexico 120 and Bencubbin produced fewer spikelets than unvernalized plants over a range of mean temperatures from 7°C to 23°C, and the number of spikelets continued to increase with temperature from 11°C to 19°C only in vernalized plants (Halse and Weir, 1974). The results have been confirmed for Mexico 120 and supplemented by responses for other varieties by Wall and Cartwright (1974) and Rawson (1970). It appears unlikely that partial vernalization was responsible for the lower spikelet number at the lower temperatures, since the time to initiation was considerably longer than at higher temperatures. It is significant that Fisher (1973) and Holmes (1973), working mainly with unvernalized spring wheats, observed striking differences in shoot apex morphogenesis between Norin 10 and standard varieties, whereas Brooking and Kirby (1981) found no consistent differences between genotypes of vernalized winter wheats. The interaction between vernalization and temperature responses merits further study, since it may provide some indication of the possible association between vernalization requirement and higher yields.

The greatest influence of temperature on floral development is exerted during the ear development phase (Warrington et al., 1977). This phase is lengthened by low temperatures; Rawson (1970) found an average increase of 90% in length in a range of varieties when the mean temperature was reduced from 18°C to 12°C. Because this is a relatively long phase, temperature has a marked effect on the length of time from sowing to heading, which is reported to be shorter at 30°C than at lower temperatures (Friend et al., 1963; Rahman and Wilson, 1978). But because all spikelets have been initiated by the start of ear development, temperature usually has less effect on spikelet number than vernalization or daylength on responsive varieties. Temperature does, however, affect the fertility of florets in different parts of the ear in ways that vary with variety (Bagga and Rawson, 1977).

In general, the number of spikelets remains closely correlated with time from sowing to ear emergence. In 14 varieties examined, Wall and Cartwright (1974) recorded an average increase of 0.27 spikelets for every day that ear emergence was delayed.

In the field, rising temperatures and increasing daylengths during the growing season generally result in faster floral development. With later sowing dates, spikelet initiation starts earlier, progresses more rapidly, and is completed sooner. Each day's delay in sowing spring wheats in Australia has been found to reduce the time to anthesis by 0.5–0.6 days (Doyle and Marcellos, 1974; French et al., 1979).

Extreme temperatures may have much more serious effects on floral development. During frosts, the air temperature 5–10 cm above ground may be 4°C–6°C cooler than at the soil surface (Marcellos and Single, 1975). Thus, when overnight temperatures in a Stephenson screen fall to about –5°C, the developing head may be killed if it has been raised above the soil surface by stem elongation (George, 1982; Doyle and Marcellos, 1974).

After the ear has emerged from the boot, it is no longer protected by surrounding tissues and florets are directly exposed to inoculation by ice nuclei in the atmosphere, with the risk of severe damage by frosts (Single and Marcellos, 1974).

High temperatures may be just as damaging. Campbell *et al.* (1981) reported that the optimum temperature for spikelet development is 22°C, and although Mangas Martin and Sanchez de la Puente (1980) found that temperatures above this merely depressed ear development, temperatures above 30°C during floret formation can cause complete sterility (Saini and Aspinall, 1982, Owen, 1971). Aitken (1966) found that the combination of high temperatures and moderate to short daylengths (14–12 h) may cause widespread death of shoot apices shortly after ears are initiated in daylength-sensitive wheats sown in spring.

The effect of temperature on floret survival and fertility may be more important than on the number of spikelets produced. Although the interest shown in spikelet number is largely based on the belief that increases in this component will lead to higher yields, Bremner and Davidson (1978), Stern and Kirby (1979) and Ford *et al.* (1981) all found little evidence to support the hypothesis that spikelet numbers have a significant influence on yield, and the same conclusion may be drawn from the results of Rawson (1970).

IV. GENETICS

A range of responses of different varieties to vernalization should be expected because the genetic control is complex. Pugsley (1972) identified a single gene, *Vrn1*, which inhibits the winter habit in WW15, an important derivative of Mexican semi-dwarf wheats. Other studies revealed that the spring habit in wheat is governed by 3 or 4 dominant genes (Pugsley, 1971; McIntosh, 1973), any of which inhibits expression of winter habit. Winter wheats with recessive genes differ among themselves because of multiple recessive alleles at these loci (Pugsley, 1971). Single chromosome substitution lines have been used to identify the chromosomes. Homoeologous group 5 chromosomes (A, B and D) have been found to be important in determining response to vernalization (Halloran, 1976; Law *et al.*, 1976). Halloran and Boydell (1976b) found that 11 chromosomes of the variety Hope influenced the effect of vernalization in Chinese Spring.

Two different types of responses have been associated with gene differences in the Triple Dirk series. The *Vrn1* gene produced a threshold type response, with no response to four weeks or less of vernalization but a 22-day response with more. Plants with the *Vrn1* gene showed a cumulative response to vernalization that varied in magnitude according to whether the gene was combined with *Vrn2* or *vrn2* (Berry *et al.*, 1980).

Salisbury *et al.* (1979) have also demonstrated that the relative effectiveness of vernalizing germinated seeds or young plants varies with these genes. They warn that data drawn from the vernalization of germinated seeds may not relate well to vernalization of seedlings in the field.

Working with Thatcher × Triple Dirk crosses, Pugsley (1966) found evidence of one major and one minor gene that controlled daylength responses and suggested that few major genes might be involved in spring wheats. Halloran and Boydell (1967a) concluded from studies based on Hope-Chinese Spring substitution lines that one major gene (on 4B) and minor genes on 1A and 6B conferred greater sensitivity to short days with minor genes on 3B and 7D being less sensitive. Law *et al.* (1978) concluded that chromosome 2B of Chinese Spring was the most important in conferring daylength insensitivity. Klaimi and Qualset (1973) found evidence of epistasis, additivity, and dominance being important in controlling daylength effects, with a high average degree of dominance for insensitivity. As with vernalization, the genetic control of response to daylength is apparently complex, involving major and minor genes on a number of chromosomes, so again a wide range of responses can be expected.

V. CONCLUSIONS

Work on the influence of vernalization, daylength and ambient temperatures on the successive phases of floral development has concentrated almost entirely on spring wheats. The conclusions reached might have been different had winter wheats also been examined. The distinction is important because winter wheats often greatly outyield spring wheats in their respective regions (Pinthus, 1967), and may replace traditional spring types where practicable (Martinic, 1975).

In the task of tailoring varieties to the requirements of particular localities and climates, enormous scope clearly exists for selection within the existing gene pool. The judicious selection of genotypes with specific vernalization and daylength responses, combined with suitable adaptation to temperature, enables us to focus attention on varieties that are capable of fully exploiting the growing season, or even of extending it by characteristics such as resistance to frost or to heat and water stresses.

However, progress towards this objective is hampered by a lack of knowledge of these characteristics, despite the large amount of research carried out. Although experiments conducted under uniform environments assist greatly in elucidating the roles of the various factors involved, treatments are often chosen for convenience rather than for their relevance to field conditions. They can therefore only be a partial guide to performance in natural environments which are continually changing throughout crop ontogeny. Furthermore, conclusions drawn from experiments confined to a limited range of genotypes often conflict with those from other sources using different material. It is evident that the spectrum of response to vernalization, daylength and temperature is not only very broad but also virtually continuous. Few worthwhile generalizations can therefore be made about the influence of the environment on flowering in wheat.

Until now, studies on flowering have consisted of empirical measurements, and a theoretical basis for the different responses has not yet been developed. It is usually supposed, for example, that vernalization requirements must be satisfied before the daylength response can be exerted, but there is little quantitative information to support this assumption. More attention to the underlying mechanisms is clearly desirable for a more rational and efficient approach to the understanding of genotypic variation.

A better knowledge of the extent of genotypic variation would be valuable for two reasons: (1) reconciliation of the existing experimental evidence would enable both the similarities and the differences between varieties to be more clearly defined; and (2) it could lead to an increased awareness of the opportunities available for selection within the gene pool. Because the plasticity of response to environment varies greatly with genotype, it is important to search for characters that show adaptation to seasonal variability and to local areas within a broader region. There is also a need to identify the sources of variation in yield components which are important in governing both yield and the stability of yield in contrasting seasons.

Progress towards the goal of predicting flowering behaviour may be aided by identifying the genes involved. Since this would require a knowledge not only of the major genes and their interactions but also of the influence of a background of minor genes which are not yet identifiable, this poses a formidable task. Nevertheless, it offers scope for planning genetic recombinations that may enable breeders to produce well-adapted varieties for specific areas.

REFERENCES

Ahrens, J. F., and Loomis, W. E. (1963). Floral induction and development in winter wheat. *Crop Science* **3**, 463–466.

Aitken, Y. (1966). Flower initiation in relation to maturity in crop plants. III. The flowering response of early and late cereal varieties to Australian environments. *Australian Journal of Agricultural Research* **17**, 1–15.

Allison, J. C. S., and Daynard, T. B. (1976). Effect of photoperiod on development and number of spikelets of a temperate and some low-latitude wheats. *Annals of Applied Biology* **83**, 93–102.

Bagga, A. K., and Rawson, H. M. (1977). Contrasting responses of morphologically similar wheat cultivars to temperatures appropriate to warm temperature climates with hot summers: A study in controlled enviroment. *Australian Journal of Plant Physiology* **4**, 877–887.

Baker, C. K., Gallagher, J. N., and Monteith, J. L. (1980). Daylength change and leaf appearance in winter wheat. *Plant, Cell and Environment* **3**, 285–287.

Berry, C. J., Salisbury, P. A., and Halloran, G. M. (1980). Expression of vernalization genes in near-isogenic wheat lines: Duration of vernalization period. *Annals of Botany* **46**, 235–241.

Bremner, P. M., and Davidson, J. L. (1978). A study of grain number in two contrasting wheat cultivars. *Australian Journal of Agricultural Research* **29**, 431–441.

Brooking, I. R. and Kirby, E. J. M. (1981). Interrelationships between stem and ear development in winter wheat: The effects of a Norin 10 dwarfing gene Gai/Rht2. *Journal of Agricultural Science* **97**, 373–381.

Campbell, C. A., Davidson, H. R., and Winkleman, G. E. (1981). Effect of nitrogen, temperature, growth stage and duration of moisture stress on yield components and protein content of Manitou spring wheat. *Canadian Journal of Plant Science* **61**, 549–563.

Chujo, H. (1966a). Difference in vernalization effect in wheat under various temperatures. *Proceedings of the Crop Science Society of Japan.* **35**, 177–186.

Chujo, H. (1966b). The effect of diurnal variation of temperature on vernalization in wheat. *Proceedings of the Crop Science Society of Japan* **35**, 187–194.

Chujo, H. (1969). Effects of fertilizer and light on vernalization of wheat plants. *Proceedings of the Crop Science Society of Japan* **38**, 234–240.

Devay, M., Paldi, E., and Kovacs, I. (1976). Thermolabile and thermostable interphases in the vernalization of winter wheat (*Tricitum aestivum* var. Bankuti 1201). *Acta Botanica Academiae Scientiarum Hungaricae* **22**, 9–16.

Doyle, A. D., and Marcellos, H. (1974). Time of sowing and wheat yield in northern New South Wales. *Australian Journal of Experimental Agriculture and Animal Husbandry* **14**, 93–102.

Evans, L. T. and Wardlaw, I. F. (1976). Aspects of the comparative physiology of grain yield in cereals. *Advances in Agronomy* **28**, 301–359.

Fischer, R. A. (1975). Yield potential in a dwarf spring wheat and the effect of shading. *Crop Science* **15**, 607–613.

Fisher, J. E. (1972). The transformation of stamens to ovaries and of ovaries to inflorescences in *Triticum aestivum* L. under short-day treatment. *Botanical Gazette* **133**, 78–85.

Fisher, J. E. (1973). Developmental morphology of the inflorescence in hexaploid wheat cultivars with and without the cultivar Norin 10 in their ancestry. *Canadian Journal of Plant Science* **53**, 7–15.

Ford, M. A., Austin, R. B., Angus, W. J., and Sage, G. C. M. (1981). Relationships between the responses of spring wheat genotypes to temperature and photoperiodic treatments and their performance in the field. *Journal of Agricultural Science* **96**, 623–634.

Forster, H. C., Tincker, M. A. H., Vasey, A. J., and Wadham, S. M. (1932). Experiments in England, Wales and Australia on the effect of length of day on various cultivated varieties of wheat. *Annals of Applied Biology* **19**, 378–412.

French, R. J., Schultz, J. E. and Rudd, C. L. (1979). Effect of time of sowing on wheat phenology in South Australia. *Australian Journal of Experimental Agriculture and Animal Husbandry* **19**, 89–96.

Friend, D. J. C. (1965). Ear length and spikelet number of wheat growth at different temperatures and light intensities. *Canadian Journal of Botany* **43**, 345–353.

Friend, D. J. C., Fisher, J. E., and Helson, V. A. (1963). The effect of light intensity and temperature on floral initiation and inflorescence development of Marquis wheat. *Canadian Journal of Botany* **41**, 1663–1674.

Friend, D. J. C., and Purvis, O. N. (1963). Studies in vernalization of cereals. 14. The thermal reactions in vernalization. *Annals of Botany* **27**, 553–579.

Garner, W. W., and Allard, H. A. (1920). Effect of the relative length of day and night and other factors of the environment on growth and reproduction in plants. *Journal of Agricultural Research* **18**, 553–606.

George, D. W. (1982). The growing point of fall-sown wheat: A useful measure of physiologic development. *Crop Science* **22**, 235–239.

Gotoh, T. (1976). Studies on varietal differences in vernalization requirement in wheat. *Japanese Journal of Breeding* **26**, 307–327.

Gott, M. B. (1957). Vernalization of green plants of a winter wheat. *Nature* (London) **180**, 714–715.

Gott, M. B. (1961). Flowering of Australian wheats and its relation to frost injury. *Australian Journal of Agricultural Research* **12**, 547–565.

Halloran, G. M. (1975). Genotype differences in photoperiodic sensitivity and vernalization response in wheat. *Annals of Botany* **39**, 845–851.

Halloran, G. M. (1976). Genes for vernalization response in homoeologous group 5 of *Triticum aestivum*. *Canadian Journal of Genetics and Cytology* **18**, 211–216.

Halloran, G. M. (1977). Developmental basis of maturity differences in spring wheat. *Agronomy Journal* **69**, 899–902.

Halloran, G. M. and Boydell, C. W. (1967a). Wheat chromosomes with genes for photoperiodic response. *Canadian Journal of Genetics and Cytology* **9**, 394–398.

Halloran, G. M. and Boydell, C. W. (1967b). Wheat chromosomes with genes for vernalization response. *Canadian Journal of Genetics and Cytology* **9**, 632–639.

Halse, N. J., and Weir, R. N. (1970). Effects of vernalization, photoperiod and temperature on phenological development and spikelet number of Australian wheat. *Australian Journal of Agricultural Research* **21**, 383–393.

Halse, N. J., and Weir, R. N. (1974). Effects of temperature on spikelet number of wheat. *Australian Journal of Agricultural Research* **25**, 687–695.

Hay, R. K. M., and Wilson, G. T. (1982). Leaf appearance and extension in field-grown winter wheat plants: The importance of soil temperature during vegetative growth. *Journal of Agricultural Science* **99**, 403–410.

Holmes, D. P. (1973). Inflorescence development of semidwarf and standard height wheat cultivars in different photoperiod and nitrogen treatments. *Canadian Journal of Botany* **51**, 941–956.

Ishihara, A. (1960). Physiological studies on the vernalization of wheat plants. II. The progress of vernalization under the field condition. *Proceedings of the Crop Science Society of Japan* **28**, 297–308.

Ishihara, A. (1961). Physiological studies on the vernalization of wheat plants. III. Direct and indirect induction by low temperature in apical and lateral buds. *Proceedings of the Crop Science Society of Japan* **30**, 88–92.

Ishihara, A. (1963). Physiological studies on the vernalization of wheat plants. IV. The effect of temperature and photoperiod on the process of vernalization in growing plant. *Proceedings of the Crop Science Society of Japan* **31**, 297–308.

Junges, W. (1959). Beeinflussung des Blühbeginns annueller landwirtschaftlicher and gärtnerischer Kulturpflanzen durch Jarowisation bei konstanten Temperaturen zwischen −10°C and +35°C. *Zeitschrift für Pflanzenzüchtung* **41**, 103–122.

Kirby, E. J. M. (1974). Ear development in spring wheat. *Journal of Agricultural Science* **82**, 437-447.

Klaimi, Y. Y., and Qualset, C. O. (1973). Genetics of heading time in wheat (*Triticum aestivum* L.). I. The inheritance of photoperiodic response. *Genetics* **74**, 139-156.

Kolderup, F. (1979). Application of different temperatures in three growth phases of wheat. II. Effects on ear size and seed setting. *Acta Agriculturae Scandinavica* **29**, 11-16.

Krekule, J. (1961). The effect of photoperiodic regime on vernalization of spring wheat. *Biologia Plantarum (Praha)* **3**, 180-191.

Langer, R. H. M. and Hanif, M. (1973). A study of floret development in wheat (*Triticum aestivum* L.). *Annals of Botany* **37**, 743-751.

Law, C. N., Worland, A. J. and Giorgi, B. (1976). The genetic control of ear-emergence time by chromosomes 5A and 5D of wheat. *Heredity* **36**, 49-58.

Law, C. N., Sutka, J. and Worland, A. J. (1978). A genetic study of daylength response in wheat. *Heredity* **41**, 185-191.

Levy, J., and Peterson, M. L. (1972). Responses of spring wheats to vernalization and photoperiod. *Crop Science* **12**, 487-490.

Lucas, D. (1972). The effect of daylength on primordia production of the wheat apex. *Australian Journal of Biological Science* **25**, 649-656.

MacDowall, F. D. H. (1977). Growth kinetics of Marquis wheat. VII. Dependence on photoperiod and light compensation point in vegetative phase. *Canadian Journal of Botany* **55**, 639-643.

McIntosh, R. A. (1973). A catalogue of gene symbols for wheat. *In* "Proceedings of the Fourth International Wheat Genetics Symposium" (Columbia, Miss.). (E. R. Sears and L. M. S. Sears, eds.), pp. 893-937. University of Missouri, Columbia.

McKinney, H. H. and Sando, W. J. (1933). Russian methods for accelerating sexual reproduction in wheat. *Journal of Heredity* **24**, 165-167.

McKinney, H. H. and Sando, W. J. (1935). Earliness of sexual reproduction in wheat as influenced by temperature and light in relation to growth phases. *Journal of Agricultural Research* **51**, 621-641.

Major, D. J. (1980). Photoperiod response characteristics controlling flowering of nine crop species. *Canadian Journal of Plant Science* **60**, 777-784.

Mangas Martin, V. J., and Sanchez de la Puente, L. (1980). Ecophysiological factors determining wheat grain production. *Agrochimica* **24**, 160-168.

Marcellos, H., and Single, W. V. (1975). Temperatures in wheat during radiation frost. *Australian Journal of experimental Agriculture and Animal Husbandry* **15**, 818-822.

Martinic, Z. (1975). Life cycle of common wheat varieties in natural environments as related to their response to shortened photoperiod. *Zeitschrift für Pflanzenzüchtung* **75**, 237-251.

Midmore, D. J., Cartwright, P. M. and Fischer, R. A. (1982). Wheat in tropical environments. I. Phasic development and spike size. *Field Crops Research* **5**, 185-200.

Mohapatra, P. K., Aspinall, D., and Jenner, C. F. (1982). The growth and development of the wheat apex: The effects of photoperiod on spikelet production and sucrose concentration in the apex. *Annals of Botany* **49**, 619-626.

Murfet, I. C. (1977). Environmental interaction and the genetics of flowering. *Annual Review of Plant Physiology* **28**, 253-278.

Owen, P. C. (1971). Responses of a semidwarf wheat to temperatures representing a tropical dry season. II. Extreme temperatures. *Experimental Agriculture* **7**, 43-47.

Pinthus, M. J. (1967). Comparative analysis of growth and grain yield components of winter wheat and spring wheat. *Israel Journal of Botany* **16**, 160-161.

Pirasteh, B., and Welsh, J. R. (1980). Effect of temperature on the heading date of wheat cultivars under a lengthening photoperiod. *Crop Science* **20**, 453-456.

Puckridge, D. W. (1968). Competition for light and its effect on leaf and spikelet development of wheat plants. *Australian Journal of Agricultural Research* **19**, 191–201.

Pugsley, A. T. (1963). The inheritance of a vernalization response in Australian spring wheats. *Australian Journal of Agricultural Research* **14**, 622–627.

Pugsley, A. T. (1966). The photoperiodic sensitivity of some spring wheats with special reference to the variety Thatcher. *Australian Journal of Agricultural Research* **17**, 591–599.

Pugsley, A. T. (1971). A genetic analysis of the spring-winter habit of growth in wheat. *Australian Journal of Agricultural Research* **22**, 21–31.

Pugsley, A. T. (1972). Additional genes inhibiting winter habit in wheat. *Euphytica* **21**, 547–552.

Rahman, M. S. (1980). Effect of photoperiod and vernalization on the rate of development and spikelet number per ear in 30 varieties of wheat. *Journal of the Australian Institute of Agricultural Science* **46**, 68–70.

Rahman, M. S., and Wilson, J. H. (1977). Determination of spikelet number in wheat. I. Effect of varying photoperiod on ear development. *Australian Journal of Agricultural Research* **28**, 565–574.

Rahman, M. S., and Wilson, J. H. (1978). Determination of spikelet number in wheat. III. Effect of varying temperature on ear development. *Australian Journal of Agricltural Research* **29**, 459–467.

Rahman, M. S., Wilson, J. H., and Aitken, Y. (1977). Determination of spikelet number in wheat. II. Effect of varying light level on ear development. *Australian Journal of Agricultural Research* **28**, 575–581.

Rawson, H. M. (1970). Spikelet number, its control and relation to yield per ear in wheat. *Australian Journal of Biological Science* **23**, 1–15.

Rawson, H. M. (1971). An upper limit for spikelet number per ear in wheat as controlled by photoperiod. *Australian Journal of Agricultural Research* **22**, 537–546.

Riddell, J. A., and Gries, G. A. (1958a). Development of spring wheat. II. The effect of temperature on responses to photoperiod. *Agronomy Journal* **50**, 739–742.

Riddell, J. A., and Gries, G. A. (1958b). Development of spring wheat. III. Temperature of maturation and age of seeds as factors influencing their response to vernalization. *Agronomy Journal* **50**, 743–746.

Saini, H. S., and Aspinall, D. (1982). Abnormal sporogenesis in wheat (*Triticum aestivum* L.) induced by short periods of high temperature. *Annals of Botany* **49**, 835–846.

Salisbury, P. A., Berry, G. J., and Halloran, G. M. (1979). Expression of vernalization genes in near-isogenic wheat lines: Methods of vernalization. *Canadian Journal of Genetics and Cytology* **21**, 429–434.

Single, W. V., and Marcellos, H. (1974). Studies on frost injury to wheat. IV. Freezing of ears after emergence from the leaf sheath. *Australian Journal of Agricultural Research* **25**, 679–686.

Stern, W. R., and Kirby, E. J. M. (1979). Primordium initiation at the shoot apex in four contrasting varieties of spring wheat in response to sowing date. *Journal of Agricultural Science* **93**, 203–215.

Tan, K.-H., Wang, W.-H., He, X.-W., and Li, S.-Q. (1981). Effect of metabolic inhibitors on vernalization in winter wheat. *Acta Botanica Sinica* **23**, 371–376.

Trione, E. J., and Metzger, R. J. (1970). Wheat and barley vernalization in a precise temperature gradient. *Crop Science* **10**, 390–392.

Wall, P. C. and Cartwright, P. M. (1974). Effects of photoperiod, temperature and vernalization on the phenology and spikelet number of spring wheats. *Annals of Applied Biology* **76**, 299–309.

Wanser, H. M. (1922). Photoperiodism in wheat, a determining factor in acclimatization. *Science* **56**, 313–315.

Warrington, I. J., Dunstone, R. L., and Green, L. M. (1977). Temperature effects at three developmental stages on the yield of the wheat ear. *Australian Journal of Agricultural Research* **28**, 11–27.

Weibel, D. K. (1958). Vernalization of immature winter wheat embryos. *Agronomy Journal* **50**, 267–270.

Wiegand, C. L., Gerbermann, A. H. and Cuellar, J. A. (1981). Development and yield of hard red winter wheats under semitropical conditions. *Agronomy Journal* **73**, 29–37.

Williams, R. F., and Williams, C. N. (1968). The physiology of growth in the wheat plant. IV. Effects of day length and light energy level. *Australian Journal of Biological Science* **21**, 835–854.

CHAPTER 9

Seed Development and Quality in Bean

P. B. GOODWIN and M. A. SIDDIQUE

I. INTRODUCTION

Seed quality is optimal when it ensures that the seed is most likely to be dispersed to a suitable environment to germinate at the most favourable time, and to give vigorous seedlings. Thus, adaptations for dispersal and appropriate dormancy systems are important elements of seed quality in non-cultivated plants. However, in cultivated plants one aspect of seed quality dominates, namely, seed vigour, recently defined as "the sum total of those properties of the seed which determine the potential level of activity and performance of the seed or seed lot during germination and seedling emergence" (Perry, 1980). Reviews on seed production are available (Hebblethwaite, 1980; Vis, 1980). This review will concentrate on common, french, snap or navy beans (*Phaseolus vulgaris* L.). The topic will be discussed under two headings; firstly, the pattern of bean seed development, and secondly, the development of seed vigour.

II. DEVELOPMENT OF THE SEED IN BEAN

Detailed reviews of the morphological, anatomical and physiological aspects of normal seed development have been presented by a number of authors

CONTROL OF CROP PRODUCTIVITY
ISBN 0 12 548280 9

(e.g. Bhatnagar and Johri, 1972; Bopp, 1979; Rubenstein, 1980; Egli, 1981). Bewley and Black (1978) reviewed the general pattern of chemical changes accompanying seed development. Although these aspects have been well studied in some of the commercially important legumes, particularly the garden pea (*Pisum sativum* L.) and field pea (*Pisum arvense* L.), fewer studies have been carried out on french bean (*Phaseolus vulgaris* L.).

Loewenberg (1955) attempted to define an index of physiological age for bean seeds. He described the development of the seeds on a cellular and organ basis from flowering to maturity. The dry weight, nitrogen and phosphorus per seed increased slowly during the first 3 weeks after flowering, reaching 17% of their final values. In the following two weeks, the rate of fresh and dry weight, nitrogen and phosphorus accumulation increased, reaching maximum rates 24 days after flowering. Older seeds continued to grow at a steadily decreasing rate up to maturity. Fresh weight reached a peak at 36 days. No satisfactory age index could be established.

In a similar study Carr and Skene (1961) found a diauxic pattern of growth. In the first phase, growth was exponential, and dry weight and fresh weight increased at about the same relative rate. In the second phase, which lasted about three days and was referred to as the lag phase, growth was much slower than during the first phase. In the third phase, growth was initially about as rapid as in the first exponential phase, but gradually declined to zero. Only slight quantitative differences were found in the growth patterns of the seeds when the beans were grown at different times of the year. In this study pod growth was completed within 16–17 days of anthesis. Seed growth began about nine days after anthesis and experienced a three day lag phase from about the 20th to the 23rd day. It was suggested that mechanical restriction caused by the surrounding structures caused the growth rate of the seed to fall during the lag phase.

While Loewenberg (1955) failed to develop an age index, Walbot *et al.* (1972) were able to construct a timetable for reproductive development in the bean cultivar Taylor's Horticultural using morphological characters of the pod, seed and embryo. They partitioned the entire period from anthesis to dry seed (36 days) into nine stages of about equal duration. Identification of each stage was based on several criteria, most notably size, colour and morphology of the pod, seed and embryo. During the first six days of seed development the embryo developed rapidly from the unicellular zygote to the multicellular form. Pod development was rapid for about 14 days, and by the end of this period, pods had reached their maximum length. Seed length began to increase immediately after fertilization, and growth continued gradually for the first two weeks. Seed development was interrupted after the 16th day and for two days there was almost no increase in length. Elongation resumed after the 18th day and the seed reached its maximum length at day 24.

The dry weight of seed increased very little during the early period of development, and it was not until the 18th day, coincident with the second period of rapid growth in seed length, that dry weight began to increase rapidly. Both seed length and dry weight increases were diauxic, but the lag phase for seed length was from day 16 to 18, whereas for dry weight it was from day 22 to 24. Fresh weight reached a maximum by day 28–30, and declined as the embryo dehydrated. Seed dry weight increased until day 32.

Hsu (1979) compared morphological development in seed of two bean cultivars, Taylor's Horticultural and PI226895. Both attained full pod length by about 15 days after anthesis, and mature seeds after 36 days. The testa developed very rapidly after anthesis, contributing the major part of the seed during the first phase of development. The embryo did not exceed the weight of the seed coat until two weeks after anthesis. Lag phases in the increase of seed length and dry weight were recorded. Seed of cv. Taylor's Horticultural grew more slowly than cv. PI226895 in the first growth phase, but more rapidly in the final phase, so that cv. Taylor's Horticultural produced larger seed within the same period of growth.

The photosynthetic contribution of the pod to the developing seed was investigated by Oliker et al. (1978). The chlorophyll content of the pod was highest in very young pods, decreased slowly at first and then disappeared at about 32 days when the seeds were at maximum fresh weight. Photosynthesis of whole pods (with seeds inside) increased slowly until about 26 days after anthesis, and then decreased rapidly as chlorophyll was destroyed. Net photosynthesis of the pod was very low, and it barely balanced carbon dioxide loss by respiration. Until 16 days of growth, photosynthesis of the pod plus seed contributed between 2.5% and 3.5% of the daily weight increment of the pod and thereafter its contribution decreased rapidly.

Oliker et al., 1978 suggested that the seed does not depend on the pod for its photosynthate, rather the growth of both depends almost exclusively on photosynthate originating outside the pod, presumably in the leaves. This is in accord with reports of Hall et al. (1972) and Crookston et al. (1974). Oliker et al. (1978) suggested that pods and seeds constitute a system of competing sinks. They proposed two possibilities for the control of sequential distribution of nutrients between the two sinks: (1) either the events that start rapid seed growth also cause the diversion of nutrients to the developing seed and so the pod stops growing, or (2) when the pod has reached its maximal size it no longer acts as a sink and the onset of rapid seed development becomes possible. The desiccation of the pod appears to serve as an additional signal which shuts off supplies to the pods and diverts them to the seeds. They do not mention the possibility of transfer of carbohydrates accumulated in the pods into the seeds during pod desiccation. However, they found no evidence of nitrogen transfer from the pods into the seeds during desiccation.

Changes in nitrogen and protein content in bean seeds have been studied by several workers. Opik (1968) found that the weight increase of bean cotyledons (cv. Belfast New Stringless) followed a sigmoid curve. The rapid phase began just after the liquid endosperm was exhausted, and slowed down when cell expansion was complete. However, the percentage of nitrogen decreased at the early stages of growth, and then (after 32 days) remained constant at 3.8% of the dry weight. Loewenberg (1955) reported similar results, where nitrogen was 6.5% of dry weight 12 days after flowering, fell to 3.5% at 32 days and then remained constant. However, nitrogenous compounds were still being translocated into the cotyledons, and total nitrogen content per cotyledon increased in a similar manner to dry weight.

Powrie et al. (1960) found that mature beans cv. Michelite contain 27%, 46% and 4.8% protein (on a dry weight basis) in the cotyledons, embryonic axis and seed coat, respectively. Ma and Bliss (1978) reported protein contents of cotyledons ranging between 25% and 28% in a range of bean genotypes at maturity. Sun et al. (1978) investigated protein synthesis and accumulation in bean cotyledons, cv. Tendergreen, during growth. The most dramatic changes in protein accumulation occurred from 10 to 31 days after flowering. Over this period elongation of the seed (from 6 mm to 21 mm) was proportional to the increase in fresh weight of the seed. Whereas only low levels of protein were extracted from the cotyledons of young seeds, after 17 days the protein content increased sharply. By the 37th day the protein content was 75 times the level at 14 days. The seeds increased in protein content by 3.3 milligrams per day per cotyledon pair over the active phase of protein accumulation. The mature cotyledons contained 20% protein, about 50% of which was storage globulin. No globulin-1 protein was detected at 10 days. The major burst of globulin-1 synthesis began at 16 days.

Although mature bean seeds contain more than 35% starch on a dry weight basis (Peterson and Churchill, 1921; Eichelberger, 1922; Powrie et al., 1960) the details of the pattern of starch accumulation in the developing bean seed are not well known. Pfenninger (1909) found that almost all starch present in the pod was gone by the end of the ripening period, and the loss per pod was only about 25% of the grain by the seeds. He suggested that the carbohydrate was translocated to the seeds from other parts of the plant, the pod acting as a temporary reservoir. Opik (1968) noted initiation of starch synthesis and subsequent enlargement of starch grains at a very early developmental stage. Yeung (1980) demonstrated the major role of the suspensor in sucrose uptake to the embryo during the heart to early cotyledon stage. Siddique (1980), working on cv. Apollo, found that the amount of both total and trichlorocetic acid (TCA) soluble nitrogen per gram of dry weight was maximal in the earliest harvested seed (19 days after anthesis), and decreased with seed development. In contrast, the percentage of starch increased markedly, from 11.1% at 19 days to 39.0% at 43 days, remaining at this level until maturity at 49 days.

Relatively few studies describe the developing bean seed at the cellular or subcellular level. Loewenberg (1955) estimated the fresh weight, dry weight, nitrogen and phosphorus content per cotyledon cell using values based on the content of the average bean seed at various stages of development, the proportion of the seed dry weight due to the cotyledons, and the number of cotyledon cells at a given stage of seed development. The fresh weight, dry weight, nitrogen and phosphorus content per cotyledon cell increased gradually during the first three weeks after flowering. At this stage cell division ceased, and the cells grew rapidly, achieving their maximal growth rate 24 days after flowering. About 36 days after flowering the cell fresh weight started to decline, and only small further gains in dry weight, nitrogen and phosphorus were made.

An elaborate study on the development of cotyledon cell structure in ripening bean seeds of the cultivar (cv.) Belfast New Stringless has been reported by Opik (1968). Observations of cell size showed that cell expansion was completed by 34 days. Cotyledon cells in the outermost three to five layers were meristematic and dividing 16 days after flowering. Vacuolation was slight and the cell's walls were thin at this stage. Plastids and mitochondria were not highly differentiated. The inner parenchyma cells were highly vacuolated. Many plastids contained starch granules with a maximum length of four microns. By 25 days all cell division had ended. Cells continued to expand, but the large vacuoles became subdivided into small ones, with flocculent contents. The vacuoles were beginning to turn into reserve protein bodies. Starch granules grew considerably, distorting the plastid membranes. With continuing growth (maturation occurred at 53–56 days) the cotyledon cells continued to expand and to accumulate reserves. Starch granules continued to enlarge, some reaching diameters of 50 microns. Growth of the cell walls proceeded in three stages. During cell division little thickening occurred. Wall thickening became apparent after cell division stopped. When cell expansion stopped, a new type of wall thickening occurred mainly at the cell corners.

III. DEVELOPMENT OF SEED VIGOUR

The preceding work is a substantial description of bean seed development in morphological, chemical, anatomical and ultrastructural terms. However, there is relatively little work linking these properties with the development of the ability of the seed to germinate and to produce a vigorous seedling.

Premature dehydration caused by unduly early harvest or any other factor can result in abnormalities such as immature embryos in carrots (e.g. Austin et al., 1969) or hollow heart (cavitation) in peas (Perry and Howell, 1965). Hollow heart has been associated with emergence failure (Scott and Close,

1976). Vigour loss in prematurely harvested bean seed has been demonstrated by a number of workers (Inoue and Suzuki, 1962; Wijandi and Copeland, 1974; Somerset, 1977; Goodwin et al., 1978). However, prematurely harvested crops also give smaller seeds.

Most work on the effect of seed size on the growth and yield of bean crops agrees that larger seeds produce larger seedlings and in some cases higher yields (Adams and Lacascio, 1968; Clark and Peck, 1968; Ries, 1971; Smittle et al., 1976). However, Hardenburg (1942) did not obtain any significant difference between the total yield of field beans grown from large seeds and those grown from small seeds. Proposed explanations for the correlation between seed size and seedling vigour include the higher photosynthetic rate of larger cotyledons from large seeds (Burris, et al., 1971), the production of seedlings with a larger initial photosynthetic area by large seeds (Black, 1957), and larger amounts of reserve food material in the larger seeds (Tomkins, 1965).

Inoue and Suzuki (1962) studied the germinability of bean seed cv. Masterpiece at different stages of development. They reported that fresh seeds harvested 15 days after anthesis did not germinate at all, but those harvested more than 25 days after anthesis showed a gradual increase in germination as the period between anthesis and seed harvest became longer. Fresh seeds harvested 35 days after anthesis showed almost 100% germination. Although the seeds harvested at 15 days did not germinate in the fresh condition, they showed high germination with 20 days after-ripening (seeds were allowed to dry on the harvested plants for 20 days). Achievement of germinability in the after-ripened seed cannot be attributed solely to drying, as a considerable increase in seed weight did occur during the after-ripening process.

Walbot et al. (1972) found that even exceedingly immature embryos retained the ability to germinate after drying. When tested on aseptic media the dried bean embryos from progressively earlier developmental stages required progressively longer periods to complete germination (production of seedlings) and germinated at reduced frequencies. The youngest embryos (eight to eleven days from anthesis; dry weight of the seed about 2.5% of the final weight) showed only about 30% germination and did not show cotyledon expansion.

Siddique (1980) attempted to relate the development of the ability to germinate to chemical, physical and ultrastructural changes in cv. Apollo. The development of ability to germinate could not be correlated with changes in total nitrogen, trichloroacetic acid soluble nitrogen, trichloroacetic acid insoluble (protein) nitrogen, starch content, soluble sugar content, thickening of cell walls or development of protein bodies. In dried seeds but not in seeds tested immediately after harvest, the ability to germinate was well correlated with seed dry weight, and with lack of ion leakage from imbibed seeds. Seeds that were tested immediately after harvest did not show 50% germination until 47 days after anthesis (assessed as per cent production of normal seedlings), by which time the seed moisture content had fallen to 30%. They showed 50% radicle emergence at 37 days. However, if the isolated seeds were allowed to dry for two to three days at 20°C, they showed 50% germination at 31 days. Seeds dried in pods, a process taking six days, gave 80% germination when the pods were collected as early as 25 days after anthesis. Even seeds dried rapidly under forced air, where the drying process was complete within 12 hours, gave 50% germination at 39 days. Thus seed drying appears to induce the development of the capacity to germinate and produce normal seedlings even in very young seeds. Seed dehydration during normal growth may play an essential role in inducing the development of seed vigour. A careful study of the changes occurring during seed drying appears warranted. Such a requirement both prevents germination during seed development and enables early seed maturation should the plant be dehydrated.

Austin (1972) and Maguire (1977) review the effects of environment on seed structure and composition and attempt to relate these effects to seed quality and performance. The environment includes factors such as mineral nutrition, temperature, rainfall and soil moisture. In bean, seedling vigour, seed weight and seed yield depend on adequate nitrogen fertilization; high vigour being associated with high protein content of the seed (Ries, 1971). Sinclair and DeWitt (1975) suggested that the high nitrogen requirement of the developing bean seeds could only be met by a rapid breakdown of proteins in the vegetative plant parts, with a consequent rapid onset of senescence. Adequate nitrogen fertilization may allow more time for seed maturation.

Hypocotyl necrosis is a serious physiological disorder in beans that are germinated in media with a low free calcium content. It is most common in cv. Tendercrop types (Clark and Kline, 1965). The disorder can be prevented by the addition of calcium salts to the germination medium (Shannon et al., 1967; Helms, 1971). Liming in low pH soils has been shown to reduce the incidence of hypocotyl necrosis in the progeny of bean plants, but not to eliminate it (Williams et al., 1966). There is a correlation between a high magnesium content in the cotyledon and enhanced resistance to mechanical damage (cotyledonary cracking) (Dickson et al., 1973; Aqil and Boe, 1975). A response in seed vigour to potassium fertilization at relatively high rates has been demonstrated by Ison (1979) in pot trials. Thus all major nutrients except phosphorus and iron have been shown to be important for the production of high-vigour bean seeds. Presumably the importance of the remaining nutrients will also be shown.

Experiments on the effects of preharvest temperature on seed development are few, and generally agree that high temperatures during seed development cause early ripening and rapid maturation of seed (lima bean, Lambeth, 1950; rice, Nagato and Ebata, 1960; pea, Robertson et al., 1962). It has been shown that low temperatures lead to the production of larger seeds in pea (Robertson et al., 1962), flax (Zhdanova, 1969), pearl millet (Fussell et al., 1980) and bean (Siddique and Goodwin, 1980a; Moss and Mullett, 1982).

The Centro Internacional de Agricultura Tropical (CIAT) annual report (1978) points out that the optimum temperature range for growing beans is very narrow. Most of the total bean yield in Central and South America is from regions with an average growing season temperature within the range of 20°C to 22.5°C. International yield trials also indicate that bean yield is maximal at mean temperatures close to 20°C.

Glasshouse investigations by Siddique and Goodwin (1980a) showed that a 24°C/19°C (day/night, 12 hours each) temperature regime during the initial growth of the plant was optimal for the development of healthy, compact and strong bean cv. Apollo plants. Seed yield was similar at seed development temperatures in the range 18°C/13°C–24°C/19°C, imposed after the plants had grown up to pod set at 25°C/18°C. However, seed size decreased with increase in temperature over this range. Plants at 27°C/22°C gave maximal pod and seed numbers per plant, but because of the small seed size, gave low seed yields. At higher temperatures (33°C/28°C and 30°C/25°C) pod set and seed development were reduced. When plants were grown to the yellow, fleshy pod stage (maximum seed fresh weight) at 25°C/18°C, and then passed to other temperatures for seed maturation, the detrimental effects of high temperature were still observed, the seed being of low vigour, and very susceptible to mechanical damage.

Further investigation (Siddique and Goodwin, 1980b) revealed a requirement for low maturation temperatures (21°C/16°C or 18°C/13°C) for optimum seed quality in ten genotypes of snap bean, including genotypes resistant to mechanical damage. Resistance to mechanical damage also was maximal in seeds of all genotypes matured at low temperature. Although in general the colour-seeded genotypes showed greater tolerance to high maturation temperatures than white-seeded types, one white-seeded genotype showed good tolerance. This suggests that breeding and selection of white-seeded lines showing improved tolerance of high seed maturation temperatures is possible.

Unpublished field studies on seed crops grown in the Burdekin, Queensland, by Somerset, Siddique and Goodwin showed that if the crop is cut in the usual manner, that is, at the stage when the leaves have fallen and all the pods are dry, or the plants are cut at any stage and allowed to dry on the ground in single rows as is commonly done, seed quality is poor. The poor quality of seed is associated with high pod temperature during seed maturation–fleshy pods exposed to the sun reached 35.6°C and dry pods reached 33.3°C (internal temperatures) at an air temperature of 30.5°C. Cutting the crop before leaf-fall, at about 50% seed moisture (20%–40% of pods dry) and windrowing immediately in 5–10 rows to 1 windrows gave high seed quality.

Moss and Mullett (1982) have suggested that there is an optimum growing temperature for bean seed quality. They grew bean plants for five generations at four temperatures (21°C/16°C, 24°C/19°C, 27°C/22°C and 30°C/25°C). For the first four generations no selection, and no seed tests were carried out. At the fifth generation, fourth generation seed from each temperature was grown at all four temperatures and its progeny assessed for vigour by both seedling evaluation and potassium ion leaching. As judged by potassium leaching, there was a clear cut response to the growing temperature during the first four generations, 27°C/22°C giving optimal seed quality. The growing temperature of the final generation had no consistent effect on ion leaching. As judged by seedling evaluation, only cv. Apollo showed a significant response to temperature during the first four generations, the optimum temperature being 27°C/22°C or 30°C/25°C. The seedling evaluation response to the growing temperature of the final generation varied between cultivars, Apollo showing poorest vigour after production at 27°C/22°C, Oregon-1604 poorest at 21°C/16°C and Goldcrop poorest at 30°C/25°. Moss and Mullett suggested that the responses to the final growing temperatures were in conflict with those of Siddique and Goodwin (1980a,b). However, the latter authors were studying the effect of maturation temperatures only.

The water regime experienced by plants with developing seeds can be damaging in two ways, firstly a low water regime can cause premature dehydration of developing seeds, usually resulting from a premature harvest, and secondly, a high and varying moisture regime can cause "weathering" of the mature seed. Moore, in a series of papers (1964, 1965a,b,c, 1966), pointed out that fracturing of bean embryos, particularly hypocotyls, cotyledons and seed coats can take place during the drying phase of seed development by causing rapid, uneven losses of moisture and the resulting unequal shrinkage of drying tissue. Similar damage is induced by rapid absorption of water into dry mature seed (Pollock and Manalo, 1970; Dickson et al., 1973). In the field, rapid drying at high pod temperatures, with moisture cycling within the dry seeds due to alternate night dews and hot dry days, greatly accentuates the problem. Even under constant conditions, the combination of high levels of seed moisture and high temperatures leads to rapid deterioration of mature seed (Justice and Bass, 1978).

Most types of seed injury, whether caused by premature harvest, field weathering, high maturation temperatures, poor mineral nutrition, or mechanical damage, lead to similar faults in the progeny seedlings. These faults include cracked, broken or necrotic cotyledons, damage to the food transport system in the embryo and loss of essential structures. Ultimately, abnormal seedlings are produced. Thus, although there are a number of vulnerable organs in the seed, no particular organ appears to show special sensitivity to a particular damaging agent, except hypocotyl sensitivity to calcium deficiency. Nor is there obvious evidence for a general failure at the cellular level, such as might be presumed by using ion leakage or respiration tests for seed vigour.

IV. CONCLUSIONS

In bean, the requirements for the production of high-vigour seed appear to be as follows. The level of mineral nutrition, particularly of nitrogen and potassium, needs to be good. The temperature regime needs to be about 24°C/22°C during plant growth and pod filling, and 21°C/16°C to 18°C/13°C during seed maturation. Maturation needs to take place in a dry environment. The ability to germinate develops only as the seeds dry. Using this information it should be possible to progress to both a better understanding of the physiology of development of seed vigour, and a more precise definition of regions, times of planting and cultural and harvest techniques most likely to lead to the production of high-quality bean seed.

REFERENCES

Adams, Z., and Lacascio, S. J. (1968). Seed size and depth of planting effects on broccoli, sweet corn and beans. *Sunshine State Agricultural Research Report,* 14–16.

Aqil, B. A., and Boe, A. A. (1975). Occurrence of cotyledonal cracking in snap beans and its relation to the nutritional status in the seed. *HortScience* **10,** 509–510.

Austin, R. B. (1972). Effects of environment before harvesting on viability. *In* "Viability of Seeds" (E. H. Roberts, ed.), pp. 114–149. Chapman and Hall, London.

Austin, R. B., Longden, P. C. and Hutchinson, J. (1969). Some effects of "hardening" carrot seed. *Annals of Botany* **33,** 883–895.

Bewley, J. D., and Black, M. (1978). "Physiology and Biochemistry of Seeds" Vol. 1. Springer-Verlag, New York.

Bhatnagar, S. P. and Johri, B. M. (1972). Development of angiosperm seeds. *In* "Seed Biology" Vol. 1 (T. T. Kozlowski, ed.), pp. 77–149. Academic Press, New York.

Black, J. N. (1957). The early vegetative growth of three strains of subterranean clover (*Trifolium subterranean* L.) in relation to seed size. *Australian Journal of Agricultural Research* **8,** 1–14.

Bopp, M. (1979). Developmental physiology. *Progress in Botany* **41,** 135–142.

Burris, J. S., Wahab, A. H., and Edge, O. T. (1971). Effect of seed size on seedling performance in soybean seedling growth and respiration in the dark. *Crop Science* **11,** 492–496.

Carr, D. J., and Skene, K. G. M. (1961). Diauxic growth curves of seeds, with special reference to French beans (*Phaseolus vulgaris* L.). *Australian Journal of Biological Sciences* **14,** 1–12.

Centro Internacional de Agricultura Tropical (CIAT). 1978. CIAT Annual Report, Bean Program, pp. 1–70. Cali, Colombia.

Clark, B. E., and Kline, D. B. (1965). Effects of water temperature, seed moisture content, mechanical injury, and calcium nitrate solution on the germination of snap bean seeds in laboratory germination tests. *Proceedings of the Association of Official Seed Analysts* **55**, 110–120.

Clark, B. E. and Peck, N. H. (1968). Relationship between the size and performance of snap bean seeds. *New York State Agricultural Experiment Station Bulletin* 819.

Crookston, R. K., O'Toole, J., and Ozbun, J. L. (1974). Characterization of the bean pod as a photosynthetic organ. *Crop Science* **14**, 708–712.

Dickson, M. H., Duczmal, K. and Shannon, S. (1973). Imbibition rate and seed composition as factors affecting transverse cotyledon cracking in bean (*Phaseolus vulgaris* L.) seed. *Journal of the American Society for Horticultural Science* **98**, 509–513.

Egli, D. B. (1981). Species differences in seed growth characteristics. *Field Crops Research* **4**, 1–12.

Eichelberger, M. (1922). The carbohydrate content of navy beans. *Journal of the American Chemical Society* **44**, 1407–1408.

Fussell, L. K., Pearson, C. J., and Norman, M. J. T. (1980). Effect of temperature during various growth stages on grain development and yield of *Pennisetum americanum. Journal of Experimental Botany* **31**, 621–633.

Goodwin, P. B., Trutza, G., Somerset, G. and Siddique, M. A. (1978). Maturation studies on bean seeds (*Phaseolus vulgaris*). *In* "Bean Improvement Workshop" (B. J. Ballantyne, ed.), pp. 29–34. NSW Department of Agriculture, Sydney.

Hall, T. C., McLeester, R. C., and Bliss, J. A. (1972). Electrophoretic analysis of protein changes during the development of French bean fruit. *Phytochemistry* **11**, 647–649.

Hardenburg, E. V. (1942). Experiments with field beans. *Cornell University Agricultural Experiment Station Bulletin* 776.

Hebblethwaite, P. D. (1980). "Seed Production" Butterworths, London.

Helms, K. (1971). Calcium deficiency of dark-grown seedlings of *Phaseolus vulgaris* L. *Plant Physiology* **47**, 799–804.

Hsu, F. C. (1979). A developmental analysis of seed size in common bean. *Crop Science* **19**, 226–230.

Inoue, Y., and Suzuki, Y. (1962). Studies on the effect of maturity and after-ripening of seeds upon the seed germination in snap bean, *Phaseolus vulgaris* L. *Journal of the Japanese Society for Horticultural Science* **31**. 146–150.

Ison, R. (1979). Seed vigour in beans in relation to potassium nutrition. *Australian Seed Science Newsletter* **5**, 66–71.

Justice, O. L., and Bass, L. N. (1978). Principles and practices of seed storage. *U.S.D.A. Agriculture Handbook* 506.

Lambeth, V. N. (1950). Some factors influencing pod set and yield of the lima bean. *Missouri Agricultural Experiment Station Research Bulletin* 466.

Loewenberg, J. R. (1955). The development of bean seeds (*Phaseolus vulgaris* L.). *Plant Physiology* **30**, 244–249.

Ma, Y., and Bliss, F. A. (1978). Seed protein in common bean. *Crop Science* **18**, 431–437.

Maguire, J. D. (1977). Seed quality and germination. *In* "The Physiology and Biochemistry of Seed Dormancy and Germination" (A. A. Khan, ed.), pp. 219–235. North-Holland Publishing Co., Amsterdam.

Moore, R. P. (1964). Natural-fracturing and crushing of snap bean seed. *Association of Official Seed Analysts Newsletter* **38**, 14–16.

Moore, R. P. (1965a). Internal seed injuries during moistening. *Association of Official Seed Analysts Newsletter* **39**, 34–35.

Moore, R. P. (1965b). Weather-fractured embryos in snap beans. *Seed Technology News* **34**, 13–14.

Moore, R. P. (1965c). Weather-fractured seed coats in bean. *Association of Official Seed Analysts Newsletter* **39**, 26–27.

Moore, R. P. (1966). Weather-fractured hypocotyls in Great Northern bean seed. *Seed Technology News* **35**, 7–8.

Moss, G. I., and Mullett, J. H. (1982). Potassium release and seed vigour in germinating bean (*Phaseolus vulgaris* L.) seed as influenced by temperature over the previous five generations. *Journal of Experimental Botany* **33**, 1147–1160.

Nagato, K., and Ebata, M. (1960). Effects of temperature in ripening periods upon development and qualities of lowland rice kernels. *Proceedings of the Crop Science Society of Japan* **28**, 275–278.

Oliker, M., Poljakoff-Mayber, A.M and Mayer, A. (1978). Changes in weight, nitrogen accumulation, respiration and photosynthesis during growth and development of seeds and pods of *Phaseolus vulgaris*. *American Journal of Botany* **65**, 366–371.

Opik, H. (1968). Development of cotyledon cell structure in ripening *Phaseolus vulgaris* seeds. *Journal of Experimental Botany* **19**, 64–76.

Perry, D. A. (1980). The concept of seed vigour and its relevance to seed production techniques. *In* "Seed Production" (P. D. Hebblethwaite, ed.), pp. 585–591. Butterworths, London.

Perry, D. A., and Howell, P. J. (1965). Symptoms and nature of "Hollow Heart" of pea seed. *Plant Pathology* **14**, 111–116.

Peterson, W. H., and Churchill, H. (1921). The carbohydrate content of the navy bean. *Journal of the American Chemical Society* **43**, 1180–1185.

Pfenninger, U. (1909). Untersuchungen der Fruchte von *Phaseolus vulgaris* L. in verschiedenen Entwicklungs-Stadien (Vorlaufige Milleilung). *Berichte der Deutschen Botanischen Gesselschaft* **27**, 227–234.

Pollock, B. M. and Manalo, J. R. (1970). Simulated mechanical damage to garden beans during germination. *Journal of the American Society for Horticultural Science* **95**, 415–417.

Powrie, W. D., Adams, M. W., and Pflung, I. J. (1960). Chemical, anatomical, and histochemical studies on the navy bean seed. *Agronomy Journal* **52**, 163–167.

Ries, S. K. (1971). The relationship of protein content and size of bean with growth and yield. *Journal of the American Society for Horticultural Science* **96**, 557–568.

Robertson, R. N., Highkin, H. R., Smydzuk, J., and Went, F. W. (1962). The effect of environmental conditions on the development of pea seeds. *Australian Journal of Biological Sciences* **15**, 1–15.

Rubenstein, I. (1980). *The Plant Seed: Development, Preservation and Germination*. Academic Press, New York.

Scott, D. J., and Close, R. C. (1976). An assessment of seed factors affecting field emergence of garden pea seed lots. *Seed Science and Technology* **4**, 287–300.

Shannon, S., Natti, J. J., and Atkin, J. D. (1967). Relation of calcium nutrition to hypocotyl necrosis of snap bean, *Phaseolus vulgaris* L. *Journal of the American Society for Horticultural Science* **90**, 180–190.

Siddique, M. A. (1980). Environmental influences on seed quality in French bean (*Phaseolus vulgaris* L.)., Ph.D. thesis, The University of Sydney, Sydney.

Siddique, M.A., and Goodwin, P. B. (1980a). Seed vigour in bean (*Phaseolus vulgaris* L. cv. Apollo) as influenced by temperature and water regime during development and maturation. *Journal of Experimental Botany* **31**, 313–323.

Siddique, M. A., and Goodwin, P. B. (1980b). Maturation temperature influences on seed quality and resistance to mechanical injury of some snap bean genotypes. *Journal of the American Society for Horticultural Science* **105**, 235–238.

Sinclair, T. R., and deWitt, C. T. (1975). Photosynthate and nitrogen requirements for seed production by various crops. *Science* **189**, 565–567.

Smittle, D. A., Williamson, R. E., and Stansell, R. J. (1976). Response of snapbean to seed separation by aerodynamic properties. *HortScience* **11**, 469–471.

Somerset, G. (1977). Seed development and effect of harvest date on seed quality in *Phaseolus vulgaris* L. B. Sc.Agr. thesis, The University of Sydney, Sydney.

Sun, S. M., Mutschler, M. A., Bliss, F. A., and Hall, T. C. (1978). Protein synthesis and accumulation in bean cotyledons during growth. *Plant Physiology* **61**, 918–923.

Tompkins, D. R. (1965). Broccoli maturity and production as influenced by seed size. *Journal of the American Society for Horticultural Science* **88**, 400–405.

Vis, C. (1980). Flower seed production. *Seed Science and Technology* **8**, 495–503.

Walbot, V., Clutter, M., and Sussex, I. M. (1972). Reproductive development and embryogeny in *Phaseolus*. *Phytomorphology* **22**, 59–68.

Wijandi, S., and Copeland, L. O. (1974). Effect of origin, moisture content, maturity and mechanical damage on seed and seedling vigour of beans. *Agronomy Journal* **66**, 546–548.

Williams, F. J., Hollis, W. L., and Day, M. H. (1966). Incidence of hypocotyl collar rot of *Phaseolus vulgaris* in the field and germination tests. *Phytopathology* **56**, 531–535.

Yeung, E. C. (1980). Embryogeny of *Phaseolus:* The role of the suspensor. *Zeitschrift für Pflanzenphysiologie* **96**, 17–28.

Zhdanova, L. P. (1969). Effect of temperature on the synthesis of fat in maturing seeds of oil bearing plants. *Fiziologiya Rastenii* **16**, 488–497.

CHAPTER 10

Maize and Pearl Millet

C. J. PEARSON and A. J. HALL

I. INTRODUCTION

Maize (*Zea mays* L.) and pearl millet (*Pennisetum americanum* (L.) Leeke) are tropical cereals. They fix carbon via the NADP–malic enzyme C_4 pathway (Hatch *et al.*, 1975) and their enzyme proteins and membrane lipids reorganize or disorganize, showing "phase changes", at about 10°C–12°C (Phillips and McWilliam, 1971; Pearson *et al.*, 1977a). They also typify a tropical crop, being erect, large-leaved, able to withstand high temperatures and capable of high growth rates. Either crop may grow in excess of 500 kg (dry matter) ha^{-1} d^{-1}. For maize in Nigeria and temperate Australia 530 kg (dry matter) ha^{-1} d^{-1} and 590 kg (dry matter) ha^{-1} d^{-1} respectively have been recorded (Kowal and Kassam, 1973; Pearson *et al.*, 1977b) and 580 kg (dry matter) ha^{-1} d^{-1} was reported for pearl millet in northern Australia (Phillips and Norman, 1967).

Maize is the C_4 cereal of greatest productive importance in the tropics (62.85 million tonne in 1978–1980, FAO, 1981) whereas pearl millet, of lesser importance (19.0 million tonne in 1978–1980) is, however, the only major C_4 cereal for which production in the tropics exceeds that from temperate

CONTROL OF CROP PRODUCTIVITY
ISBN 0 12 548280 9

latitudes. The physiology of maize has been intensively researched and reviewed, but almost always from the viewpoint of, and using data from temperate maize, most often "dent" hybrids in North America (e.g. Duncan, 1975). Pearl millet has received relatively little attention although some physiological reviews have recently been written based wholly or in part on performance of plants grown in the tropics (Alagarswamy et al., 1977; Pearson, 1984). This chapter is not a comprehensive review of the physiology of these tropical crops but aims to discuss the physiological controls and limitations to their productivity.

II. DEVELOPMENT PATTERNS AND GROWTH

A. Development

The developmental pattern of maize has been described in terms of appearance of leaf pairs by Hanway (1963) with amplification by Swan et al. (1981). The total duration of maize growth and the timing of developmental events within this mainly depend on temperature and genotype. Since maize in the tropics is grown predominantly in the wet tropics from lowlands to over 3500 m altitude, there is great variation in rates of development even for the same genotype. Duration to maturity increased at a mean rate of 7.6 days per 100 m when one cultivar (H6302) was grown at three elevations from 1270 m to 2250 m at the equator in Kenya (Cooper, 1979). Considering genotype differences in longevity, CIMMYT (1980) have divided the tropical maize gene pool into three; early, medium and late. These three types take up to 50 days, 50–59 days and over 60 days to reach silking in the lowland tropics at 0–1600 m altitude where the average growing temperatures are 25°C–28°C, and take up to 70 days, 70–94 days and 95–120 days to reach silking in the highland or cool tropics where average growing temperatures are 15°C–17°C.

The development pattern of pearl millet has been divided into nine stages by Alagarswamy et al. (1977). An Indian hybrid (HB3) and a West African landrace (Mel Zengo) reach anthesis 40 days and 53 days respectively after emergence, and reach maturity at 65 days and 75 days after emergence at Hyderabad (lat. 17°N) (Alagarswamy et al., 1977). In West Africa, where pearl millet is thought to have originated, genotypes range in maturity from 60 days to 180 days and two maturity groups are recognized, the "gero" and "maiwa" types. Much of the pearl millet grown today is of "gero" or short-duration type; the "maiwa" types are more strongly sensitive to daylength and are of longer growth duration (a large "maiwa" sample showed a normal distribution about a mean of 110 days to anthesis, Pearson, 1984). Pearl millet is grown in the wet-and-dry tropics, largely within 200 mm to 800 mm annual average rainfall (Bidinger *et al.*, 1981). As a consequence of this wide range in duration of growing season, long season or "maiwa" types tend to be grown in wetter regions, usually at lower latitude (7°N–9°N) and the short season types are grown in drier areas, or as opportunistic crops or relay crops towards the end of the wet season.

B. Vegetative Growth and Carbon Gain

Generally, the growth of tropical maize (i.e. of landraces and cultivars which have evolved and are grown within the tropics in distinction to the growth of short-stature temperate hybrids which are now sometimes grown in the tropics) is not seriously constrained by either emergence or establishment problems. By contrast, establishment of pearl millet may be a serious constraint on subsequent productivity. The caryopsis of pearl millet is small (3–4 mm or 5–12 mg in comparison with 200–400 mg in maize). Small size is associated with lack of vigour, as in other cereals (Pearson, 1975; ICRISAT, 1977). Seed weight and seedling vigour are also relatively low when the seed is matured rapidly under high temperatures (Fussell and Pearson, 1980). High temperature and rapid seed maturation occur in the wet-and-dry tropics. Moreover, pearl millet, since it is reputedly tolerant to drought and light, sandy soils, is often sown on soils which are liable to wind erosion, as in the deep sands of the sub-Saharan wet-and-dry tropics of Africa. Seedlings are also susceptible to soil surface crusting, as may happen in soils of almost any texture (even a Hissar sandy loam containing (69% sand, Agrawal and Sharma, 1980). Finally, problems of small seed size, wind and soil are exacerbated by high topsoil temperature. During the planting season, topsoil temperatures commonly exceed 50°C in Niger, Africa (L. K. Fussell. pers. comm.) in contrast to temperatures of 30°C–35°C which are common where maize is sown in the wet tropics. A disadvantageous thermal environment has a cumulative effect on seedling performance: percentage germination may be relatively unaffected (although, of course, temperature influences the rate of seed germination) whereas emergence is reduced and mortality becomes progressively greater as the seedling begins expansion of embryonic leaves (Pearson, 1975).

Once (or if) a reasonable leaf area is established, growth rates may be high as a result of potentially high photosynthetic rates, 1.75 mg (CO_2) m^{-2} s^{-1} has been measured in laboratories (McPherson and Slatyer, 1973) and 3.0 mg (CO_2) m^{-2} s^{-1} in the field (J. D. Hesketh and C. J. Pearson, unpublished). High photosynthetic rates and tall canopy structure after stem elongation lead to maize and pearl millet having accentuated canopy profiles of carbon dioxide, and of net photosynthesis. Fluxes of heat, humidity and carbon dioxide at different times during a day are presented in Figure 10.1. These commonly

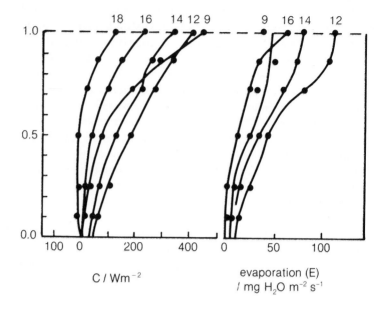

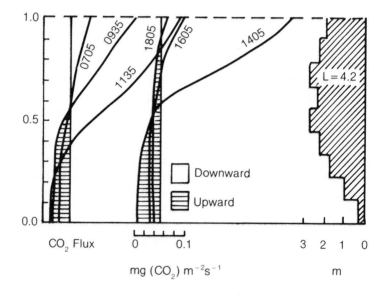

Fig. 10.1 Vertical profiles of air temperature (°C) and profiles of fluxes of evaporation and carbon dioxide (CO_2) within a maize canopy. Leaf area profile (L) is shown as a hatched histogram. Times refer to local times. (Data from various authors, figures adapted from Uchijima, 1976.)

show upward fluxes of sensible heat and water from all layers within the canopy whereas carbon dioxide fluxes are downward in the top half of the canopy (Uchijima, 1976). Naturally, the extent and direction of these fluxes depends on solar angle and atmospheric conditions. Diurnal oscillations in carbon dioxide fluxes are most pronounced when the canopy is dense and solar radiation high; the position of the profile minimum moves downwards as solar angle approaches local noon (Lemon and Wright, 1969).

Net photosynthesis of the total canopy and the contribution of upper leaves to total dry matter gain depend on leaf architecture and on the density of the canopy, that is, leaf area. Computer simulations suggest that canopies that have the upper strata comprised of erect leaves and the lower strata of more horizontal leaves are more efficient than canopies with other leaf orientations (references in Uchijima, 1976). Temperate cultivars that are high yielding tend to have this leaf arrangement. Furthermore, Uchijima *et al.* (1968) predicted that the angle of leaf inclination has relatively little effect on radiation profile (mean extinction coefficient) where the solar angle is 30°–50°, but that erect leaves confer a very considerable advantage—a low extinction coefficient—at high solar angles, such as those encountered in the tropics (Fig. 10.2).

Canopy and leaf photosynthesis have been measured in old (circa 1942) and modern (1975) temperate, hybrid maize cultivars; the latter have canopies of erect upper leaves as suggested may be desirable by Uchijima (1976) and others. Individual leaf photosynthesis rates were about 15% higher in the modern than in the old cultivars. The ratio of photosynthesis to evaporation was the same in both cultivars, ranging linearly from 0.02 at the base of the canopy to 0.055 in uppermost leaves. This suggests that there may be some, albeit small, advantage in "desirable" canopy structure while leaf area is a constraint upon photosynthesis, that is, until canopy closure. However, canopy exchange measurements after canopy closure showed no difference in photosynthetic rates per unit ground area. Furthermore, the canopy photosynthesis–irradiance responses (computed from short-term measurements throughout the day at intervals from stem elongation until near maturity) were the same in old and modern maize genotypes until after anthesis (Fig. 10.3; C. J. Pearson *et al.*, unpublished).

It is not clear to what extent canopy structure, as varied by plant population, affects photosynthesis and growth rates of maize and pearl millet in the tropics. For maize, high growth rates are achieved at plant populations of 50 000 plants ha^{-1} (e.g. Kowal and Kassam, 1973) and total above-ground yields of experimental crops in the tropics are commonly between 12 and 20 t ha^{-1} at maturity. Since maize is grown in the wet tropics and it is relatively sensitive to water deficits, the effectiveness of increasing growth rate through an increase in plant population (which accelerates the attainment of canopy closure) depends largely on water availability. Babalola and Oputa (1981)

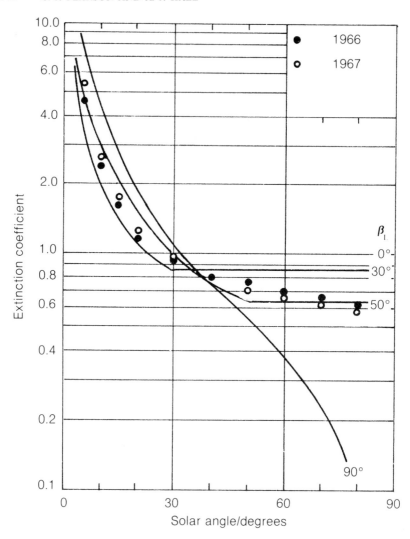

Fig. 10.2 Effects of solar angle and angle of leaf inclination (β_L) on the extinction coefficient (k) of maize canopies (from Uchijima *et al.*, 1968).

have shown that total dry matter and grain yield of early planted maize, which would grow under adequate soil water, responds linearly to increasing plant population to 72 000 plants ha^{-1} whereas late crops show no grain yield response to increasing population. For pearl millet, plant populations are commonly much lower than for maize. In Niger, populations are about 5 000–10 000 plants ha^{-1} and even on soils having high water-holding capacity (Vertisols) in India, populations commonly seem to be about 30 000 ha^{-1}. Above-ground growth rates are linearly related to intercepted radiation

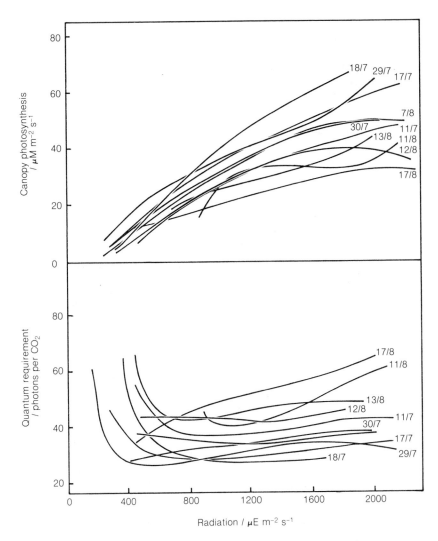

Fig. 10.3 Canopy photosynthetic rates, presented as irradiance responses from measurements throughout days at various intervals during grain filling of maize at Illinois. The date of maximum rates (18/7: 18 July 1981) coincided with anthesis, when rates were the same in cultivar Mo17 × B73 (a relatively new cultivar) as in L289 × Os420 (commercialized in 1947).

from emergence until the time of canopy closure and complete light interception (ICRISAT, 1978). Low plant populations, resulting in low leaf area indices, may maintain the growth rate–radiation relationship throughout the life of pearl millet in the semi-arid tropics. Thus, if some event (water stress, low temperatures, insects or disease) retards development, this retardation of growth will persist throughout the crop life instead of the more usual situation where a retardation primarily affects the timing of canopy closure, and has little influence on subsequent growth. For example, cool temperatures may retard leaf area development during the summer (dry) growing season so that maximum leaf area indices of pearl millet in summer were only 1.7 compared to 2.9 in the wet season in India (ICRISAT, 1978). Maximum leaf area is reached at inflorescence emergence and leaf area declines throughout grain development (e.g. Kassam and Kowal, 1975).

C. Water Use and Nitrogen Uptake

Crop water use has been studied by Kassam and Kowal (1975) and others. The ratio of evaporation to pan evaporation averages about 0.8 throughout the life of both crops. Crop water use by pearl millet is relatively more efficient than by maize, being 300 mm or 148 g water per g (top dry weight) for millet and 253 g water per g (top dry weight) for maize in Northern Nigeria (Kassam and Kowal, 1975). However, because of low partitioning of dry matter into grain in pearl millet, water use per g (grain weight) was 863 g per g in pearl millet and 747 g per g in maize.

Maize in particular, and pearl millet to a much smaller extent have been used as "type specimens" for studies of the mechanisms of mineral uptake in grasses. Both species are capable of high rates of net nitrogen uptake (28–42 nmol s^{-1} per g (dry weight/root) arising from high rates of nitrate influx, and appreciable but in the case of pearl millet, very variable, rates of efflux from the root (Pearson et al., 1981). About half the nitrogen which is taken up is reduced in the root at normal temperatures for growth (30°C/25°C) so that, after some reduced nitrogen is retained for root metabolism, export to the plant top through the xylem consists of 2:1 nitrate:reduced nitrogen (Theodorides and Pearson, 1982).

However, although physiological studies such as these suggest that nitrogen uptake and distribution are dynamic and relatively efficient, tropical grasses in field situations are widely recognized as having low tissue concentrations of nitrogen (e.g. Brown, 1978). Concentrations of 1%–3% are common, given the tendency not to apply nitrogen fertilizer on dryland cereals in the tropics. When highly fertilized, the crops maintain reasonable concentrations of nitrogen (about 4% on tissue dry weight basis) and constant ratios of nitrogen uptake to dry weight increment. In pearl millet these ratios were found to be virtually the same throughout vegetative growth for plants in controlled environments and in crops in the field (Theodorides and Pearson, 1981).

Crop yield (dry matter or nitrogen) responses to fertilizer nitrogen generally show a long period of near linearity; for example, in Brazil

$$Y = 0.68 + 4.49 \text{ (applied nitrogen)} \qquad (10.1)$$

where Y is tonne protein ha[-1] and nitrogen is applied up to 300 kg ha[-1] (Medeiros et al., 1978). Fertilization increases the ratio of grain to stover (leaf, stem and tassel) and it increases grain yield in maize but in pearl millet, which has a low ratio of grain to stover at maturity (e.g. 0.15), fertilization may have a negligible effect on grain production (e.g. Singh, 1976). The responsiveness of maize and unresponsiveness of millet cannot be attributed to environment (e.g. longer period of water availability in maize-growing areas) because it occurs when both crops are grown at the same location. This response to nitrogen fertilization is most plausibly related to the more determinate pattern of development of maize whereas pearl millet may respond to applied nitrogen by initiating more, and later tillers which may not bear grain.

D. Reproductive Growth

Reproductive growth can be viewed as a set of four sequential interrelated steps: (1) inflorescence development; (2) pollen shedding and fertilization; (3) seed setting; and (4) seed filling. These phases can be distinguished clearly in a tiller but there is blurring of boundaries for each phase within a plant due to non-synchronous tillering and effects of position of the inflorescence; that is, the site of ear on the shoot in maize and the order of inflorescence tiller (primary, basal or high-order axillary) in millet. There is still greater blurring of reproductive phases in a crop.

Floral initiation in both maize and pearl millet is accelerated by short days and the time from initiation to inflorescence emergence is shortened by high temperature (e.g. Aitken, 1977, 1980; Pearson, 1984).

Unlike other cereals, maize is diclinous, with the female inflorescences (ears) initiated on lateral shoots up to twelve days after the tassel. Temporal differences, albeit much smaller ones, also exist in the initiation of ears on the axillary shoots of the same plant. These temporal differences in initiation of reproductive structures tend to be maintained throughout the initiation to the flowering period, and are amplified by environmental stresses and defoliation, with the latest formed structure suffering the greatest delay (Edmeades and Daynard 1979a; Hall et al. 1980; 1981; Trapani and Hall, 1980); thus the heirarchy of reproductive structures and its modulation has important implications for the number of ears which develop, extrude silks, are pollinated and set kernels.

Increased tolerance to density has played an important part in raising the yield of maize (Duvick, 1977), and it appears that the low yields still found in some tropical countries (e.g. 1 t ha[-1] in India and 1.5 t ha[-1] in Brazil and

Mexico; FAO, 1981) may be partly related to the intolerance of tropical landraces to increased plant population. Tolerance to increased population is associated with reduced tassel size, coincidence of pollen shedding and silk extrusion on the same plant and prolificacy when measured at reduced density (Buren *et al.*, 1974). Prolificacy, in turn, is associated with greater synchrony in silk extrusion of the second and third ears with respect to the first (uppermost) one (Harris *et al.*, 1976).

By contrast, pearl millet yields in the tropics are low because the crop is grown most widely at low populations in semi-arid conditions. Pearl millet landraces have retained the ability to produce a variable and sometimes large number of both basal and axillary tillers. Basal tillering commences at or before floral initiation of the main stem-apex, and up to 40 tillers may form before first anthesis. Axillary tillers form from upper nodes in flushes under intermittent drought and may contribute up to 50% of the total yield of pearl millet under rain-fed conditions. Thus, "high" yields are associated with asynchronous flowering and plasticity in yield components.

In maize a single tassel will shed pollen during seven to ten days; there is a large range among genotypes in number of pollen grains produced per tassel (Hall *et al.*, 1982). Because pollen production is skewed towards the start of anthesis and tends to be delayed less by stress than silking, stress can markedly increase the likelihood that there will be insufficient pollen to fertilize the last ears to extrude silks (Fig. 10.4).

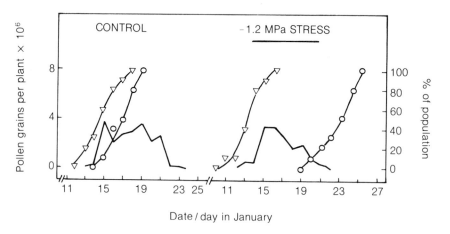

Date / day in January

Fig. 10.4 Water deficits cause asynchronous flowering and reduce pollen production in maize. Percentage of populations with extruded anthers (∇) and silks (O) and temporal distribution of pollen production (–) in control (left) and stressed plants (right) of cultivar Dekalb 2F10. Horizontal bar shows period of intense stress, when mean nightfall leaf water potential was –1.2 MPa (after Hall *et al.*, 1982).

Extrusion of silks on a single ear takes place, in the absence of stress, over four to five days in temperate cultivars; most silks appear within the first two days (Tollenaar and Daynard, 1978a). Stress imposed by high plant density (Edmeades and Daynard, 1979; Wilson and Allison, 1978) and water deficits (Herrero and Johnson, 1981) slows down silk growth and presumably increases the time taken by a single ear to extrude such silks as do grow. The transition from ear development to flowering (silking) is particularly sensitive to stress: ears that develop under stress can form a close to normal number of spikelets and then fail to silk (Edmeades and Daynard 1979) and set kernels. Water stresses capable of delaying silking of the uppermost ear can inhibit silking and kernel set of the second ear (Table 10.1).

High temperatures and tissue-atmosphere vapour pressure deficits (VPD) reduce pollen viability and seed set (Lonnquist and Jugenheimer, 1943; Tatum and Kehr, 1951). It is not clear whether the latter effect is due to loss of pollen viability or the failure of pollen to germinate successfully on the silks or both; nor has the relative importance of high temperatures and VPD (which usually co-vary under field conditions) as causal factors been established, notwithstanding a recent attempt to do so for pollen viability (Herrero and Johnson, 1980). Plant water deficits, in the absence of extreme temperature and VPD conditions, do not seem to affect pollen viability (Herrero and Johnson 1981; Hall *et al.*, 1982) although there is some evidence that water deficit may cause embryo sac abnormalities in maize (Moss and Downey, 1971). It is not at all clear how the various possible mechanisms (e.g. embryo sac abnormalities, failure of pollen to germinate, post-fertilization abortion) contribute to the reduction in kernel set in ears of stressed plants adequately supplied with viable pollen (Table 10.1).

Because of the negative association between the initiation hierarchy of inflorescences and sensitivity to stress conditions, it seems likely that some traits associated with density tolerance and prolificacy (smaller tassels, greater flowering synchrony within and between plants) may not always ensure greater yield stability, in spite of some limited evidence to the contrary. Although the relationship between yield and silk delay has been recognized, there have been few attempts to separate the effects of pollen availability from other effects of stress. A simple model of pollen production and fertilization that modulates kernel set (number of kernels: number of visible spikelets) of the uppermost ear as a function of pollen availability and a fertility factor (a variable which integrates all effects of stress other than pollen availability) explained 86% of the variability in kernel set found in nine sets of Argentinian data. This data was obtained using seven different genotypes stressed in up to three different periods close to, or during, flowering. The fertility factor was used to estimate the reduced kernel set in those ovules receiving pollen while under stress, and the proportion of these ovules

Table 10.1

The effect of drought (withholding water until 50% or more of the population showed signs of wilting at 8.30 a.m.) on some determinants of kernel number in maize. Some plants were rewatered after drought (D1-1, D11-1, D111-1) whereas others were subjected to two drought cycles. (From Hall et al., 1981.)

| | Control | Time of droughting | | | | | | LSD |
		D_{1-1} (before tasselling)	D_{1-2}	D_{11-1} (before pollen shedding)	D_{11-2}	D_{111-1} (before silking)	D_{111-2}	(P = 0.05)
Plants with basal ears that silked (% of total)	100	73[a]	40[a]	86	60[a]	80[a]	73[a]	NA[c]
Basal ears that set kernels (% of ears with silks)	64	27	0[a]	10[a]	27	27	0[a]	NA
Spikelets per apical ear	561	415	378	492	442	534	516	86
Kernel set of apical ear[b] (arc. sin (p)$^{1/2}$)	73.5	72.4	63.9	68.8	65.3	58.3	61.6	9.8

[a] Differs from control (P = 0.05) as determined by χ^2 test
[b] In apical ears of plants with sufficient pollen and no basal ear
[c] NA = Not applicable.

varied between 0 and 1 with a mean of 0.4 according to genotype and timing of stress (Sadras *et al.*, unpublished).

Stress imposed by density, defoliation, shading and drought during flowering and seed set is largely manifested in changes in kernel number (e.g. Edmeades and Daynard, 1979; Fischer and Palmer, 1980). Assimilate availability has been advanced as the likely reason for some of these effects but how assimilate availability is sensed is far from clear. Whether or not assimilate availability is the over-riding factor, it would appear that the developmental shifts brought about by stress may be sufficient to account for an important fraction of the total effect. In maize, fertilization may take place in only some of the ovules that silk due to lack of pollen brought about by silk delay, but even where fertilization is successful the asynchrony of fertilization may be enough to allow the hierarchical mechanism that adjusts kernel number between (Harris *et al.*, 1976) and, possibly, within ears to act on an increasing proportion of the total number of fertilized ovules. Direct effects of stress (e.g. tassel blasting, loss of pollen or embryo sac viability, etc.) on kernel number may be important under conditions of extreme stress.

There is a lag between fertilization and initiation of the linear phase of grain filling. Some evidence suggests that the lag phase in maize coincides with the period in which the degree of kernel set is fixed (Egharevba *et al.*, 1976; Tollenaar and Daynard, 1978b). The lag phase appears to be shorter under tropical conditions; this may indicate a shorter period during which kernel set is susceptible to stress.

Mathematically, final grain size is determined by the actual grain filling period (AGFP) multiplied by rate of filling (RF). The RF is usually calculated from samples taken throughout the linear phase of grain filling, which stops at a time close to that of attainment of maximum grain weight, when a black closing layer appears at the base of the embryo (in maize, Daynard and Duncan, 1969; in pearl millet, Fussell and Pearson, 1978). Large genotypic variations in both AGFP and RF have been demonstrated in maize (Poneleit *et al.*, 1980 and references herein) or can be inferred from published data (Aitken, 1977); in millet, genotypic differences so far reported have been due mostly to RF (Fussell and Pearson, 1978). Variations in AGFP and RF are not correlated, suggesting that yield improvements may be obtained through appropriate combinations of these characteristics. Environmental effects on grain filling can be quite important, although it is not always possible to resolve these effects into AGFP and RF components from published data. Fischer and Palmer (1980) report a regression equation, derived from data on 12 populations of tropical maize grown in 15 environments, that suggests that a 10°C rise in mean temperature from 16°C to 26°C almost triples RF while decreasing AGFP by about one-quarter, which results in a doubling of kernel weight. In pearl millet, increasing temperature during grain filling reduces the AGFP by about 0.2 days per degree Celsius from 21°C

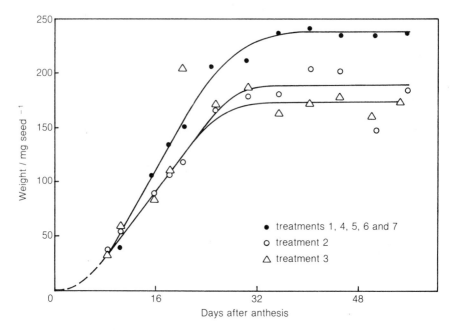

Fig. 10.5 Effect of decapitation at various times after anthesis on growth of individual grains of maize cultivar Kretek in East Java. Treatments 1 to 7 were control, decapitation 5, 15 and 30 d after anthesis and defoliation (removal of all leaves below the ear) at 5, 15 and 30 d after anthesis, respectively.

to 33°C while having virtually no effect on RF (Fussell *et al.*, 1980). Stress imposed by defoliation (Egharevba *et al.*, 1976, Tollenaar and Daynard, 1978b) and water deficit (Classen and Shaw, 1970; Jurgens *et al.*, 1978) can affect grain filling. AGFP and RF may both be reduced by defoliation (Fig. 10.5; Tollenaar and Daynard 1978b).

III. MANAGEMENT CONSTRAINTS

Management obviously exerts considerable control over crop productivity. Some constraints of major importance, such as weeds and diseases, low soil fertility and inadequate supply or control of water, are well known in tropical agriculture and a discussion of these is beyond the scope and size of this book. We wish to mention two practices that impinge closely on the physiology and performance of maize and pearl millet in the tropics. These are the complex of low plant populations–poor seed quality–multiple seeds per hill, and the practice of cutting stover for forage during grain development.

High temperatures during seed maturation shorten the AGFP and cause smaller seeds. High temperatures and humidity during seed storage over the wet season might be expected to reduce seed size further through elevating post-maturation respiration (see Fussell and Pearson, 1980) and pathogen attack. All these factors contribute to poor seed viability and low seedling vigour (see Chapter 9, Section III). Farmers commonly compensate for this by sowing several seeds per hill. (Many seeds per hill also increase the likelihood of overcoming impedance caused by surface crusting.) Naturally, if more than one plant survives there is increased interplant competition. In East Java, traditional planting (2 or 3 plants per hill) results in lower leaf area per plant at anthesis and lower average cob weight (e.g. 71 g compared to 85 g) due to fewer seeds and a slightly reduced individual seed weight compared to one plant per hill.

A variable number of plants per hill is commonly coupled with low plant populations. Low populations have direct effects on vegetative productivity, as already mentioned. They may also constrain grain production since the current viewpoint admits that grain growth may be limited by both total photosynthate production after anthesis and "sink" size (Tollenaar, 1977).

Farmers throughout the tropics may remove upper leaves or decapitate maize plants above the ear at or after anthesis. This provides forage for livestock and may mitigate the effects of end-of-season drought. Decapitation of maize at anthesis markedly reduces the growth rate of individual grains and thereby reduces final grain size and total yield; progressively later decapitations have progressively smaller, or no effect on yield (Fig. 10.5). In contrast to decapitation or removal of upper leaves, removal of a comparable area of leaves below the ear has an insignificant effect on yield. Such effects of "source" manipulation on sink growth may be interpreted from the viewpoint of the low sink capacity of tropical maize in comparison of high sink activity in hybrid, temperate maize (review by Fischer and Palmer, 1980).

IV. CONCLUSIONS

Tropical maize and pearl millet are characterized by high photosynthetic rates and relatively high temperature optima for development and growth processes. We have argued that photosynthetic rates *per se* establish high potential for vegetative productivity (and that this potential is higher in new cultivars than in old by virtue of leaf arrangement, not rate per area of leaf). Potential productivity is constrained in the tropics by low leaf areas, environmental effects particularly upon the determination of yield components, and management practices. Determination of the relative weighting of these constraints is a prerequisite for the development of optimal strategies for breeding and management in each environment. These strategies may often differ from those which have been found useful in temperate areas.

REFERENCES

Aitken, Y. (1977). Evaluation of maturity genotype-climate interactions in maize (*Zea mays* L.) *Zeitschrift für Pflanzenzüchtung* **78**, 216–237.

Aitken, Y. (1980). The early maturing character in maize (*Zea mays* L.) in relation to temperature and photoperiod. *Zeitschrift für Acker-und Pflanzenbau* **149**, 89–106.

Agrawal, R. P., and Sharma, D. P. (1980). Management practices for improving seedling emergence of pearl millet (*Pennisetum glacum* L.) under crusting conditions. *Zeitschrift für Acker-und Pflanzenbau* **149**, 398–405.

Alagarswamy, G., Maiti, R. K., and Bidinger, F. R. (1977). Physiology Program. International Pearl Millet Workshop, ICRISAT, International Crops Research Institute for the Semi-Arid Tropics, Hyderabad, India.

Babalola, O., and Oputa, C. (1981). Effects of planting patterns and population on water relations of maize. *Experimental Agriculture* **17**, 97–104.

Bidinger, F. R., Mahalakshmi, V., Talukdar, B. S., and Alagarswamy, G. (1981). Improvement of drought resistance in pearl millet. ICRISAT Conference Paper No. 44. International Crops Research Institute for the Semi-Arid Tropics, Hyderabad, India.

Brown, R. H. (1978). A difference in N use efficiency in C_3 and C_4 plants and its implications in adaptation and evolution. *Crop Science* **18**, 93–98.

Buren, I. L., Mock, J. J., and Anderson, I. C. (1974). Morphological and physiological traits in maize associated with tolerance to high plant density. *Crop Science* **14**, 426–429.

CIMMYT (1980). CIMMYT Review 1980. Centro International de Majoramiento de Maiz y Trigo, El Batan, Mexico.

Claassen, M. M., and Shaw, R. H. (1970). Water deficit effects on corn. II. Grain components. *Agronomy Journal* **62**, 652–655.

Cooper, P. J. M. (1979). The association between altitude, environmental variables, maize growth and yield in Kenya. *Journal of Agricultural Science* **93**, 635–649.

Daynard, T. B., and Duncan, W. G. (1969). The black layer and grain maturity in corn. *Crop Science* **9**, 473–476.

Duncan, W. G. (1975). Maize. *In* "Crop Physiology, Some Case Histories" (L. T. Evans, ed.), pp. 25–50. Cambridge University Press, Cambridge.

Duvick, D. N. (1977). Genetic rates of gain in hybrid maize yields during the past 40 years. *Maydica* **22**, 187–196.

Edmeades, G. O., and Daynard, T. B. (1979). The development of plant-to-plant variability in maize at different planting densities. *Canadian Journal of Plant Science* **59**, 561–576.

Egharevba, P. N., Horrocks, R. D., and Zuber, M. S. (1976). Dry matter accumulation in maize in response to defoliation. *Agronomy Journal* **68**, 40–43.

FAO (1981). Production Yearbook, 1980. Vol. 34. Food and Agriculture Organization, Rome.

Fischer, K. S., and Palmer, A. F. E. (1980). Yield efficiency in tropical maize. Symposium on potential productivity of field crops under different environments, 22–26 September 1980. International Rice Research Institute, Los Baños, Philippines.

Fussell, L. K., and Pearson, C. J. (1978). Course of grain development and its relationship to black region appearance in *Pennisetum americanum*. *Field Crops Research* **1**, 21–31.

Fussell, L. K., and Pearson, C. J. (1980). Effects of grain development and thermal history on grain maturation and seed vigour of *Pennisetum americanum*. *Journal of Experimental Botany* **31**, 635–643.

Fussell, L. K., Pearson, C. J., and Norman, M. J. T. (1980). Effect of temperature during various growth stages on grain development and yield of *Pennisetum americanum*. *Journal of Experimental Botany* **31**, 621–633.

Hall, A. J., Lemcoff, J. W., and Trapani, N. (1981). Water stress before and during flowering in maize and its effects on yield, its components, and their determinants. *Maydica* **26**, 19–38.

Hall, A. J., Ginzo, H. D., Lemcoff, J. H., and Soriano, A. (1980). Influence of drought during pollen-shedding on flowering, growth and yield of maize. *Zeitschreift für Pflanzenbau* **149**, 287–298.

Hall, A. J., Vilella, F., Trapani, N., and Chimenti, C. (1982). The effects of water stress, its timing, and genotype on the dynamics of anthesis and on pollen production in maize. *Field Crops Research* **5**, 349–363.

Hanway, J.J. (1963) Growth stages of corn. *Agronomy Journal* **55**, 487–492.

Harris, R. E., Moll, R. H. and Stuber, C. W. (1976). Control and inheritance of prolificacy in maize. *Crop Science* **16**, 843–850.

Hatch, M. D., Kagawa, T. and Craig, S. (1975). Subdivision of C_4-pathway species based on differing C_4 acid carboxylating systems and ultra-structural features. *Australian Journal of Plant Physiology* **2**, 111–128.

Herrero, M. P., and Johnson, R. R. (1980). High temperature stress and pollen viability of maize. *Crop Science* **20**, 796–800.

Herrero, M. P., and Johnson, R. R. (1981). Drought stress and its effects on maize reproductive systems. *Crop Science* **21**, 105–110.

ICRISAT (1977). Pearl millet. Annual Report 1975-76. International Crops Research Institute for the Semi-Arid Tropics, Hyderabad, India.

ICRISAT (1978). Pearl millet. Annual Report 1977-78. International Crops Research Institute for the Semi-Arid Tropics, Hyderabad, India.

Jurgens, S. K., Johnson, R. R., and Boyer, J. S. (1978). Dry matter production and translocation in maize subjected to drought during grain fill. *Agronomy Journal* **70**, 678–682.

Kassam, A. H., and Kowal, J. M. (1975). Water use, energy balance and growth of Gero millet at Samaru, Northern Nigeria. *Agricultural Meteorology* **15**, 333–342.

Kowal, J. M., and Kassam, A. H. (1973). Water use, energy balance and growth of maize at Samaru, Northern Nigeria. *Agricultural Meteorology* **12**, 391–406.

Lemon, R. R., and Wright, J. L. (1969). Photosynthesis under field conditions. XA. Assessing sources and sinks of carbon dioxide in corn (*Zea mays* L.) crop using a momentum balance approach. *Agronomy Journal* **61**, 405–411.

Lonnquist, J. H., and Jugenheimer, R. W. (1943). Factors affecting the success of pollination in corn. *Journal of the American Society of Agronomy* **35**, 923–933.

Medeiros, R. B., Saibro, J. C. de, and Jacques, A. V. (1978). Efeito do nitrogenio e da populacao de plantas no rendimento e equilidade do milheto (*Pennisetum americanum* Schum). *Revista da Sociedade Brasileira de Zootecnia* **7**, 276–285.

McPherson, H. G., and Slatyer, R. O. (1973). Mechanisms regulating photosynthesis in *Pennisetum typhoides. Australian Journal of Biological Sciences* **26**, 329–339.

Monteith, J. L. (1980). Microclimatology in tropical agriculture report No. 4. Sutton Bonington, U.K.: University of Nottingham School of Agriculture (Mimeo, 123 pp).

Moss, C. I. & Downey, L. A. (1971). Influence of drought stress on female gametophyte development in corn (*Zea mays* L.) and subsequent grain yield. *Crop Science* **11**, 368–371.

Pearson, C. J. (1975). Thermal adaptation of *Pennisetum*: Seedling development. *Australian Journal of Plant Physiology* **2**, 413–424.

Pearson, C. J. (in press). Pennisetum millets. *In* "The Physiology of Tropical Field Crops (P. R. Goldsworthy and N. Fischer, eds.), Wiley, London.

Pearson, C. J., Bishop, D. G., and Vesk, M. (1977a). Thermal adaptation of *Pennisetum*: Leaf structure and composition. *Australian Journal of Plant Physiology.* **4**, 541–554.

Pearson, C. J., Dawbin, K. W., Muldoon, D. K., and Campbell, L. C. (1977b). Growth and quality of tropical forages in a temperate environment. *Australian Journal of Experimental Agriculture and Animal Husbandry* **17**, 991–994.

Pearson, C. J., Volk, R. J. and Jackson, W. A. (1981). Daily changes in nitrate influx, efflux and metabolism in maize and pearl millet. *Planta* **152**, 319–324.

Phillips, P. J., and McWilliam, J. R. (1971). Thermal responses of the primary carboxylating enzymes from C3 and C4 plants adapted to contrasting temperature environments. *In* "Photosynthesis and Photorespiration" (M. D. Hatch, C. B. Osmond and R. O. Slatyer, eds.), pp. 82–88. Wiley, New York.

Phillips, L. J., and Norman, M. J. T. (1967). A comparison of two varieties of bullrush millet (*Pennisetum typhoides*) at Katherine, N.T. Division of Land Research Technical Memorandum 67/18. Commonwealth Scientific and Industrial Research Organization, Australia.

Poneleit, C. G., Egli, D. B., Cornelius, P. L., and Reicosky, D. A. (1980). Variation and associations of kernel growth characteristics in maize populations. *Crop Science* **20**, 766–770.

Singh, R. P. (1976). Fertilizer use in dryland crops: Effect of levels and methods of nitrogen application on the yield and yield attributes of Kharif cereals. *Indian Journal of Agronomy* **21**, 27–32.

Swan, D., Brown, D. M., and Coligado, M. C. (1981). Leaf emergence rates of corn (*Zea mays* L.) as affected by temperature and photoperiod. *Agricultural Meteorology* **24**, 51–73.

Tatum, L. A., and Kehr, W. R. (1951). Observations on factors affecting seed-set with in bred strains of dent corn. *Agronomy Journal* **43**, 270–275.

Theodorides, T. N., and Pearson, C. J. (1981). Effect of temperature in controlled environments and planting date and nitrogen application in the field on total nitrogen uptake and distribution in pearl millet. Department of Agronomy and Horticultural Science, University of Sydney, Research Report No. **9**, 12.

Theodorides, T. N., and Pearson, C. J. (1982). Effect of temperature on nitrate uptake, translocation and metabolism in *Pennisetum americanum*. *Australian Journal of Plant Physiology,* **9**, 309–320.

Tollenaar, M. (1977). Sink–source relationships during reproductive development in maize. A review. *Maydica* **22**, 49–75.

Tollenaar, M., and Daynard, T. B. (1978a). Kernel growth and development at two positions on the ear of maize (*Zea mays*). *Canadian Journal of Plant Science* **58**, 189–197.

Tollenaar, M., and Daynard, T. B. (1978b). Effect of defoliation on kernel development in maize. *Canadian Journal of Plant Science* **58**, 207–212.

Trapani, N., and Hall, A. J. (1980). Influencia de la eliminación de hojas basales sobre el rendimiento y sus componentes en el maíz. *Revista de Facultad de Agronomia* (Bueno Aires) **1**, 77–85.

Uchijima, Z. (1976). Maize and rice. *In* "Vegetation and the Atomophere Vol. 2" (J. L. Monteith, ed.), pp. 33–64. Academic Press, London.

Uchijima, Z., Udagawa, T., Horie, T., and Kobayashi, K. (1968). Studies of energy and gas exchange within crop canopies 4. The penetration of direct solar radiation into corn canopy and the intensity of direct radiation on the foliage surface. *Journal of Agricultural Meteorology* (Tokyo). **23**, 141–151.

Wilson, J. H., and Allison, J. C. S. (1978). Effect of plant population on ear differentiation and growth in maize. *Annals of Applied Biology* **90**, 127–132.

Temperate Pastures

D. R. KEMP

I. INTRODUCTION

Temperate pastures comprise species that have the C_3 pathway of photosynthesis and are considered chilling and frost tolerant. The grasses are usually members of the tribes Agrosteae, Poeae and Phalarideae, whereas the legumes are from the family Fabaceae (terminology of Jacobs and Pickard, 1981). Temperate pastures predominate in temperate zones; they are also found at higher latitudes such as the subarctic and at lower latitudes in montane environments with mild climates. In the tropics temperate species are used as short-term forages.

A major aim of research on temperate pastures has been to identify and overcome those factors that limit productivity. There has been more research on temperate crops than on temperate pastures but as many similarities exist between them it is possible to infer some of the responses of pasture species

CONTROL OF CROP PRODUCTIVITY
ISBN   0 12 548280 9

from data on crops. Also, more is known about European temperate species than others as these provide most of our cultivated species. In this brief review emphasis will be largely on controls on plant productivity in established swards and to a lesser extent on the influences of man and his grazing animals.

II. SEASONAL PRODUCTION

The extent of control of pasture productivity can be appreciated by considering the seasonal growth patterns of *Lolium perenne* pastures in the southern and northern hemispheres (Fig. 11.1). In temperate areas growth rates vary from zero in winter to the full genetic potential of the species, usually in spring, coincident with flowering (Langer, 1959; Silsbury, 1965). After the spring peak there is often a dip in production in mid-summer; a second, a lesser peak may occur in autumn, then growth rates decline to a minimum in winter. This sequence is modified by local rainfall patterns and also depends on factors such as temperature, light leaf area and frequency and intensity of defoliation. At lower latitudes where the range in temperatures is less, growth patterns reflect more the rainfall pattern of wet and dry seasons. The carrying capacity of the grazed sward is usually limited by productivity at the time of minimum growth rates. Where winter growth is negligible, forage may be stored from spring surpluses for feeding to stock in winter.

Differences between species and cultivars in factors such as the time of flowering cause small differences in season production (Anslow and Green, 1967). The peak of production in *L. perenne–Trifolium repens* pastures is broader than in *L. perenne* alone (Fig. 11.1) as *T. repens* has a later peak of production than *L. perenne* (Brougham, 1959). However, it is only by incorporating contrasting species such as tropical grasses or legumes such as *Medicago sativa*, that the pattern of seasonal growth changes significantly. Even then, winter temperatures usually set the minimum for growth rates. Tropical grasses only introduce a second peak or expand an existing peak in summer production. For an ungrazed or uncut pasture the rates of sward growth reflect the interaction of temperature, light and moisture on growth. This interaction is significantly modified by the changes in plant development that occur within the sward.

III. PLANT DEVELOPMENT

The major developments that occur during the seasonal growth cycle are associated with the change from a vegetative to a reproductive state. This

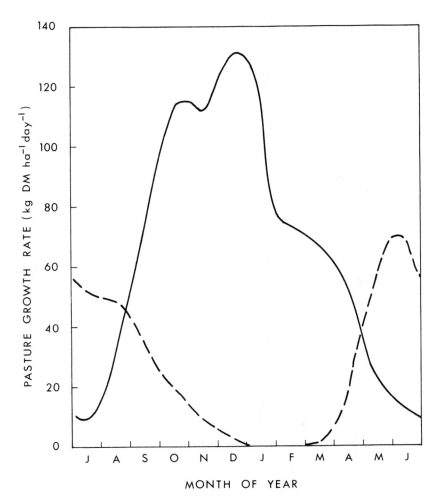

Fig. 11.1 Seasonal growth patterns for a *Lolium perenne/Trifolium repens* sward at Palmerston North (lat. 40°S) in New Zealand (————; Brougham, 1959) and a *L. perenne* sward at Hurley (lat. 51°N) in England (- - -; Anslow and Green, 1967).

change is initiated in response to temperature and day length and causes the stem apex to switch from leaf to flower production. Temperate pasture species are generally long day plants (Cooper, 1960; Evans, 1964) that require a critical day length of either 12 hours or more or a period of increasing day length to initiate flowering. The day length requirement often needs to be supplemented by a period of low temperatures (-5°C to 10°C) to vernalize the plants (Cooper, 1956). Day length requirements may in some instances be replaced by low temperatures (McWilliam and Jewiss, 1973).

The required conditions vary between genotypes and depend on the eco-type origin. As a general rule the higher the latitude of origin, the longer the critical day length (e.g. Cooper, 1960) and the colder the climate of origin the greater the vernalization requirement (e.g. Aitken, 1974). These controls prevent flowering under extreme conditions of cold and frost. Many ecotypes are fully vernalized before the end of winter but do not flower until spring when the required day length occurs. Vernalization thus prevents flowering in autumn when day lengths are the same as in spring (Evans, 1964). Ver-nalized plants can also be devernalized by a period of warm temperatures greater than 17°C (e.g. Evans, 1964), whereas some species such as *Phleum pratense* form vegetative proliferations on the apex and revert to leaf produc-tion when transferred from long to short days after being induced to flower (Langer and Ryle, 1958). During normally warm weather, such a mechanism prevents short, cold periods from inducing flowering and prevents flowering when water stress is likely.

The change from vegetative to reproductive states has a major effect on the productivity of the whole sward, and results in greater pasture growth rates in spring than in autumn, although temperatures are the same (Fig. 11.2). Light levels are not sufficiently different between spring and autumn

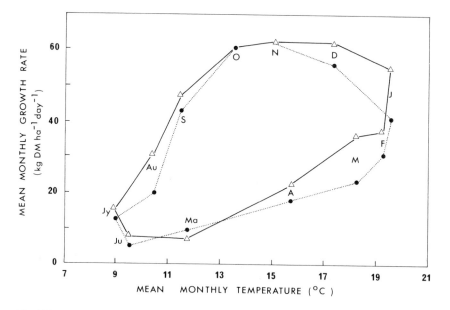

Fig. 11.2 Mean monthly pasture growth rates of farms at Awaroa (————) and Omeheu (. . . .) in the Bay of Plenty district in New Zealand in relation to mean monthly temperatures (after Field, 1980).

to explain this difference. The increased productivity recorded in reproductive swards is largely in vegetative components such as leaves. For pasture grasses once a stem apex switches to spikelet production no more leaves are initiated on that tiller, although usually several leaves are still growing. Growth rates of these remaining leaves increase considerably over those on vegetative plants (Fig. 11.3a) and this increases the growth rate of the whole tiller (Hunt and Field, 1979). Not only do leaf growth rates increase but so also does the unit leaf photosynthetic rate (Fig. 11.3b), while the proportion of assimilated carbon translocated to the roots is decreased (Fig. 11.3c). An association between increased photosynthetic rates and flowering in long day plants has often been observed (Bodson et al., 1977). Changes in leaf extension rates and unit leaf photosynthetic rates can be detected before reproductive development is clearly visible at the shoot apex (Fig. 11.3). Thus, the increased productivity recorded in reproductive swards is largely in vegetative components such as leaves and only later, as flower heads emerge, do stems and flowers contribute significantly to the biomass.

The timing of the change to reproductive development and its implications for sward productivity can be important in explaining differences in seasonal production patterns between species. For instance, the winter growth rates of a range of ecotypes can be related to the timing of reproductive development (Fig. 11.4; Williams and Biddiscombe, 1964; Thomas and Norris, 1979; Thomas, 1981).

After inflorescences emerge, the production of new leaves in perennial grass swards depends on the production of new tillers. During spring the standing biomass of the sward can reach several tonnes per hectare so that new tillers and leaves emerge in shade at the bottom of the sward. These shaded leaves have a lower potential for photosynthesis than leaves produced at other times of the year (Woledge, 1979), so that the photosynthetic capacity of the sward after flowering is depressed (e.g. Parsons and Robson, 1982); this partly explains the observed decrease in sward productivity (Fig. 11.1).

In grasses the tiller rather than the whole plant is considered the unit of production. The rate of tiller production normally declines or ceases at the time of stem elongation and flowering, although it may be resumed after inflorescences have emerged (Cooper and Saeed, 1949; Robson, 1968) provided no summer droughts intervene. The number of tillers per unit area thus declines around flowering as tillers that die are not immediately replaced (Langer, 1958). Tiller production appears to be under hormonal control because the hormones that stimulate stem elongation or maintain apical dominance also inhibit tiller bud growth (Jewiss, 1972; Clifford, 1977). Tiller production is also reduced when the base of plants is shaded as occurs in spring (e.g. Alberda and Sibma, 1982).

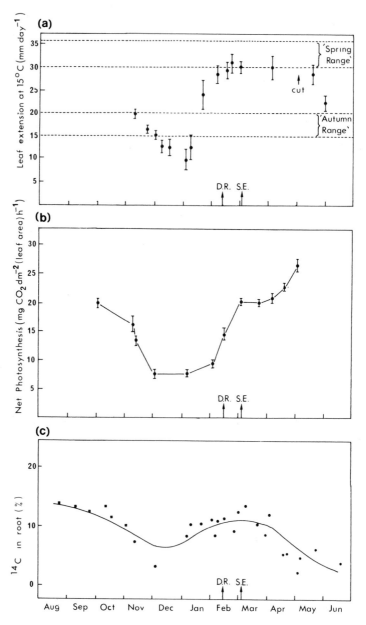

Fig. 11.3 Seasonal changes in (a) leaf extension rates at 15°C and (b) net photosynthetic rates at 250 J m^{-2} s^{-1} of successive young leaves and (c) percentage of ^{14}C translocated to the roots for *L. perenne* swards growing at Hurley in England. The times of first formation of double ridges (D.R.) and stem elongation (S.E.) are indicated by arrows (data taken from Parsons and Robson, 1980, 1981a,b).

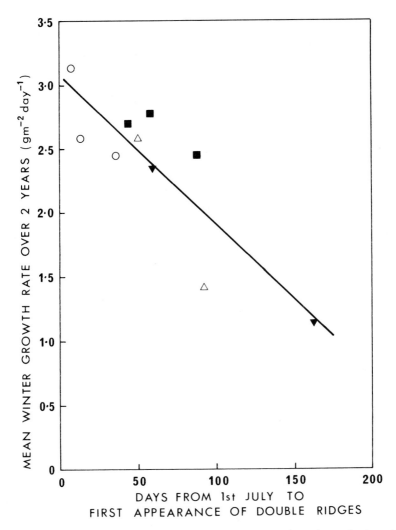

Fig. 11.4 Relationship between timing of the change between vegetative and reproductive apices, as discerned by double ridge formation and mean winter growth rates of microswards in the field at Orange, NSW, Australia. Data are for *L. perenne* (o), *P. aquatica* (■), *F. arundinacea* (△) and *D. glomerata* (▼). Swards were irrigated and received 150 kg N ha^{-1} y^{-1}.

Most studies on the seasonal changes in development in temperate grasses have been conducted on perennial swards in the UK. Little information is available about grass swards elsewhere or about developmental changes in legumes. With more information, phenomena such as apparent summer dormancy of mediterranean ecotypes could be analysed and more productive ecotypes could be selected for different climates.

IV. CLIMATE AND PASTURE GROWTH

Within a species genetic potential for growth, the actual growth rate depends very much on climate, as is shown by seasonal growth curves (Fig. 11.1). The main components of climate are light, temperature and water. Effects of light and temperature on plant development have been considered in Section III; here I consider their effects on pasture growth.

A. Light

In pastures constraints set by light on photosynthetic rate and hence growth can be considered at two levels: (1) light saturation and the maximum photosynthetic rates of single leaves; and (2) the distribution of light within the canopy. Within a sward, when 95% of incident light is intercepted, only those leaves at the top of the canopy usually reach maximum photosynthetic rates and then for only part of the day (Leafe, 1972). Light saturation of leaves limits productivity of the sward where leaves have a low capacity for photosynthesis (Robson, 1980). This occurs for leaves that develop at the bottom of dense swards as in spring and, as mentioned in Section III, partly explains the post-flowering decrease in sward growth rates. As leaves age their photosynthetic potential declines. Mutual shading within the canopy limits sward productivity further by preventing the sward from operating at maximum photosynthetic rates.

In swards, canopy structure influences light penetration. Simple models show that pasture productivity is limited more by canopy structure than by light levels (Monteith, 1981). It is possible to improve light penetration by selecting for longer and more rigid leaves on erect tillers; this results in increased growth rates in dense swards under infrequent cutting (Rhodes, 1975). Plants selected for improved canopy structure have outyielded the original populations by 30% (Wilson et al., 1980). In spring, greater growth rates of flowering swards can partly be attributed to both improved light interception within the elongating sward and higher maximum rates of leaf photosynthesis for reproductive plants (Parsons and Robson, 1982). Changes in proportions of components within a canopy will also influence growth rates.

The ideal grass canopy is composed of young green leaves that are prostrate initially to better intercept light and become more erect as the leaf area index increases (Robson, 1980). This is an ideal that cannot be maintained for long.

Where a canopy intercepts less than 95% of incident light the growth rate of the canopy is often related to the amount of intercepted light (e.g.

Brougham and Glenday, 1967). It is, however, unlikely that average leaf photosynthetic rates (or net assimilation rates) increase as the amount of light intercepted increases. The reverse is more likely: as leaf area increases, the amount of light intercepted increases as does the amount of shading within the canopy.

Shading within swards can also affect the development of components of the plant. Shading causes a reduction in tillering rate (Ryle, 1961), longer and thinner grass leaves with decreased leaf area ratios and longer legume petioles (Cooper and Tainton, 1968; Hunt and Halligan, 1981). Studies in *Triticum aestivum* (Kemp, 1981a) showed that shading increased the length of the elongation zone at the leaf base and that this allowed leaf extension rates to equal or exceed that of unshaded plants, although shading reduced dry weight growth rates. The elongation mechanism allows shaded leaves to reach the top of a sward.

B. Temperature

Temperature controls plant growth more through regulating the rates of reactions that utilize the products of photosynthesis than through direct effects on photosynthesis. This view is supported by several observations that when temperatures decline, carbohydrates accumulate (e.g. Eagles, 1967). It is therefore useful to consider how the growth of pasture plants respond to temperature in terms of the processes for utilizing photosynthate.

Utilization of products of photosynthesis may be grouped into two processes; (1) synthesis of new tissue; and (2) maintenance of grown and mature tissues. These two processes are often assessed as growth (or synthesis) and maintenance respiration respectively. Interpretation of temperature responses in terms of growth and maintenance components explains, for example, the shift in temperature optima as pastures age. For example, *Trifolium subterraneum* seedlings have a distinct optimum for whole plant growth of around 25°C but mature swards show a broad optimum of 10°C–25°C (Davidson *et al.*, 1970; Greenwood *et al.*, 1976; Fukai and Silsbury, 1976); this difference can be attributed to the higher maintenance component (which increases dramatically as temperature increases) of mature plants (e.g. Fukai and Silsbury, 1977). Leaf extension rates of *L. perenne* and *T. aestivum* have an optimum around 25°C and 28°C (Peacock, 1975; Kemp and Blacklow, 1982) whereas whole plant growth rates even in young plants, have an optimum 5°C lower (Friend and Helson, 1976; Hunt and Halligan, 1981).

Temperature responsiveness is often described by the Q_{10}, that is, the increase in rate for a 10°C rise in temperature. The Q_{10} is a measure of the efficiency of growth similar to relative growth rates (Kemp, 1981b). Processes with a high Q_{10} show a greater increase for a few degrees rise in temperature than processes with a low Q_{10} and this can explain some of the differences in sward productivity. Leaf appearance rates are more responsive to temperature—have higher Q_{10} values—in reproductive than vegetative swards (Thomas and Norris, 1977, 1979) and are more responsive than leaf death rates (Fig. 11.5). Similarly, leaf extension rates have higher Q_{10} values in reproductive than vegetative plants of *L. perenne* (Fig. 11.5). The Q_{10} values for leaf extension rates, which measure growth without any maintenance component, can exceed the commonly accepted value of 2 for biological reactions (Fig. 11.5). Q_{10} values for maintenance respiration rates also vary, though they tend to be closer to 2 (Fig. 11.5; Robson 1981).

Maintenance respiration consumes a large part of the products of photosynthesis. *Lolium perenne* lines selected for slow rates of dark (mainly maintenance) respiration produce more dry matter than either the parent lines (6%–13% more) or a fast respiration line (up to 28% more), especially in warm weather and under infrequent cutting (Wilson, 1982). The advantage of the slow respiration line may be as a result of an overall more efficient utilization of substrates principally in the production of more tillers (Robson, 1982a, 1982b). No data is as yet available on the growth and maintenance respiration components in these fast and slow respiring lines.

The minimum temperature for growth of temperate pastures is close to 0°C (e.g. Cooper and Tainton, 1968; Peacock, 1975). Most legumes produce little forage below 5°C as nodulation and nitrogen fixation are severely retarded at these temperatues (Gibson, 1963; Roughley *et al.*, 1970). Grass ecotypes show some variation in growth at low temperatures. Ecotypes from the mediterranean often produce more growth at low temperatures than ecotypes from northern Europe (Robson and Jewiss, 1968). This is apparently related to a better ability to utilize photosynthate at low temperatures in mediterranean ecotypes (Cooper, 1964). However, an inverse relationship can sometimes occur between the ability to grow at low temperatures and frost resistance (e.g. Cooper, 1964).

Temperate pasture species are regarded as resistant to chilling temperatures of around 10°C but variable in frost resistance ($< 0°C$) (McWilliam, 1978). Frosts can either kill the leaves, limiting productivity by reducing leaf areas or kill the whole plant. Frost hardening depends to some extent upon the accumulation of solutes within the plant (Breese and Foster, 1971; Levitt, 1980). Hardening is therefore enhanced by exposure to light (Lorenzetti *et al.*, 1971; Lawrence *et al.*, 1973) and by drought stress (Gusta *et al.*, 1980). The utilization of solutes for growth can increase a plant's susceptibility to frost damage as can application of nitrogen or irrigation (Breese and Foster, 1971). New leaves are more susceptible to frost damage than old leaves

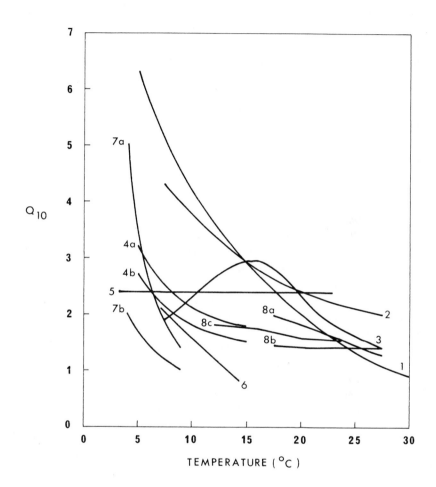

Fig. 11.5 Responsiveness of different plant processes to temperature as measured by Q_{10}. Values were calculated from curves of best fit to rate data or from tables. (1) Leaf extension rates of *T. aestivum* (Kemp and Blacklow, 1982). (2) Maintenance respiration rates for *T. repens* (McCree, 1974). (3) Maintenance respiration rates for *L. perenne* (Robson, 1981). (4) Leaf extension rates of *L. perenne* in (a) spring, (b) autumn (Peacock, 1972). (5) Dark respiration rates for *T. repens* leaves (Woledge and Dennis, 1982). (6) Relative leaf growth rates of *D. glomerata* (Broué *et al.*, 1967). (7) Rates for *L. perenne* in the field of (a) leaf appearance, (b) leaf death (Thomas and Norris, 1977). (8) *T. subterraneum* (a) leaf appearance rate, (b) leaf death rate, (c) dark respiration rate (from Fukai and Silsbury, 1976, 1977).

especially as they emerge at the top of the canopy where temperatures are lowest (Marcellos and Single, 1975).

High temperature stress limits sward productivity and prolonged temperatures above 35°C can be lethal (Andrew, 1965; Cooper and McWilliam, 1966). Temperature of 43°C–44°C for one hour can severely reduce respiration rates of *Trifolium* sp. (Golovko, 1978). High temperatures increase maintenance respiration rates, which is generally considered as causing a decline in growth rate at high temperatures, although Duff and Beard (1974) found otherwise. Many temperate grasses, particularly those adapted to mediterranean environments, survive hot summers by going into a state of dormancy.

C. Water

There have been many exhaustive reviews on the effects of water on plants. Here the main concern is with how plants regulate water fluxes and water potentials within their tissues for maintenance and growth. Temperate pasture species are able to adjust to moderate water stress from plant potentials of 0 to –2 MPa and survive severe water stress (below –2 MPa plant potentials) for short periods.

The flux (E) of water from the soil through the plant to the atmosphere depends on water potential gradients ($\Delta\Psi$) and the conductance (g) of the pathway as given in the general eqn (Van den Honert, 1948)

$$E = g \, \Delta\Psi \qquad (11.1)$$

In this eqn the conductance term is not constant but can vary with E and $\Delta\Psi$, especially when the pathway is complex. There are several points along the hydraulic pathway from the soil to the atmosphere where conductances can vary and can regulate flux rates. Conductance of water through the soil to the roots is usually higher than plant conductances and consequently rarely limits water flow. At the soil level the main factor that limits pasture growth is the ability of a species to exploit the soil volume with roots (see also Chapter 6, Section II). The greater productivity of *Festuca arundinacea* and *Phalaris aquatica* over *L. perenne* through periods of intermittent drought has been attributed to their greater quantity of roots, particularly at depth (Pook and Costin, 1970; Johns and Lazenby, 1973; Garwood *et al.*, 1979).

Where plants have many roots, the limiting hydraulic conductance is probably the diffusive conductance of water flow in the vapour phase at the stomata (Cowan and Milthorpe, 1968). One of the more significant arguments for the method of regulation of stomatal conductances is advanced in Chapter 2.

Decreases in stomatal conductance as a result of fewer stomata, smaller stomata or reduced epidermal ridging have been shown to reduce water use in *L. perenne* swards, allowing a longer period of growth during a drying cycle (Wilson, 1975). Improved water use efficiency in *Festuca* sp. and *Dactylis* sp. is also associated with reduced stomatal conductance and stomata that respond quickly to changes in ambient conditions (Silcock and Wilson, 1981).

The maintenance of turgor (or pressure) potentials within plant tissue is required for the expansion of leaves and other organs. At zero turgor no expansion growth occurs. Turgor potentials depend on the total water potential and osmotic potential. Pastures minimize the effects of water stress by osmoregulation that may maintain a positive turgor for expansion and allow stomata to remain open for carbon dioxide uptake by increasing the concentration of solutes within the cell sap. The initial increase in solutes results from the non-utilization of substrates for growth (Munns and Weir, 1981). Solute concentrations also increase when plants transpire water at a faster rate than it is being taken up by the roots. With increasing water stress and further decreases in total water potential, solutes continue to accumulate. Growing leaves and buds can osmotically adjust better than mature leaves as solutes continue to be translocated to growing regions during the water stress, whereas mature leaves are net exporters of solutes (Munns *et al.*, 1979). Shading and high nitrogen will depress solute concentrations as will cutting or grazing. In such circumstances plants lose turgor faster than might be expected and growth rates are reduced earlier. Osmoregulation may also lag behind changes in total water potentials and this produces hysteresis in the relationship between leaf growth rates and water potentials (Biscoe and Gallagher, 1977).

Pasture species also use other mechanisms to cope with water stress. These include change of leaf orientation (e.g. *Trifolium* sp.) to minimize radiation load, passive leaf flagging and leaf rolling (e.g. *Fescue* sp.) and early leaf senescence.

Maximizing the efficiency of water use by temperate pastures is an objective in many regions where soil moisture deficits occur. Water use efficiency can be considered at a leaf or sward level. Species differences occur in leaf water use efficiency but the most efficient species are not necessarily the most productive (Silcock and Wilson, 1981). Water use efficiency of a sward encompasses many relationships and it is necessary to resolve what these relationships are so that progress can be made in improving pasture productivity.

Legumes such as *T. repens* tend to have lower water use efficiencies than grasses, whereas *F. arundinacea* can have a greater efficiency of water use than *L. perenne* (Johns and Lazenby, 1973; Garwood *et. al.*, 1979). The greater efficiency of *F. arundinacea* may simply be a reflection of its more extensive root system (Garwood and Sinclair, 1979) whereas the poorer efficiency of *T. repens* could be due to both reduced leaf area and less extensive roots (Johns and Lazenby, 1973). *Trifolium repens* under irrigation increases its leaf area and achieves water use efficiencies equal to the grasses (Johns and Lazenby, 1973). It is also suggested that the poorer water use efficiency of *T. repens* under dryland is because of a less efficient stomatal mechanism than the grasses (Johns and Lazenby, 1973) but comparative studies on this are lacking. No single factor is likely to explain why some species are more efficient than others and the relevant factors are also likely to differ between species and environments.

V. PLANT INTERACTIONS

Pastures are dynamic and comprise several species, some sown, others not. Under grazing, pastures tend towards diversity both in the number of species and genetic diversity within species (Harper, 1978). The basis of many temperate pastures is a grass and a legume, typically *L. perenne* and *T. repens*. Few studies have traced in detail the interactions between legumes and grasses in a grazed sward but general experiences would support the model, developed in Wales, that is shown in Fig. 11.6. Species may change between regions, for example *Agrostis tenuis* may invade swards in New Zealand (Harris, 1974) and *Bothriochloa macra* swards on the Northern Tablelands of New South Wales (Cook *et al.*, 1978a), but the general cycle would follow Fig. 11.6. Note that there is no stable climax association in this cycle.

The regeneration cycle (Fig. 11.6) depends upon environmental and edaphic factors. Where soil phosphorus is low the pasture may simply degenerate to domination by weed species (Cook *et al.*, 1978b) and *T. repens* and *L. perenne* become insignificant until soil fertility is restored by the application of phosphate. The continual application of nitrogen to a grass–clover sward can result in grass dominance and exclusion of clover (e.g. Reid, 1972). Species may also be displaced as a result of seasonal changes in temperature (e.g. Cook *et al.*, 1978a) and soil moisture (Pook and Costin, 1970). Changes in botanical composition will also occur in response to cutting or grazing and this will be dealt with in Section VI.

Plant structure is important in determining the outcome of interactions between species and ecotypes. Long-leaved cultivars of *L. perenne* tend to dominate short-leaved and prostrate cultivars when cutting frequency is low (Hill and Shimamato, 1973) whereas grasses such as *Bromus catharticus* with few erect tillers provide ample space for clover and weed growth in contrast

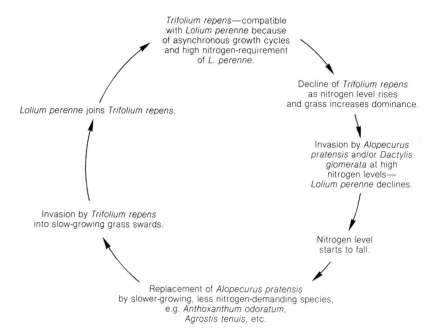

Fig. 11.6 Postulated regeneration cycle within an area of permanent pasture in Wales (from Turkington and Harper, 1979a).

to *L. multiflorum* (Pineiro and Harris, 1978b). Cultivars of *T. subterraneum* with long petioles may dominate those with short petioles (Black, 1960) and long petiole types of *T. repens* are better able to compete with *L. perenne* than short-statured cultivars (Wilson *et al.*, 1980). Results with a wide range of species indicate that under conditions of adequate soil fertility and moisture a species must be able to display most of its leaves at the top of the canopy to compete successfully in mixtures. Where soil moisture becomes a limiting factor differences in root development can affect the competitiveness of species, for example, where the deeper rooted *P. aquatica* replaces *L. perenne* during drought (Pook and Costin, 1970).

There has been a continuing debate as to whether mixed swards are more productive than monocultures. Where experiments have measured the productivity of species with similar growth patterns mixtures have not yielded more than monocultures but where species with complementary growth patterns are used then mixtures have been more productive. For example, an increase in pasture productivity has been achieved by mixtures of C_3 and C_4 grasses (Harris and Lazenby, 1974). The productivity of grass-clover mixtures is often higher than monocultures and this is partly because of the different times of peak growth and the nitrogen supplied by the legume. When species have complementary growth rhythms competition between them can be minimized. The yield of species in mixtures will also depend on the aggressiveness of the individual species. *Lolium perenne* is usually more aggressive than *D. glomerata*, which in turn is more aggressive than *Phleum pratense*. When sown in mixtures total sward yields appear inversely related to aggressiveness. Swards of *Phleum pratense* and *D. glomerata* yielded more than their respective monocultures whereas *L. perenne* and *D. glomerata* yielded less. Mixtures of *Phleum pratense* and *L. perenne* were intermediate.

We are only starting to appreciate that evolutionary pressures within a pasture can exert profound changes in genotype (Harper, 1978; Snaydon, 1978). Such changes occur rapidly as shown by the 40% change in genotypic composition of *L. perenne* swards that can occur within four months of sowing (Brougham and Harris, 1967). Studies from a one hectare pasture in Wales have shown significant differences in genotype and growth rates among *T. repens* clones taken from the field (Burdon, 1980). Clones of *T. repens* differed in growth rates by a factor of two, depending upon the species they were growing with and were most productive in mixtures with the original associate species (Turkington and Harper, 1979b). These results support data for *L. perenne* that the productivity of a genotype is often best under the conditions where that genotype developed (Charles, 1972). The stability of mixtures is also greater for co-adapted species (Snaydon, 1978).

VI. MANAGEMENT

A. Grazing and Cutting

The main function of pastures is to provide forage for grazing animals. Sometimes forage is cut and taken to stock but mostly forage is harvested by animals; they also deposit dung and urine and frequently sit, lie, scratch and paw on the pasture as well as walking, running and jumping on it. Each of these activities has its own effect upon plant performance and needs to be considered when assessing the regulation of pasture productivity. It is not always possible to standardize grazing effects in experiments and therefore cutting techniques have frequently been used to simulate what seems to be the main effect of grazing, that is, herbage removal. Cutting effects are defined by the frequency, intensity, uniformity and timing of defoliation in relation to the development of plants or swards. Most of our knowledge about pasture productivity has been learnt from cutting experiments and cutting treatments in themselves are a valuable tool. However, cutting does not simulate grazing exactly, as forage yields under cutting frequently exceed those under grazing (e.g. up to 68% more, Bryant and Blaser, 1968), though the reverse can also be found.

Before considering grazing and cutting influences it is worthwhile remembering that from an evolutionary point of view pasture plants depend on being grazed for their persistence. Unpalatable or toxic genotypes would be liable to invasion by the later shrub and tree phases of a plant succession, whereas grazing animals would limit any shrub or tree growth when grazing more palatable genotypes. Very palatable genotypes are likely to be grazed out and hence the successful genotype is a compromise. From this point of view one expects pasture species to have evolved strategies to replace the grazed parts and to be adaptable to a range of grazing pressures. Some variation in nutritive value would also be expected. Pasture plants are also adapted to predators and pathogens to which they have evolved a range of defence mechanisms. Unlike crops, pasture plants are rarely selected by man for disease resistance. The extent of these adaptations would, of course, depend upon the area of origin of the ecotype.

The ability to produce many branches, tillers, stolons or rhizomes, is an important feature of a grazed plant. Tillers are produced from the base of a grass. The base is shaded in ungrazed swards and as previously mentioned, shading limits tiller production. When swards are grazed, however, and especially when apices are removed, tillering and branching can be stimulated (e.g. Clifford, 1977). A common observation is that ungrazed grass swards have relatively few and erect tillers whereas grazed swards have many prostrate tillers. Increased tillering could be a response to the removal of apical dominance, although more than the stem apex needs to be removed as the presence of young foliage leaves also seems to limit tillering (Laude, 1975; Laidlow and Berrie, 1976). Defoliation can also stimulate flowering in some species (e.g. in *T. repens*, Thomas, 1981).

Leaf area is an important determinant of pasture productivity yet grazing mainly removes leaves. Donald and Black (1958) suggested that grazing should be controlled to maintain maximum sward productivity by keeping the leaf area index around the optimum. However, this strategy has not proved practical (e.g. Brougham, 1970) and it has also been criticized on the basis that it is a gross simplification of the factors that determine productivity in a pasture. Highest sward productivities have often been obtained by allowing swards to grow until light interception is complete and leaf area optimal but then defoliate them to near ground level (e.g. Korte *et al.*, 1982).

The regulation of rates of regrowth has been the subject of many investigations particularly with respect to carbohydrate reserves. However, much of the evidence indicates that the main determinants of regrowth rates are the amount of leaf area on the shoot and the rate of leaf area expansion. Reserves play a secondary role and are only important for the first few days after defoliation (Milthorpe and Davidson, 1966). The notable exception to this is *M. sativa* where reserves provide as much as 60% of new growth for up to two weeks after defoliation (Leach, 1978).

The frequency that a plant is grazed in a sward will affect its ability to regrow. Very frequent defoliation limits the number of tiller buds and leaves to the point where a plant is unable to compete. The grazing frequency is inversely related to the quantity of forage on offer per grazing animal. This means that when plant growth is slow, for example during low temperatures or seasonal water shortages, plants are heavily grazed. Under these conditions plant mortalities are high as plants are small and more vulnerable to desiccation, frosting and treading. *Medicago sativa* is particularly susceptible to frequent cutting. A depletion of reserves eventually limits the plant's ability to regrow and plant death then occurs (Leach, 1978). Studies on grazing habits have shown a higher frequency than is used in cutting experiments, for example, once every seven days (Hodgson, 1966). This difference is part of the reason why yields under cutting and grazing differ.

Cutting and grazing practices markedly influence the botanical composition of a sward. Experience in New Zealand (Harris, 1978) showed that rotational defoliation with a high level of pasture utilization generally leads to a simple mixture dominated by *L. perenne* and *T. repens*. A low level of utilization allows ingress of taller species like *D. glomerata, Holcus lanatus* and *B. catharticus*. Close continuous defoliation leads to a more species-rich association dominated by plants with prostrate, rhizomatous, stoloniferous or basal rosette habits. Where defoliation is especially severe and disturbance is accentuated by treading, annuals such as *Poa annua* can dominate. Similar patterns of response are likely in other climates. Where a sward contains palatable and unpalatable species grazing often results in the unpalatable species dominating. Grazing practices can effect the botanical composition more than the composition of species sown and these changes may have a bigger effect on production than any immediate effect of defoliation on herbage growth (Morley *et al.*, 1969). Grazing management can also be used to aid the replacement of competitive, undesirable cultivars by less competitive but desirable cultivars.

Treading influences pasture productivity severely when stocking rates are high and soils wet. Growth reductions of 20% to 30% have been recorded as a response to treading (Brown and Evans, 1973). Treading physically damages the plant destroying leaves and buds, although after treading tillering rates can increase (Edmond, 1970). Some species are more susceptible to treading than others and treading alone can change the botanical composition, *L. perenne* being one of the most resistant species (Edmond, 1964). Although eventually returning nutrients to the soil, the deposition of dung and urine on pastures can shade plants by covering them or burn leaves by dousing them in solutions of high solute concentration. The leaf area of plants is reduced and sward productivity declines. Urine returns more nitrogen and potassium than dung and this can stimulate grass growth at the expense of clover (Drysdale, 1965). Dung returns phosphorus, calcium and magnesium as well as nitrogen and potassium. The rapid return of nutrients in dung to the soil requires coprophagous activity by soil fauna (Watkins and Clements, 1978). Dung pats can shade and kill plants while those around the periphery are stimulated and as peripheral plants are often rejected by stock the botanical composition can be changed (Watkins and Clements, 1978). Rejection periods can last up to 18 months (Castle and Mac-Daid, 1972). Estimates of the reduction in herbage intake resulting from fouling of pasture by cattle vary from 11% to 47% (Greenhalgh and Reid, 1969; Taylor and Large, 1955).

B. Nutrition

The fertility of soils can have a major impact on the productivity of pastures. The interactions of nutrition, species, soil types and climate are complex but some general points have emerged from the extensive literature.

Species are often selected for use in pastures based on their responses to nutrients. *Medicago* sp. are used in neutral to alkaline soils whereas *Trifolium* sp. will tolerate mild acidity. *Lolium perenne* will tolerate more acid soils than *P. aquatica* but *P. aquatica* will grow better on a zinc deficient soil than *M. sativa* (Anderson, 1946). Where intensive, high-producing pastures are required, grasses are used in preference to legumes because of their greater productivity under high rates of nitrogen fertilization (Reid, 1972), but where nitrogen fertilizer cannot be used, legumes become the basis of pastures. The sensitivity of legumes to some soil conditions may reflect sensitivity of the *Rhizobium* symbiosis more than sensitivity of the host plant (Robson and Loneragan, 1978). Similarly it would be expected that inorganic nutrition of other nitrogen-fixing micro-organisms and mycorrhizas would be important in determining productivity of other pasture species.

Pasture species show a wide range of responses to nutrients and these responses are conditioned by the origin of the plant. Ecotypes of *L. perenne* have been shown to differ in their response to nitrogen and in each case the response was directly proportional to the nitrogen status of the soils from where the plant originated (Antonovics *et al.*, 1967; Charles, 1972). For intensive forage systems *L. perenne* has since been selected for cultivars that use nitrogen more efficiently (Goodman, 1977).

The nutrient requirement of pastures depends on the stage of plant development, age of tissue, climate and level of other nutrients. The amount of nutrient required is often related to the plants growth rate and to the mobility of the element within the plant (Robson and Loneragan, 1978). Defoliated plants have lower requirements than undefoliated plants. Thus the nutrient levels required by a sward vary throughout the year and are associated with seasonal growth patterns and with varying requirements of component species.

Nutrients are frequently in the top soil from where they are extracted by the roots. However, the top soil can frequently dry out and nutrient absorption then ceases, even though the soil at depth remains moist. This causes major reductions in pasture growth (e.g. Garwood and Williams, 1967). Injection of nutrients (mainly nitrogen) 45 cm below the surface, where the soil was still moist, significantly increased grass growth under conditions of a dry soil surface. Similarly it was demonstrated that the productivity of *Medicago* pasture increased with depth of placement of phosphate in a semi-arid environment (B. Scott, pers. comm., 1982).

A major part of pasture management has been to match species to soils and ensure adequate nutrition to optimize production. With increasing cost of fertilizer and use of marginal lands more attention is being given to selecting species and cultivars that can utilize nutrients more efficiently and tolerate toxic or adverse sites. There is considerable scope for these aims to be achieved (Robson and Loneragan, 1978).

VII. CONCLUSIONS

The productivity of pastures is determined in a complex way. Unlike crops that exist in monoculture and where predators are minimized, pastures are dynamic ecosystems where plants, predators and herbivores interact and where survival may be more important than maximizing production. It is therefore essential that pasture cultivars are selected from those growing in a pasture and managed in the way they will ultimately be used. For improving productivity or in some instances for just maintaining existing levels, it is necessary to understand not only what regulates the development and productivity of pasture components but to appreciate how such regulation is modified within the pasture ecosystem. Knowledge from physiology, ecology and agronomy needs to be blended to reach this end. The limits of pasture productivity have not been reached and future research will undoubtedly continue to provide insights into the regulation of these complex systems.

REFERENCES

Aitken, Y. (1974). "Flowering time, Climate and Genotype". Melbourne University Press, Melbourne.

Alberda, Th., and Sibma, L. (1982). The influence of length of growing period, nitrogen fertilization and shading on tillering of perennial ryegrass (*Lolium perenne* L.) *Netherlands Journal of Agricultural Science* **30**, 127–135.

Anderson, A. J. (1946). Molybdenum in relation to pasture improvement in South Australia. *Journal of Council of Scientific and Industrial Research of Australia* **19**, 1–15.

Andrew, W. D. (1965). Moisture and temperature requirements for germination of three annual species of Medicago. *Australian Journal of Experimental Agriculture and Animal Husbandry* **5**, 450–452.

Anslow, R. C., and Green, J. O. (1967). The seasonal growth of pasture grasses. *Journal of Agricultural Science* **68**, 109–122.

Antonovics, J., Lovett, J. and Bradshaw, A. D. (1967). The evaluation of adaptation to nutritional factors in populations of herbage plants. *In* "Isotopes in Plant Nutrition and Physiology". pp. 549–567. International Atomic Energy Agency, Vienna.

Biscoe, P. V. and Gallagher, J. N. (1977). Weather, dry matter production and yield. *In* "Environmental effects on crop physiology" (J. J. Landsberg and C. V. Cutting, ed.), pp. 75–100. Academic Press, London.

Black, J. N. (1960). The significance of petiole length, leaf area and light interception in competition between strains of subterranean clover (*Trifolium subterraneum* L.) grown in swards. *Australian Journal of Agricultural Research* **11**, 277–291.

Bodson, M., King, R. W., Evans, L. T., and Bernier, G. (1977). The role of photosynthesis in flowering of the long-day plant *Sinapis alba*. *Australian Journal of Plant Physiology* **4**, 467–478.

Breese, E. L., and Foster, C. A. (1971). Breeding for increased winter hardiness in perennial ryegrass. *Report Welsh Plant Breeding Station for 1970* pp. 77–86.

Brouè, P., Williams, C. N., Neal-Smith, C. A., and Albrecht, L. (1967). Temperature and day length response of some cocksfoot populations. *Australian Journal of Agricultural Research* **18**, 1–13.

Brougham, R. W. (1959). The effects of season and weather on the growth rate of a ryegrass and clover pasture. *New Zealand Journal of Agricultural Research* **2**, 283–296.

Brougham, R. W. (1970). Agricultural research and farming practice. Proceedings XIth International Grassland Congress, Surfers Paradise, April 1970. pp. A120–126.

Brougham, R. W., and Glenday, A. C. (1967). Grass growth in mid-summer—a re-interpretation of published data. *Journal of British Grassland Society* **22**, 100–107.

Brougham, R. W., and Harris, W. (1967). Rapidity and extent of changes in genotypic structure induced by grazing in a ryegrass population. *New Zealand Journal of Agricultural Research* **10**, 56–65.

Brown, K. R., and Evans, P. S. (1973). Animal treading: A review of the work of the late D. B. Edmond. *New Zealand Journal of Experimental Agriculture* **1**, 217–226.

Bryant, N. T., and Blaser, R. E. (1968). Effects of clipping compared to grazing of Ladino clover-orchard grass and alfalfa–orchard grass mixtures. *Agronomy Journal* **60**, 165–166.

Burdon, J. J. (1980). Intra-specific diversity in a natural population of *Trifolium repens*. *Journal of Ecology* **68**, 717–735.

Castle, M. E., and MacDaid, E. (1972). The decomposition of cattle dung and its effect on pasture. *Journal of British Grassland Society* **27**, 133–137.

Charles, A. H. (1972). Ryegrass populations from intensively managed leys. III. Reaction to management, nitrogen application and *Poa trivialis* L. on field trials. *Journal of Agricultural Science* **79**, 205-215.

Clifford, P. E. (1977). Tiller bud suppression in reproductive plants of *Lolium multiflorum* Lam. cv. Westerwoldicum. *Annals of Botany* **41**, 605-615.

Cook, S. J., Blair, G. J., and Lazenby, A. (1978a) Pasture degeneration II. The importance of superphosphate, nitrogen and grazing management. *Australian Journal of Agricultural Research* **29**, 19-29.

Cook, S. J., Lazenby, A., and Blair, G. J. (1978b). Pasture degeneration. I. Effect on total and seasonal pasture production. *Australian Journal of Agricultural Research* **29**, 9-18.

Cooper, J. P. (1956). Developmental analysis of populations in the cereals and herbage grasses. I. Methods and techniques. *Journal of Agricultural Science* **47**, 262-279.

Cooper, J. P. (1960). The use of controlled life-cycles in the forage grasses and legumes. *Herbage Abstracts* **30**, 71-79.

Cooper, J. P. (1964). Climatic variation in forage grasses. 1. Leaf development in climatic races of Lolium and Dactylis. *Journal of Applied Ecology* **1**, 45-61.

Cooper, J. P., and McWilliam, J. R. (1966). Climatic variation in forage grasses. 2. Germination, flowering and leaf development in mediterranean populations of *Phalaris tuberosa. Journal of Applied Ecology* **3**, 191-212.

Cooper, J. P., and Saeed, S. W. (1949). Studies on growth and development in Lolium. 1. Relation of the annual habit to head production under various systems of cutting. *Journal of Ecology* **37**, 233-259.

Cooper, J. P., and Tainton, N. M. (1968). Light and temperature requirements for the growth of tropical and temperate grasses. *Herbage Abstracts* **38**, 167-176.

Cowan, I. R., and Milthorpe, F. L. (1968). Plant factors influencing the water status of plant tissues. *In* "Water Deficits and Plant Growth (T. T. Kozlowski, ed.), Vol. 1 pp. 137-193. Academic Press, New York.

Davidson, J. L., Gibson, A. H., and Birch, J. W. (1970). Effects of temperature and defoliation on growth and nitrogen fixation in subterranean clover. Proceedings XIth International Grasslands Congress, Surfers Paradise, 1970, 542-545.

Donald, C. M. and Black, J. N. (1958). The significance of leaf area in pasture growth. *Herbage Abstracts* **28**, 1-6.

Drysdale, A. D. (1965). Liquid manure as a grassland fertilizer. III. The effect of liquid manure on the yield and botanical composition of pasture, and its interaction with nitrogen, phosphate and potash fertilizers. *Journal of Agricultural Science* **65**, 333-340.

Duff, D. T., and Beard, J. B. (1974). Supra optimal temperature effects upon *Agrostis palustris.* II. Influence on carbohydrate levels, photosynthetic rate and respiration rate. *Physiologia Plantarum* **32**, 18-22.

Eagles, C. F. (1967). The effect of temperature on vegetative growth in climatic races of *Dactylis glomerata* in controlled environments. *Annals of Botany* **31**, 31-39.

Edmond, D. B. (1964). Some effects of sheep treading on the growth of ten pasture species. *New Zealand Journal of Agricultural Research* **6**, 265-276.

Edmond, D. B. (1970). Effects of treading on pastures, using different animals and soils. Proceedings of the XIth International Grassland Congress, Surfers Paradise. pp. 604-608.

Evans, L. T. (1964). Reproduction. *In* "Grasses and Grassland" (C. Barnard, ed.), pp. 126-153. MacMillan, London.

Frield, T. R. O. (1980). Can we manipulate the annual pattern of pasture growth? *Proceedings New Zealand Grassland Society* **41**, 80-88.

Friend, D. J. C., and Helson, V. A. (1976). Thermoperiodic effects on the growth and photosynthesis of wheat and other crop plants. *Botanical Gazette* **137**, 75-84.

Fukai, S. and Silsbury, J. H. (1976). Responses of subterranean clover communities to temperature. I. Dry matter production and plant morphogenesis. *Australian Journal of Plant Physiology* **3**, 527–543.

Fukai, S. and Silsbury, J. H. (1977). Responses of subterranean clover communities to temperature. II. Effects of temperature on dark respiration rate. *Australian Journal of Plant Physiology* **4**, 159–167.

Garwood, E. A. and Sinclair, J. (1979). Use of water by six grass species. 2. Root distribution and use of soil water. *Journal of Agricultural Science* **93**, 25–35.

Garwood, E. A., Tyson, K. C., and Sinclair, J. (1979). Use of water by six grass species. 1. Dry-matter yields and response to irrigation. *Journal of Agricultural Science* **93**, 13–24.

Garwood, E. A., and Williams. T. E. (1967). Growth, water use and nutrient uptake from the subsoil by grass swards. *Journal of Agricultural Science* **69**, 125–130.

Gibson, A. H. (1963). Physical environment and symbiotic nitrogen fixation. I. The effect of root temperature on recently nodulated *Trifolium subterraneum* L. plants. *Australian Journal of Biological Science* **16**, 28–42.

Golovko, T. K. (1978). Critical respiration temperature of clover leaves. *Soviet Journal of Ecology* **9**, 553–555.

Goodman, P. J. (1977). Selection for nitrogen responses in *Lolium*. *Annuals of Botany* **41**, 243–256.

Greenhalgh, J. F. D., and Reid, G. W. (1969). The effects of grazing intensity on herbage consumption and animal production. III. Dairy cows grazed at two intensities on clean or contaminated pasture. *Journal of Agricultural Science* **72**, 223–228.

Greenwood, E. A. N., Carbon, B. A., Rossiter, R. C., and Beresford, J. D. (1976). The response of defoliated swards of subterranean clover to temperature. *Australian Journal of Agricultural Research* **27**, 593–610.

Gusta, L. V., Butler, J. D., Rajashekar, C. and Burke, M. J. (1980). Freezing resistance of perennial turf grass. *Hort. Science* **15**, 494–496.

Harper, J. L. (1978). Plant relations in pastures. *In* "Plant Relations in Pastures" (J. R. Wilson, ed.), pp. 3–16. CSIRO, Melbourne.

Harris, W. (1974). Competition among pasture plants. V. Effects of frequency and height of cutting on competition between *Agrostis tenuis* and *Trifolium repens*. *New Zealand Journal of Agricultural Research* **17**, 251–256.

Harris, W. (1978). Defoliation as a determinant of the growth, persistence and composition of pasture. *In* "Plant Relations in Pastures" J. R. Wilson, ed.), pp. 67–85. CSIRO, Melbourne.

Harris, W. and Lazenby, A. (1974). Competitive interaction of grasses with contrasting temperature responses and water stress tolerances. *Australian Journal of Agricultural Research* **25**, 227–246.

Hill, J., and Shimamato, Y. (1973). Methods of analysing competition with special reference to herbage plants. I. Establishment. *Journal of Agricultural Science* **81**, 77–89.

Hodgson, J. (1966). The frequency of defoliation of individual tillers in a set-stocked sward. *Journal of British Grassland Society* **21**, 258–263.

Hunt, W. F., and Field. T. R. O. (1979). Growth characteristics of perennial ryegrass. *Proceedings of the New Zealand Grassland Association*, **40**, 104–113.

Hunt, W. F., and Halligan, G. (1981). Growth and developmental responses of perennial ryegrass grown at constant temperature. I. Influence of light and temperature on growth and net assimilation. *Australian Journal of Plant Physiology* **8**, 181–190.

Jacobs, S. W. L., and Pickard, J. (1981). *Plants of New South Wales. A census of the Cycads, Conifers and Angiosperms*. National Herbarium of N.S.W. Royal Botanic Gardens, Sydney.

Jewiss, O. R. (1972). Tillering in grasses—its significance and control. *Journal of British Grassland Society* **27**, 65–82.

Johns, G. G., and Lazenby, A. (1973). Effect of irrigation and defoliation on the herbage production and water use efficiency of four temperate pasture species. *Australian Journal of Agricultural Research* **24**, 797–808.

Kemp, D. R. (1981a). The growth rate of wheat leaves in relation to the extension zone sugar concentration manipulated by shading. *Journal of Experimental Botany* **32**, 141–150.

Kemp, D. R. (1981b). Comparison of growth rates and sugar and protein concentrations of the extension zone of main shoot and tiller leaves of wheat. *Journal of Experimental Botany* **32**, 151–158.

Kemp, D. R., and Blacklow, W. M. (1982). The responsiveness to temperature of the extension rate of leaves of wheat growing in the field under different levels of nitrogen fertilizer. *Journal of Experimental Botany* **33**, 29–36.

Korte, C. J., Watkin, B. R., and Harris, W. (1982). Use of residual leaf area method and light interception as criteria for spring-grazing management of a ryegrass dominant pasture. *New Zealand Journal of Agricultural Research* **25**, 309–320.

Laidlow, A. S., and Berrie, A. M. M. (1976). The influence of expanding leaves and the reproductive stem apex on apical dominance of *Lolium multiflorum. Annals of Applied Biology* **78**, 75–82.

Langer, R. H. M. (1958). Changes in the tiller population of grass swards. *Nature* (London) **182**, 1817–1818.

Langer, R. H. M. (1959). A study of growth in swards of timothy and meadow fescue. II. The effects of cutting treatments. *Journal of Agricultural Science* **52**, 273–281.

Langer, R. H. M., and Ryle, G. J. A. (1958). Vegetative proliferations in herbage grasses. *Journal of British Grassland Society* **13**, 29–33.

Lawrence, T., Cooper, J. P., and Breese, E. L. (1973). Cold tolerance and winter hardiness in *Lolium perenne.* II. Influence of light and temperature during growth and hardening. *Journal of Agricultural Science* **80**, 341–348.

Laude, H. M. (1975). Tiller bud inhibition by young foliage leaves in jointing wheat. *Crop Science* **15**, 621–624.

Leach, G. J. (1978). The ecology of lucerne pastures. In "Plant Relations in Pastures" (J. R. Wilson, ed.), pp. 290–308. CSIRO, Melbourne.

Leafe, E. L. (1972). Micro-environment, carbon dioxide exchange and growth in grass swards. *In* "Crop Processes in Controlled Environments" (A. R. Rees, K. E. Cockshull, D. W. Hand and R. G. Hurd, eds.), pp. 157–174. Academic Press, London.

Levitt, J. (1980). "Responses of Plants to Environmental Stresses. Vol. 1, *Chilling, Freezing and High Temperature Stresses.*" Academic Press, New York.

Lorenzetti, F., Tyler, B. F., Cooper, J. P., and Breese, E. L. (1971). Cold tolerance and winter hardiness in *Lolium perenne.* 1. Development of screening techniques for cold tolerance and survey of geographical variation. *Journal of Agricultural Science* **76**, 199–209.

McCree, K. J. (1974). Equations for the rate of dark respiration of white clover and grain sorghum, as functions of dry weight, photosynthetic rate and temperature. *Crop Science* **14**, 509–514.

McWilliam, J. R. (1978). Response of pasture plants to temperature. *In "Plant Relation in Pastures"* (J. R. Wilson, ed.), pp. 17–34. CSIRO, Melbourne.

McWilliam, J. R., and Jewiss, O. R. (1973). Flowering of S.24 Perennial Ryegrass (*Lolium perenne* L.) in short days at low temperature. *Annals of Botany* **37**, 263–265.

Marcellos, H., and Single, W. V. (1975). Temperatures in wheat during radiation frost. *Australian Journal of Experimental Agriculture and Animal Husbandry* **15**, 818–822.

Milthorpe, F. L. and Davidson, J. L. (1966). Physiological aspects of regrowth in Grasses. *In* "The Growth of Cereals and Grasses" (F. L. Milthorpe and J. D. Ivins, eds.), pp. 241–255. Butterworths, London.

Monteith, J. L. (1981). Does light limit crop production. *In* "Physiological Processes Limiting Plant Productivity (C. B. Johnson, ed.), pp. 23–38. Butterworths, London.

Morley, F. H. W., Bennett, D., and McKinney, G. T. (1969). The effect of intensity of rotational grazing with breeding ewes on phalaris–subterranean clover pastures. *Australian Journal of Experimental Agricultural and Animal Husbandry* **9**, 74–84.

Munns, R., Brady, C. J., and Barlow, E. W. R. (1979). Solute accumulation in the apex and leaves of wheat during water stress. *Australian Journal of Plant Physiology* **6**, 379–389.

Munns, R., and Weir, R. (1981). Contribution of sugars to osmotic adjustment in elongating and expanded zones of wheat leaves during moderate water deficits at two light levels. *Australian Journal of Plant Physiology* **8**, 93–105.

Parsons, A. J., and Robson, M. J. (1980). Seasonal changes in the physiology of S.24 perennial ryegrass (*Lolium perenne* L.). 1. Response of leaf extension to temperature during the transition from vegetative to reproductive growth. *Annals of Botany* **46**, 435–444.

Parsons, A. J., and Robson, M. J. (1981a). Seasonal changes in the physiology of S.24 perennial ryegrass (*Lolium perenne* L.) 2. Potential leaf and canopy photosynthesis during the transition from vegetative to reproductive growth. *Annals of Botany* **46**, 249–258.

Parsons, A. J., and Robson, M. J. (1981b). Seasonal changes in the physiology of S.24 perennial ryegrass (*Lolium perenne* L.) 3. Partition of assimilates between root and shoot during the transition from vegetative to reproductive growth. *Annals of Botany* **48**, 733–744.

Parsons, A. J., and Robson, M. J. (1982). Seasonal changes in the physiology of S.24 perennial ryegrass (*Lolium perenne* L.) 4. Comparison of the carbon balance of the reproductive crop in spring and the vegetative crop in autumn. *Annals of Botany* **50**, 167–177.

Peacock. J. M. (1972). Interaction between sward and the environment in the field, *Annual Report Grassland Research Institute, Hurley, for 1971*, 47–49.

Peacock, J. M. (1975). Temperature and leaf growth in *Lolium perenne*. II. The site of temperature perception. *Journal of Applied Ecology* **12**, 115–123.

Pineiro, J. and Harris, W. (1978b). Performance of mixtures of ryegrass cultivars and prairie grass with red clover cultivars under two grazing frequencies. II. Shoot populations and natural reseeding of a prairie grass. *New Zealand Journal of Agricultural Research* **21**, 665–673.

Pook, E. W., and Costin, A. B. (1970). Changes in pattern and density of perennial grasses in an intensively grazed sown pasture influenced by drought in southern New South Wales. *Australian Journal of Experimental Agriculture and Animal Husbandry* **10**, 286–292.

Reid, D. (1972). The effects of the long term application of a wide range of nitrogen rates on the yields from perennial ryegrass swards with and without white clover. *Journal of Agricultural Science* **79**, 291–301.

Rhodes, I. (1975). The relationship between productivity and some components of canopy structure in ryegrass (*Lolium* sp.) 4. Canopy characters and their relationship with sward yields in some intra-population selections. *Journal of Agricultural Science* **84**, 345–351.

Robson, A. D., and Loneragan, J. F. (1978). Responses of pasture plants to soil chemical factors other than nitrogen and phosphorus, with particular emphasis on the legume symbiosis. *In* "Plant Relations in Pastures" (J. R. Wilson, ed.), pp. 128–144. CSIRO, Melbourne.

Robson, M. J. (1968). The changing tiller population of spaced plants of S.170 Tall Fescue (*Festuca arundinacea*). *Journal of Applied Ecology* **5**, 575–590.

Robson, M. J. (1980). A physiologist's approach to raising the potential yield of the grass crop through breeding. *In* "Opportunities for Increasing Crop Yields" (R. G. Hurd, P. V. Biscoe and C. Dennis, eds.), pp. 33–49. Pitman, London.

Robson, M. J. (1981). Respiratory efflux in relation to temperature of simulated swards of perennial ryegrass with contrasting soluble carbohydrate contents. *Annals of Botany* **48**, 269–273.

Robson, M. J. (1982a). The growth and carbon economy of selection lines of *Lolium perenne* cv. S23 with differing rates of dark respiration. 1. Grown as simulated swards during a regrowth period. *Annals of Botany* **49**, 321–329.

Robson, M. J. (1982b). The growth and carbon economy of selection lines of *Lolium perenne* cv. S23 with differing rates of dark respiration. 1. Grown as young plants from seed. *Annals of Botany* **49**, 331–339.

Robson, M. J., and Jewiss, O. R. (1968). A comparison of British and North African varieties of Tall Fescue (*Festuca arundinacea*). II. Growth during winter and survival at low temperatures. *Journal of Applied Ecology* **5**, 179–190.

Roughley, R. J., Dart, P. J., Nutman, P. S., and Roderiquez-Barrueco, C. (1970). The influence of root temperature on root-hair infection of *Trifolium subterraneum* L. by *Rhizobium trifolii* Dang. Proceedings XIth International Grassland Congress, Surfers Paradise, April 1970. 451–455.

Ryle, G. J. A. (1961). Effects of light intensity on reproduction in S.48 Timothy (*Phleum pratense* L.). *Nature* (London) **191**, 196–197.

Silcock, R. G., and Wilson, D. (1981). Effect of watering regime on yield, water use and leaf conductance of seven *Festuca* species with contrasting leaf ridging. *New Phytologist* **89**, 569–580.

Silsbury, J. H. (1965). Inter-relation in the growth and development of *Lolium*. I. Some effects of vernalization on growth and development. *Australian Journal of Agricultural Research* **16**, 903–913.

Snaydon, R. W. (1978). Genetic changes in pasture populations, *In* "Plant Relations in Pastures" (J. R. Wilson, ed.), pp. 253–272. CSIRO, Melbourne.

Tayler, J. C., and Large, R. V. (1955). The comparative output of two seeds mixtures. *Journal of British Grassland Society* **10**, 341–351.

Thomas, H. and Norris, I. B. (1977). The growth responses of *Lolium perenne* to the weather during winter and spring at various altitudes in mid-Wales. *Journal of Applied Ecology* **14**, 949–964.

Thomas, H., and Norris, I. B. (1979). Winter growth of contrasting ryegrass varieties at two altitudes in mid-Wales. *Journal of Applied Ecology* **16**, 553–565.

Thomas. R. G. (1981). Flower initiation in relation to cool season growth of four lines of white clover. *New Zealand Journal of Agricultural Research* **24**, 37–41.

Turkington, R. and Harper, J. L. (1979a). The growth, distribution and neighbour relationships of *Trifolium repens* in a permanent pasture. *Journal of Ecology* **67**, 201–218.

Turkington, R., and Harper, J. L. (1979b). The growth, distribution and neighbour relationships of *Trifolium repens* in a permanent pasture. IV. Fine-scale biotic differentiation. *Journal of Ecology* **67**, 245–254.

Van den Honert, T. H. (1948). Water transport in plants as a catenary process. *Discussions Faraday Society* **3**, 146–153.

Watkins, B. R., and Clements, R. J. (1978). The effects of grazing animals on pastures. *In* "Plant Relations in Pastures" (J. R. Wilson, ed.), pp. 273–289. CSIRO, Melbourne.

Williams, C. N., and Biddiscombe, E. F. (1964). The winter growth of selected pasture grasses. *Australian Journal of Experimental Agriculture and Animal Husbandry* **4**, 357–362.

Wilson, D. (1975). Stomatal diffusion resistances and leaf growth during droughting of *Lolium perenne* plants selected for contrasting epidermal ridging. *Annals of Applied Biology* **79**, 83–94.

Wilson, D. (1982). Response to selection for dark respiration rate of mature leaves in *Lolium perenne* and its effects on growth of young plants and simulated swards. *Annals of Botany* **49**, 303–312.

Wilson, D., Eagles, C. F., and Rhodes, I. (1980). The herbage crop and its environment— exploiting physiological and morphological variations to improve yields. *In* "Opportunities for Increasing Crop Yields" (R. G. Hurd, P. V. Biscoe, and C. Dennis, eds.), pp. 21–32. Pitman, London.

Woledge, J. (1979). Effect of flowering on the photosynthetic capacity of ryegrass leaves grown with and without natural shading. *Annals of Botany* **44**, 197–207.

Woledge, J., and Dennis, W. D. (1982). The effect of temperature on photosynthesis of ryegrass and white clover leaves. *Annals of Botany* **50**, 25–35.

Woledge, J. and Leafe, E. L. (1976) Single leaf and canopy photosynthesis in a ryegrass sward. *Annals of Botany* **40**, 733–783.

Tropical Pastures

J. R. WILSON

I. INTRODUCTION

Sown pastures within the subtropics and tropics (30°N to 30°S) are largely confined to areas with annual precipitation greater than 600 mm ranging from the semi-arid subtropics and monsoonal tropics, with a 4–6 months' growing period, to the humid tropics with a year-long growing season. Hutton (1971) listed about 40 grasses and 23 legumes used as sown species for these areas. In Queensland, Australia, Scattini (1981) indicated that 5% of the area of sown pasture is sown to legume alone (i.e. an introduced legume surface-sown or sod-seeded into a native grass pasture), 73% to grass alone, and 22% to a legume–grass mixture. In other major areas where sown tropical pastures are widely used, such as South and Central America, and the southern USA, there is an even greater emphasis on pure grass pastures. Panicoid tropical grasses compared to tropical legumes represent a major difference in physiology, not only because of the nitrogen fixing capacity of the legumes, but also because the tropical grasses have the C_4 photosynthetic pathway and the legumes have the C_3 pathway. These differences are important to any discussion of the control of productivity of tropical pastures.

185

Table 12.1

Comparison of dry matter, beef and milk production from sown temperate and tropical farm pastures. (After Dirven, 1977).

		Temperate	Tropical
Dry matter (t ha^{-1})	N fertilized grass	10	25
Liveweight gain (kg ha^{-1})	grass/legume pastures	600	600
	N fertilized grass	1000	1650
Milk production (kg ha^{-1})	grass/legume pastures	4000	5000
	N fertilized grass	9000	9000

Conventionally, plant physiologists reviewing tropical and temperate pasture species, or comparing C_4 and C_3 plants, have been exclusively concerned with dry matter production; its level (Cooper, 1970) and the efficiency of carbon assimilation (Ludlow, 1984). However, Dirven (1977) has pointed out that while dry matter production from sown tropical pastures is more than double the production from temperate pastures, there is little difference in the beef and especially the milk production from these pastures per given area (Table 12.1). There is thus often a poor correlation between standing dry matter yield and animal production because the quality of tropical pastures, in terms of intake of digestible energy and protein, is a severe limitation to obtaining maximum animal production from the high amounts of dry matter produced (Henzell, 1968; Cooper, 1970). Consequently this review, although selective because of space limitations, attempts a wider consideration of productivity. It discusses not only dry matter production but also gives a physiologist's view of the interactions between the consequences of that productivity, the physiological characteristics of the species groups involved, and the quality of the herbage for animal production. Most emphasis is given to grasses because of their importance to the yield of sown pastures.

II. CHARACTERISTICS OF TROPICAL GRASSES AND LEGUMES

A. Dry Matter Production

Ludlow (1984) recently reviewed the photosynthesis and dry matter productivity of C_4 tropical grasses compared to C_3 legumes. This review and others have dealt at length with the comparative biochemical and physiological efficiency of the C_4 and C_3 photosynthetic pathways and its implication for yield, and I will not reiterate in detail here. Despite some dissent the consensus reached by Ludlow (1984) and others (e.g. Monteith, 1978) is that the high photosynthetic efficiency of the C_4 pathway is translated into maximum growth rates of C_4 grasses which are 1.5–2 times higher than that of C_3 species (temperate grasses and all legumes), and an annual pasture production of tropical grasses which is at least twice that of tropical legumes. The assimilation efficiency is related to a higher efficiency of light utilization at lower light levels, with the C_4 grass superior to the C_3 legume at all temperatures within the normal growth range of 15°C–40°C (Ludlow, 1981). Legumes also differ from grasses in that nitrogen (N) is usually provided from symbiotic fixation. However, the biological nitrogen fixation process *per se* does not appear to impose any large, additional penalty on potential legume productivity because the energy cost is similar to that of nitrate reduction (Gibson *et al.*, 1977).

High photosynthetic efficiency does not necessarily lead to higher plant growth rate or higher production in C_4 species (as found, for example, in comparisons of C_4 and C_3 *Panicum* spp.; Wilson and Brown, 1983). Effective development of leaf area and light interception is often a prime determinant of yield. Data comparing the physiological differences in leaf development of tropical grasses and tropical legumes are scant or non-existent, but the tall-growing tussock-type tropical grasses, despite a low tiller density, intercept light better than many tropical legumes (Ludlow, 1978). Trailing or climbing legumes are commonly used in mixed tropical pastures in an attempt to overcome this problem; also, these legumes, such as Siratro, may produce new leaves on their stem at twice the rate of a companion grass (J. R. Wilson, unpublished). However, the legumes tend to dispose their leaves horizontally compared to the erect grass leaves and the resulting higher light extinction coefficient (e.g. Wong and Wilson, 1980) means a lower leaf area index of legume swards at full light interception (Ludlow, 1984). This would be another factor contributing to the lower potential of maximum yield of many legumes. The sown tropical grasses also gain a yield advantage over the legumes because in general, the grasses grow actively over a wider temperature range (McWilliam, 1978) and usually start growth earlier in the spring than do most of the commonly used legumes.

While maximum dry matter yield of tropical grasses may be as high as 85 t ha⁻¹ y⁻¹ with high fertilizer inputs, irrigation and a year-long growing season in the humid tropics (see Cooper, 1970), the potential is less than half that value in the monsoonal tropics or subtropics. In these environments growth of both grasses and legumes can be severely curtailed for about six months of the year by cool temperatures, frosts, or water stress.

It has been suggested that the inability of tropical grasses to grow at low temperature is associated with excessive starch accumulation in chloroplasts that causes damage and loss of photosynthetic capacity (West, 1973). This conclusion was derived from experiments in which grass was grown under summer temperature and light conditions but exposed to 10°C nights. Other experiments with more normal day temperature/night temperature for a cool climate regime do not support West's conclusion and indicate that little starch accumulates in tropical grasses under low temperature conditions (Forde *et al.*, 1975; Pearson *et al.*, 1977). In fact, tropical grasses in general seem to accumulate less total non-structural carbohydrate at low temperatures than do temperate grasses and the limited data available suggest that this is because photosynthesis activity and assimilate utilization (leaf area growth) of tropical grasses are similarly affected by low temperature (Wilson, 1975a).

Ivory and Whiteman (1978a,b) found that growth of tropical grasses ceased at mean daily temperatures of 8°C–12°C and that there was some thermal adaptation with the critical night temperature for growth decreasing with lower day temperatures. Similar analyses of temperature response with tropical legumes ('t Mannetje and Pritchard, 1974; Sweeney and Hopkinson, 1975) suggest that their growth is more sensitive to a decrease in temperature than growth of the tropical grasses, although considerable variation in response was evident. Some "tropical" species originating in the high altitude regions in Kenya, such as *Setaria anceps* cv. Narok and *Trifolium semipilosum*, have reasonable winter growth.

The specific problem of chilling injury at 10°C–15°C shown by some tropical graminaceous crops as a result of chlorophyll loss when low temperatures are experienced during the day and to phase changes in membrane lipids (McWilliam, 1978) is not in my opinion of particular significance for the winter performance of sown tropical grasses in subtropical areas. Frost injury, however, is a major limitation to the productivity of both grasses and legumes, especially in the subtropics (McWilliam, 1978). Both groups of species show little capacity to harden to winter temperatures and are consequently intolerant of mild frost. Severe top damage occurs at –2°C to –4°C, compared to –25°C for temperate grasses (Ludlow, 1980); although most tropical species will perennate by regrowth from underground organs. Intolerance to severe frosts and to temperatures that can freeze the soil are probably the major factors that determine the latitudinal distribution of these tropical species, and although C_4 grasses are present at high latitudes in North America, they grow there as summer annuals.

Nitrogen (N) deficiency seriously limits the dry matter production of tropical pastures through its influence on grass growth. Even in soils of high total nitrogen content the rate of mineralization of N is insufficient to provide for maximum growth of tropical grasses (Henzell, 1968). Tropical legumes once considered inefficient as fixers of atmospheric nitrogen are now known to fix amounts of nitrogen equivalent to most temperate legumes, but these amounts of nitrogen (40–260 kg (N) ha^{-1} y^{-1}) are usually well below the 300–1000 kg (N) ha^{-1} y^{-1} needed to sustain the high growth of the grasses (Henzell, 1968). Consequently, sown tropical grasses are frequently deficient in nitrogen and their crude protein contents are low.

Brown (1978) postulated that the C_4 photosynthetic pathway developed as an adaptation to cope with low nitrogen in tropical soils. Certainly C_4 grasses appear able to maintain a higher photosynthetic activity than C_3 grasses over a wide range of plant nitrogen status (Fig. 12.1). This ability may be associated with a lower investment of N in the soluble proteins involved in photosynthesis, especially Rubisco [ribulose-bisphosphate (RuP_2) carboxylase–oxygenase] which comprises 29%–43% of the soluble protein in temperate C_3 grasses but only 11%–16% in C_4 grasses (Pheloung and Brady, 1979). Moore and Black (1979) have further postulated that the compartmentation of carbon dioxide (CO_2) fixing activity between bundle sheath and mesophyll cells also contributes to higher nitrogen use efficiency (NUE) of the C_4 system because the nitrate and nitrite reducing enzymes located in the mesophyll are separated from the final CO_2 fixation process in the bundle sheath and thus do not compete directly for photosynthetically-produced reductant.

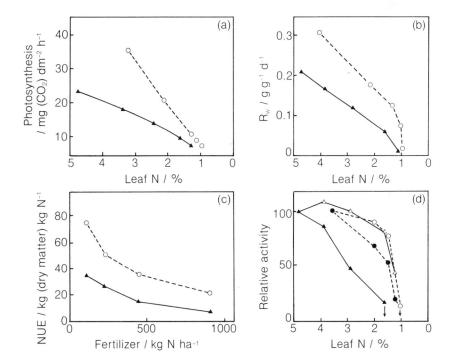

Fig. 12.1 Nitrogen use efficiency (NUE) of tropical (- - - - -) and temperate (————) grasses. Comparison of C_4 *Panicum maximum* (○, ●) with C_3 *Lolium perenne* (△, ▲) for (a) photosynthesis, (b) relative growth rate (R_W) and (d) net assimilation rate (open symbols) and relative leaf area growth rate (closed symbols) (after Wilson, 1975b); (c) dry matter production per unit of applied nitrogen (after Norton, 1982).

However, recent comparisons between C_4 and C_3 *Panicum* species (Brown and Wilson, 1983; Wilson and Brown, 1983) suggest that the C_4 pathway *per se* does not necessarily confer high NUE in terms of mg CO_2 fixed per mg of leaf nitrogen or account for relative growth rates at different leaf nitrogen concentrations. Also, other studies (Fig. 12.1) have indicated that response of leaf area development to low N may be more significant to differences in NUE between C_4 and C_3 grasses than differences in photosynthetic efficiency. This is supported by the generally lower accumulation of soluble carbohydrate in nitrogen-deficient tropical grasses compared to temperate grasses (see Wilson, 1975b). Possibly, C_4 grasses have a lower nitrogen requirement for leaf development than C_3 grasses (Wilson and Brown, 1983) but more study is needed to confirm this. Although the physiological basis is not yet well understood, C_4 grasses appear to have a high NUE in relation to relative growth rate at varying nitrogen status and high dry matter production per unit nitrogen (Fig. 12.1).

Other sources of N, such as from root association with *Azospirillum* and nitrogen fixing organisms in the phyllosphere (leaf sheaths), appear to be widespread in tropical grasses. However, it seems that the nitrogen contribution from these sources may allow some small extra yield with the plants still essentially nitrogen deficient.

Water stress is another limitation to yield in many regions where tropical pastures are sown. The C_4 pathway is thought to have evolved as an adaptation to hot, arid environments, but any general benefits in adaptation of the tropical grasses over the C_3 tropical legumes, apart from a higher water-use efficiency because of the C_4 photosynthetic efficiency, are not readily apparent. Sown tropical grasses which are adapted to semi-arid environments show osmotic and stomatal adaptation (Turner and Jones, 1980) and a tolerance of high leaf-water deficits (Ludlow, 1976). Similar attributes have been shown by some tropical legumes such as *Stylosanthes* (Ludlow, M. M., unpublished data) and *Centrosema* (Ludlow *et al.*, 1983). By contrast, *Macroptilium*, which is equally adapted to dry areas, has completely different attributes of little osmotic adaptation, tight stomatal control and intolerance of high leaf-water deficits, as well as a deep-rooting taproot and leaf movement. These various adaptations are perhaps more related to survival under stress than growth capacity, because leaf growth of both legumes and grasses is usually rapidly curtailed as water stress develops.

B. Nutritive quality

The high yield of tropical grasses creates problems with herbage quality in relation to digestible energy and protein as reviewed by Wilson and Minson (1980). Studies with single plants have indicated a negative relationship between growth rate and digestibility (Ivory *et al.*, 1974). The conservative stocking rates of most tropical pastures leads to excessive accumulation during the growing season of mature, stemmy grass with a low digestibility and low intake (Chacon *et al.*, 1978). Even with plenty of available feed, cattle may gain only 0.5–0.7 kg day^{-1} on tropical pastures compared with 1–1.2 kg on temperature pastures (see Wilson and Minson, 1980). The problem is compounded because the high NUE of tropical grasses leads to excessive dilution of N and to crude protein levels of less than 6%–8% which cause severe reduction in herbage intake. The digestibility of the grass, especially stems, falls off rapidly as the ungrazed material ages.

The low tiller density of tropical grasses and the tall, elongated culms of many of the commonly used species lead to a low leaf bulk density and consequently to a low intake of leaf per animal bite (as the leaves are widely dispersed vertically and horizontally) compared with temperate pasture grasses (Stobbs and Hutton, 1974). Interestingly, recent work by Ludlow *et al.* (1982) showed that preventing stem extension with hormone application increased leaf area density (i.e. leaf area per unit sward height per unit ground area) and increased intake per bite by grazing cattle without reducing canopy photosynthetic rate. They concluded that there would be a negligible reduction in sward yield of these grasses if their leaf area densities per unit ground area could be increased up to a value of 40 per metre of canopy height, which is higher than values for temperate grasses (33 per metre of canopy height).

Mature stem is avoided by grazing animals but even when stem is young and of similar digestibility to a leaf, its intake by grazing animals is lower than that of leaf (Laredo and Minson, 1973). The problem of high proportion of stem is accentuated in tropical grasses, because unlike temperate grasses, most of them have no specific environmental requirements for flowering, and elongated culms are produced continually over the growing season. Grass quality and animal liveweight gain are highest early in the growing season when stem development is minimal and leaf digestibility is high. Quality of grass pastures, particularly those fertilized with high N, can only be maintained by sustained heavy grazing, which is feasible only in well-watered regions with a long growing season (e.g. Puerto Rico or the humid coastal areas of Queensland). By contrast with the grasses, legume digestibility and nitrogen content are much more stable with season and plant maturity.

The potential animal production from tropical grasses is limited by the inherently higher cell wall content of their tissues (see Norton, 1982) and their lower digestibility (13% units) and intake compared to temperate grasses as summarized by Minson (1981), whereas the difference in mean digestibility between tropical and temperate legumes is only 4% units (Minson and Wilson, 1980). Leaves of tropical grasses have a lower proportion of the readily-digested mesophyll tissue and a higher proportion of the less

Table 12.2

Leaf attributes[a] of C_4, intermediate ($C_{3/4}$), and C_3 photosynthetic types of *Panicum* species. (After Wilson *et al.*, 1983.)

Type of *Panicum*	Dry matter digestibility (%)	Cell wall content (%)	Tissue proportion in leaf cross-section[b]				
			MES (%)	BS (%)	VAS (%)	EPI (%)	SCL (%)
C_4	69	50	43	20	8	27	1.7
$C_{3/4}$	70	42	48	18	6	26	1.7
C_3	76	33	66	10	3	22	0.5

[a] Means of 18 (C_4), 3 ($C_{3/4}$) and 6 (C_3) species.
[b] MES, mesophyll; BS, bundle sheath; VAS, vascular bundle; EPI, epidermis; SCL, sclerenchyma.

digestible and indigestible epidermal, vascular, sclerenchyma and bundle sheath tissues. Kranz anatomy of the C_4 tropical grasses explains these differences in tissue proportions and could account for as much as 17% units difference in cell wall content and 7% units difference in digestibility between C_4 and C_3 species (Table 12.2). The C_4 grasses have a high frequency of vascular bundles per unit leaf width and hence many potential sites for lignification and sclerenchyma strands exist. Tropical legumes have the C_3 type of anatomy and hence are little different from temperate legumes (Minson and Wilson, 1980). The interaction of the consequences of anatomy with rate of breakdown of fibre in the rumen and hence herbage intake is an important area for future research.

The higher temperatures of the tropics are another major limitation to herbage quality. They accelerate stem development and maturation, increase cell wall content and lignification, and hence lower herbage digestibility. The higher temperatures also accelerate the rates of senescence and decline in digestibility of leaves. These aspects are summarized by Dirven and Deinum (1977) and Wilson (1982) and it has been calculated from growth room studies that digestibility of grasses decreases by about 0.6% units per 1°C rise in growth temperature which could account for 5%–10% units difference in digestibility between tropical and temperate grass pastures (Wilson and Minson, 1980). The tropical legumes decrease by only about 0.28% units per 1°C increase in temperature.

Thus C_4 anatomy and high growth temperatures explain much of the low digestibility of tropical grasses. The high irradiance and the short-term water stresses common in tropical regions do not seem to be detrimental to forage quality (Wilson, 1982). In fact, water stress reduces stem development and can delay leaf ageing which may result in a slower than usual decline in digestibility and decline in protein content of the material with time (Wilson and Ng, 1975; Wilson, 1982). Provided that growth is sufficient to satisfy animal intake, substantially higher liveweight gains per beast have been recorded from tropical pastures in dry years than in wet years (Mannetje, 1982).

To the extent that total soluble carbohydrate (TSC) levels influence digestibility and palatability (intake) of grass herbage, tropical grass quality is disadvantaged because TSC concentrations are low, usually less than 10%, compared to greater than 10% for most temperate grasses. The low percentage TSC in tropical grasses may partly reflect high assimilate usage to support rapid growth—the rate of translocation of assimilate from leaves is higher for tropical than temperate grasses—though whether this is cause or effect in relation to growth is not clear. Another contributing factor is the greater night respiratory load because of the longer, higher temperature nights of the tropics during the growing season compared to temperate regions. Even when growth is slowed because of low N or low temperature, extra TSC accumulation is only modest and it appears that in these conditions photosynthesis and assimilate usage (leaf growth) decline in closer concert than in temperate grasses in which leaf growth is more sensitive than photosynthesis (Wilson, 1975a,b).

III. THE GRAZED PASTURE

The physiology of the grazed pasture has received much less attention than that of the individual species. Tropical grass pastures that are well watered and heavily fertilized with N can have carrying capacities of up to 12 beasts ha^{-1}. Extremely high gains in cattle liveweight of 1300–2760 kg ha^{-1} y^{-1} have been achieved (Henzell and 't Mannetje, 1980), although generally liveweight gain or milk production per animal is restricted by the quality of the herbage. Irrespective of these high production figures, heavily fertilized grass pastures are uneconomic at current beef prices. Most sown tropical grass pastures in Australia are grown on moderately fertile heavy soils without nitrogen fertilizer.

Grass–legume mixtures generally support much lower levels of annual liveweight gain of 180–500 kg ha^{-1} (Henzell and 't Mannetje, 1980). Many grazing studies have shown that the higher the proportion of legume in a tropical pasture, the greater the animal liveweight gain, and from pen studies, the higher the proportion of legume, the higher is the animal's intake of herbage (Fig. 12.2). However, the combination of a relatively slow-growing C_3 legume with a tall, fast-growing C_4 grass poses special problems for management of tropical pastures to maintain a desirable legume content of about 30%. The high NUE of the tropical grass compounds this problem because even when N is relatively low the grass can still manage a reasonable

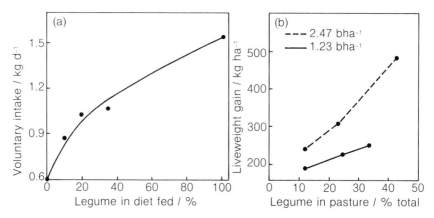

Fig. 12.2 (a) Effect of proportion of legume in the diet on the voluntary intake of sheep fed pangola grass, and (b) effect of proportion of legume in sown tropical grass–legume pastures on animal productivity at two stocking rates ((a) after Minson and Milford, 1967 and (b) after Evans, 1970).

yield, whereas in temperate pastures the grass growth and hence grass–legume balance is very sensitive to pasture nitrogen status. Competition for light is a problem (Ludlow, 1978) overcome in part by selection of legumes with a climbing habit which allows them to reach the top of the canopy. Such legumes are, however, susceptible to frequent grazing because the stems and potential growing points are removed and must be replaced by dormant axillary buds in a shaded position lower in the canopy. Selection of shorter tropical grasses for combination with legumes of bushy or stoloniferous habit, which are more tolerant of grazing, may lessen the problem of light competition (see Ludlow *et al.*, 1982). However, some short mat-forming grasses such as pangola and kikuyu have not proven particularly compatible with legumes. The other alternative, to reverse the order of shading by using shrub legumes which are taller than the grasses, e.g. *Leucaena leucocephala* has proven productive in this respect. Complex management systems that can integrate different grass–legume mixtures and balance growth against herbage quality are often needed to maximize animal production from tropical pastures (e.g. Roberts, 1979).

There is an important need for a wider range of legumes better adapted to close grazing so that mixed pastures can be more heavily defoliated to control grass quality. Most of the commonly used tropical legumes originate from Central and South America where they have been exposed to low grazing pressure, whereas the grasses come from Africa and are adapted to grazing by bovines. The search for better adapted tropical legumes is continuing actively and the numerical classifactory techniques developed by Burt and Williams (1979) may lead to more effective separation of taxonomic and agronomic groups and help to narrow the areas of search.

IV. CONCLUSIONS

The C_4 photosynthetic pathway and other physiological attributes of tropical grasses result in a high dry matter yield of sown tropical pastures and a potentially high carrying capacity of animals per hectare. However, these same physiological features and the high temperature environment, impose considerable penalties on the quality of the grass and prevent the full expression of this yield in terms of production per animal. Productive legumes better adapted to grazing are needed to increase the yield and quality of mixed grass–legume pastures and hence production per animal but the consequences of the different physiology of grasses and legumes, partly associated with the C_4 versus C_3 photosynthetic pathway, requires better understanding to develop suitable management techniques to maximize production from such mixed swards.

REFERENCES

Brown, R. H. (1978). A difference in N use efficiency in C_3 and C_4 plants and its implications in adaptation and evolution. *Crop Science* **18**, 93–98.

Brown, R. H. and Wilson, J. R. (1983). Nitrogen response of *Panicum* species differing in CO_2 fixation pathways. II. Carbon dioxide exchange characteristics. *Crop Science* **23**.

Burt, R. L., and Williams, W. T. (1979). Strategy of evaluation of a collection of tropical herbaceous legumes from Brazil and Venezuela. III. The use of ordination techniques in evaluation. *Agro-Ecosystems* **5**, 135–146.

Chacon, E. A., Stobbs, T. H., and Dale, M. B. (1978). Influence of sward characteristics on grazing behaviour and growth of Hereford steers grazing tropical grass pastures. *Australian Journal of Agricultural Research* **29**, 89–102.

Cooper, J. P. (1970). Potential production and energy conversion in temperate and tropical grasses. *Herbage Abstracts* **40**, 1–15.

Dirven, J. G. P. (1977). Beef and milk production from cultivated tropical pastures. A comparison with temperate pastures. *Stikstof* **20**, 2–15.

Dirven, J. G. P., and Deinum, B. (1977). The effect of temperature on the digestibility of grasses. An analysis. *Forage Research* **3**, 1–17.

Evans, T. R. (1970). Some factors affecting beef production from subtropical pastures in the coastal lowlands of southeast Queensland. *In* "Proceedings 11th International Grasslands Congress, Surfers Paradise, 1970" (M. J. T. Norman, ed.), pp. 803–807. University of Queensland Press, Brisbane.

Forde, B. J., Whitehead, H. C. M., and Rowley, J. A. (1975). Effect of light intensity and temperature on photosynthetic rate, leaf starch content and ultrastructure of Paspalum dilatatum. *Australian Journal of Plant Physiology* **2**, 185–195.

Gibson, A. H., Scowcroft, W. R., and Pagan, J. D. (1977). Nitrogen fixation in plants: an expanding horizon. *In* "Recent Developments in Nitrogen Fixation" (W. Newton, J. R. Postgate and C. Rodriguez-Barrueco, eds.), pp. 387–417. Academic Press, London.

Henzell, E. F. (1968). Sources of nitrogen for Queensland pastures. *Tropical Grasslands*. **2**, 1–17.

Henzell, E. F., and Mannetje, L.'t (1980). Grassland and forage research in tropical and subtropical climates. *In* "Perspectives in World Agriculture" pp. 485–532. Commonwealth Agricultural Bureau, London.

Hutton, E. M. (1971). Plant improvement for increased animal production. *Journal of the Australian Institute of Agricultural Science* **37**, 212–225.

Ivory, D. A., Stobbs, T. H., McLeod, M. N., and Whiteman, P. C. (1974). Effect of day and night temperatures on estimated dry matter digestibility of *Cenchrus ciliaris* and *Pennisetum clandestinum. Journal of the Australian Institute of Agricultural Science* **40**, 156–158.

Ivory, D. A., and Whiteman, P. C. (1978a). Effect of temperature on growth of five subtropical grasses. I. Effect of day and night temperature on growth and morphological development. *Australian Journal of Plant Physiology* **5**, 131–148.

Ivory, D. A., and Whiteman, P. C. (1978b). Effect of temperature on growth of five subtropical grasses. II. Effect of low night temperature. *Australian Journal of Plant Physiology* **5**, 149–157.

Laredo, M. A., and Minson, D. J. (1973). The voluntary intake, digestibility and retention time by sheep of leaf and stem fractions of five grasses. *Australian Journal of Agricultural Research* **24**, 875–888.

Ludlow, M. M. (1976). Ecophysiology of C_4 grasses. *In* "Ecological Studies Analysis & Synthesis, Water and Plant Life." Vol. 19 (O. L. Lange, L. Kappen, and E-D. Schulze, eds.), pp. 364–386. Springer-Verlag, Berlin.

Ludlow, M. M. (1978) Light relations of pasture plants. *In* "Plant Relations in Pastures" (J. R. Wilson, ed.), pp. 35–49. CSIRO, Melbourne.

Ludlow, M. M. (1980). Stress physiology of tropical pasture plants. *Tropical Grasslands* **14**, 136–145.

Ludlow, M. M. (1981) Effect of temperature on light utilization efficiency of leaves in C_3 legumes and C_4 grasses. *Photosynthesis Research* **1**, 243–249.

Ludlow, M. M. (1984). Photosynthesis and dry matter production in C_3 and C_4 pasture plants, with special emphasis on tropical C_3 legumes and C_4 grasses. *Photosynthesis Research.*

Ludlow, M. M., Chu, A. C. P., Clements, R. J., and Kerslake, R. G. (1983). Adaptation of accessions of *Centrosema* to water stress. *Australian Journal of Plant Physiology.*

Ludlow, M. M., Stobbs, T. H., Davis, R., and Charles-Edwards, D. A. (1982). Effect of sward structure in two tropical grasses with contrasting canopies on light distribution, net photosynthesis and the size of bite harvested by grazing cattle. *Australian Journal of Agricultural Reseach* **33**, 187–201.

McWilliam, J. R. (1978). Response of pasture plants to temperature. *In* "Plant Relations in Pastures" (J. R. Wilson, ed.), pp. 17–34. CSIRO, Melbourne.

Mannetje, L.'t (1982). Problems with tropical pastures. *In* "Nutritional Limits to Animal Production from Pastures" (J. B. Hacker, ed.), pp. 67–85. Commonwealth Agricultural Bureau, London.

Mannetje, L.'t, and Pritchard, A. J. (1974). The effect of daylength and temperature on introduced legumes and grasses for the tropics and sub-tropics of coastal Australia. 1. Dry matter production, tillering and leaf area. *Australian Journal of Experimental Agriculture and Animal Husbandry*, **14**, 173–181.

Minson, D. J. (1981). Nutritional differences between tropical and temperate pastures. *In* "Grazing Animals" (F. H. W. Morley, ed.), pp. 143–157. Elsevier, Amsterdam.

Minson, D. J., and Milford, R. (1967). The voluntary intake and digestibility of diets containing different proportions of legume and mature Pangola grass (*Digitaria decumbens*). *Australian Journal of Experimental Agriculture and Animal Husbandry* **7**, 546–551.

Minson, D. J., and Wilson, J. R. (1980). Comparative digestibility of tropical and temperate forage—a contrast between grasses and legumes. *Journal of the Australian Institute of Agricultural Science* **46**, 247–249.

Monteith, J. L. (1978). Reassessment of maximum growth rates for C_4 and C_3 crops. *Experimental Agriculture* **14**, 1–5.

Moore, R., and Black, C. C. (1979). Nitrogen assimilation pathways in leaf mesophyll and bundle sheath cells of C_4 photosynthesis plants formulated from comparative studies with *Digitaria sanguinalis* (L.) Scop. *Plant Physiology* **64**, 309–313.

Norton. B. W. (1982). Differences between species in forage quality. *In* "Nutritional Limits to Animal Production from Pastures" (J. B. Hacker, ed.), pp. 89–110. Commonwealth Agricultural Bureau, London.

Pearson, C. J., Bishop, D. G., and Wesk, M. (1977). Thermal adaptation of *Pennisetum*: Leaf structure and composition. *Australian Journal of Plant Physiology* **4**, 541–554.

Pheloung, P., and Brady, C. J. (1979). Soluble and fraction 1 protein in leaves of C_3 and C_4 grasses. *Journal of Science Food and Agriculture* **30**, 246–250.

Roberts, C. R. (1979). Grazing management of tall tropical legume based pastures. Australian Society of Animal Production Meeting, Wollongbar, 1979. J. H. Williams and Sons, Murwillumbah, NSW. Australia.

Scattini, W. J. (1981). Queensland's pastures: Past, present and future—a perspective. *Tropical Grasslands* **15**, 65–71.

Stobbs, T. H., and Hutton, E. M. (1974). Variation in canopy structures of tropical pastures and their effect on the grazing behaviour of cattle. *In* "Proceedings 12th International Grasslands Congress, Moscow" Vol. 4, pp. 680–687.

Sweeney, F. C., and Hopkinson, J. M. (1975). Vegetative growth of nineteen tropical and subtropical pasture grasses and legumes in relation to temperature. *Tropical Grasslands* **9**, 209–217.

Turner, N. C., and Jones, M. M. (1980). Turgor maintenance by osmotic adjustment: A review and evaluation. *In* "Adaptation of Plants to High Temperature Stress" (N. C. Turner and P. J. Kramer, eds.), pp. 87–103. Wiley, New York.

West, S. H. (1973). Carbohydrate metabolism and photosynthesis of tropical grasses subjected to low temperatures. *In* "Plant Responses to Climatic Factors, Ecology and Conservation" Vol. 5 (R. O. Slatyer, ed.), pp. 165–168. UNESCO, Paris.

Wilson, J. R. (1975a). Influence of temperature and nitrogen on growth, photosynthesis and accumulation of non-structural carbohydrate in a tropical grass, *Panicum maximum* var. *trichoglume*. *Netherlands Journal of Agricultural Science* **23**, 48–61.

Wilson, J. R. (1975b). Comparative response to nitrogen deficiency of a tropical and temperate grass in the interrelation between photosynthesis, growth and the accumulation of non-structural carbohydrates. *Netherlands Journal of Agricultural Science* **23**, 104–112.

Wilson, J. R. (1982). Environmental and nutritional factors affecting herbage quality. *In* "Nutritional Limits to Animal Production from Pastures) (J. B. Hacker, ed.), pp. 111–131. Commonwealth Agricultural Bureau, London.

Wilson, J. R., and Brown, R. H. (1983). Nitrogen response of *Panicum* species differing in carbon dioxide fixation pathways. 1. Growth analysis, and carbohydrate accumulation. *Crop Science.*

Wilson, J. R., Brown, R. H., and Windham, W. R. (1983). Influence of leak anatomy on the dry matter digestibility of C_4, C_3 and C_3/C_4 intermediate types of *Panicum* species. *Crop Science* **23**, 141–146.

Wilson, J. R., and Minson, D. J. (1980). Prospects for improving the digestibility and intake of tropical grasses. *Tropical Grasslands* **14**, 253–259.

Wilson, J. R., and Ng, T. T. (1975). Influence of water stress on parameters associated with herbage quality of *Panicum maximum* var. *trichoglume*. *Australian Journal of Agricultural Research* **26**, 127–136.

Wong, C. C., and Wilson, J. R. (1980). Effects of shading on the growth and nitrogen content of green panic and Siratro in pure and mixed swards defoliated at two frequencies. *Australian Journal of Agricultural Research* **31**, 269–285.

CHAPTER 13

Natural Grasslands

E. K. CHRISTIE

I. INTRODUCTION

A grassland may be best defined as a plant community with a low-growing plant cover of non-woody species (Milner and Hughes, 1968). Climax grasslands occur on all the earth's major land masses and represent the potential vegetation for around 25% of the earth's land area (Shantz, 1954). Grasslands possess a marked resemblance and constancy in climates, in growth form and in physiognomy. The use of the term "grassland" is generally applied to such communities as the prairies of North America, the pampas of South America, the veldt of South Africa and the steppes of the USSR, as well as other equivalent stands throughout the world. The occurrence of grasslands is largely controlled by climate and climatically determined grasslands occur in environments where mean annual precipitation varies from 250 mm to 1000 mm and mean annual temperature from 0°C–26°C (Leith, 1975). These climates are mostly found between deserts

199

and forests or in the rainshadow of major mountain ranges. Furthermore, grasslands occur in areas where, for some period in the year, soil water availability falls below the requirement for forests, yet sufficient precipitation is received during some period of the year to sustain grasses as the dominant, or at least, major vegetation component (Lauenroth, 1979). Most grasslands have a unimodal pattern of increase and decrease in green biomass (Sims *et al.*, 1978), but in arid and semi-arid environments a much more irregular response pattern of green biomass production may occur. Several peaks in live biomass are the result of randomly distributed precipitation (Noy-Meir, 1973).

This chapter will be concerned only with climatically determined grasslands and some successional types that result from removal of the original vegetation (e.g. scrub) and are maintained by management practices such as grazing and fire. The systems discussed rely on the soil as the sole source of minerals for plant growth; no inputs or fertilizer are applied. Lauenroth (1979) has presented above-ground net primary production values for 52 climatically determined grassland sites throughout the world. Mean annual precipitation varied from 136 mm to 1810 mm and above-ground net production rate from 42 g m^{-2} y^{-1} to 1387 g m^{-2} y^{-1}.

As grasslands are a major source of animal protein for human consumption, producing around 15% of the total protein available to man (Pimentel *et al.*, 1976), it is not surprising that a considerable amount of research has focused on the effects of defoliation on grassland growth and production. Overgrazing of grasslands characteristically leads to a decrease in herbage biomass production and photosynthetic surface area. When grazing pressure is reduced or completely removed, yield response is related to processes such as carbohydrate availability, photosynthesis and, at later stages, rate of nutrient uptake by the roots (May, 1960; Davidson and Milthorpe, 1966). Equally as well documented are the effects of frequency and intensity of defoliation on herbage biomass production and stability of native grasslands (e.g. Trlica, 1977). Although these aspects are very relevant in grassland management, equally important is the study of the limiting resources of the environment such as water, minerals and light and grassland growth and production. An understanding of physiological and ecological aspects of yield is therefore fundamental to the study of the functioning of grassland systems.

Most research has focused on the climate–grassland production interrelationships and until recently, mineral nutrition and cycling had received less attention. Herbage biomass production of grasslands is closely related to both the interactions between precipitation and available soil water and precipitation and organic mineral mineralization. For example, mineral nitrogen is considered to be the principal nutrient controlling herbage biomass in many grassland systems (Woodmansee, 1978). In addition, as other organic-matter fractions are recognizable biological entities that are significant in cycling organically bound nutrients such as phosphorus and sulphur (McGill *et al.*, 1975) these nutrients may also influence biomass production on some soils.

Representatives of both the C_3 and C_4 photosynthetic pathway are found in grassland species throughout the world. In some systems such as the North American prairies and the Australian *Acacia aneura* grazing lands, C_3 and C_4 species co-occur (Williams and Markley, 1973; Christie, 1975). Studies comparing growth between C_3 and C_4 pathway grasses will form the basis of this chapter. An analysis of the physiological determination of yield will consider environmental influences on growth and production and competition between species for limited resources.

II. PRODUCTIVITY

A. Mineral Nutrition

Grasses that occur naturally on infertile soils may often be characterized by their slow growth rates and low yield response as nutrient supply increases. Christie and Moorby (1975) examined the effects of increasing the supply of phosphorus on the growth of two native grasses that occur on either very low phosphate soils, mulga grass (*Thyridolepis mitchelliana*) or moderately high phosphate soils, Mitchell grass (*Astrebla lappacea*) and also buffel grass, (*Cenchrus ciliaris*), a species of wide edaphic tolerance. After six weeks' growth at a very low external concentration of phosphorus, mulga grass produced 25% of its maximum yield, compared with 5% and 1% for Mitchell and buffel grasses, respectively (Fig. 13.1). The higher relative yield of mulga grass at low phosphorus concentrations was related to its superior ability to absorb phosphorus, especially over the vegetative growth stages (Fig. 13.2a) and, as well, in transporting phosphorus to the shoots (Christie and Moorby, 1975). These processes were more important than the capacity to function at lower tissue concentration, since the concentration of phosphorus in the shoots of mulga grass was greater than for the other two grasses, at all external concentrations (Christie and Moorby, 1975). However, the superiority in absorption and transport of phosphorus was not related to yield, yield being related more to the photosynthetic than to the nutritional characteristics of the plants. Mulga grass had a much lower growth rate than the two other grasses. This was related to its relatively less efficient photosynthetic system as was evident by higher values for leaf area ratio and lower values for net assimilation rate, relative to, for example, buffel grass (Fig. 13.2).

Mulga grass is a C_3 pathway species whereas Mitchell and buffel grasses are C_4 pathway plants. Rates of net photosynthesis were found to be a major

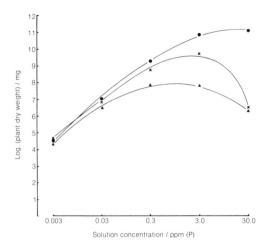

Fig. 13.1 Dry weight achieved in six weeks by mulga (▲), Mitchell (X) and buffel (●) grasses over a solution concentration range 0.003–30.0 ppm phosphorus (from Christie and Moorby, 1975).

determinant of yield in these species (Christie, 1975). C_4 plants exhibit superior growth and efficiency of nitrogen use compared with C_3 plants, particularly under conditions of low soil nitrogen supply (Wilson and Haydock, 1971). This apparent advantage in nitrogen use may result from a lower investment in photosynthetic carboxylating enzymes than C_3 plants (Brown, 1978). Because soil nitrogen is frequently a limiting resource, the interrelationships between nitrogen supply and growth of co-occurring C_3 and C_4 grasses would be studied best in replacement series mixtures. One such study reported by Christie and Detling (1982) examined the effects of nitrogen and temperature on competitive ability of *Bouteloua curtipendula* (a C_4 prairie grass) and *Agropyron smithii* (a C_3 prairie grass). At 20°C/12°C day/night temperatures, direct competition for soil nitrogen resulted in lowered growth of *B. curtipendula* when grown in mixtures with *A. smithii*. However, at 30°C/15°C day/night temperature the situation reversed in that the growth of *A. smithii* was restricted when grown in the presence of *B. curtipendula*, through competition for nitrogen (Fig. 13.3). These results suggest that yield and competitive ability is more closely related to temperature and photosynthetic pathway type than soil nitrogen supply. Moreover, although differences in nitrogen uptake and use occurred between the two grasses, the growth differential at two temperatures would result in uneven utilization of the soil nitrogen on a seasonal basis. There was little evidence from this study to support the postulate that differences in nitrogen use efficiency between C_3 and C_4 pathway plants may result in yield advantages for C_4 plants only.

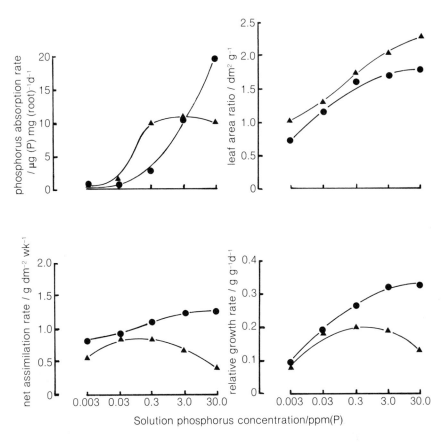

Fig. 13.2 Influence of phosphorus supply on relative growth rate, net assimilation rate, leaf area ratio and phosphorus absorption rate of mulga (▲) and buffel (●) grasses during vegetative growth (days 12–32) (after Christie and Moorby, 1975).

B. Water Relations

In a controlled study with the semi-arid C_4 species, Mitchell grass, in which a water deficit was imposed slowly, Doley and Trivett (1974) found there was an approximately linear relationship between the rates of carbon dioxide assimilation and transpiration as leaf water potential declined. Assimilation approached zero at –6 MPa potential. The decrease in assimilation was attributed to stomatal closure and a marked direct inhibition of the photosynthetic system. Although assimilation is reduced during periods of water deficit, Detling *et al.* (1978) have estimated that under optimal light and temperature, but at a soil water potential of –5 MPa in the root zone, assimilation

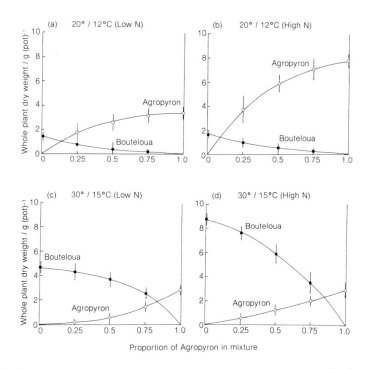

Fig. 13.3 Replacement series (substitutive) experiments with grass mixtures. The diagrams show the influence of temperature and soil nitrogen supply (Low N : 20 ppm mineral N; High N : 80 ppm mineral N) on growth of *Agropyron* (C_3 species) and *Bouteloua* (C_4) species at various frequencies. Values shown are mean whole plant weight $\pm$ 1 S.E. (from Christie and Detling, 1982).

rates of the North American C_4 prairie grass *Bouteloua gracilis* are comparable to values recorded for many C_3 species from more productive systems. They suggest that the low above-ground production values recorded for *B. gracilis* (Table 13.1) result from the allocation of a large proportion of the photosynthetically fixed carbon to below-ground organs, rather than to the production of additional photosynthetic tissue. Such a pattern of carbon partitioning leads to wide variation in root/shoot ratio. This pattern of partitioning may be significant in a grass species response to defoliation rather than as some general index of drought tolerance. In drought situations, absolute root biomass and depth of root penetration, rather than root/shoot ratio, is perhaps a better indicator of a species capacity to withstand periods of prolonged water deficit.

Under conditions of non-limiting water supply C_4 plants have greater values for efficiency of water use (net photosynthetic rate: transpiration rate)

Table 13.1
A comparison of net primary production studies into ungrazed temperate and subtropical grasslands

Location	Dominant grass species (Photosynthetic pathway[a])	Mean annual temperature (°C)	Annual precipitation (mm)	Annual evapotranspiration (mm)	Annual green shoot biomass production (g.m⁻².y⁻¹)	Maximum short-term green shoot biomass production rate (g.m⁻².d⁻¹)	Litter biomass crop (g.m⁻²)	Maximum root biomass crop (g.m⁻²)	Annual root biomass production (g.m⁻².y⁻¹)	Source
TEMPERATE GRASSLANDS										
1. Colorado, U.S.A.	*Bouteloua gracilis* (C₄) *Buchloe dactyloides* (C₄)	7.6	232	230	101	2.69	251	1598	422	b
2. New Mexico, U.S.A.	*Bouteloua eriopoda* (C₄) *Sporobolus flexuosus* (C₄)	14.1	224	224	125	1.95	71	207		b
3. Texas, U.S.A.	*Bouteloua gracilis* (C₄) *Aristida longiseta* (C₄)	13.7	355	351	176	3.04	331	859	436	b
4. South Dakota, U.S.A.	*Agropyron smithii* (C₃) *Stipa viridula* (C₃)	7.2	360	346	188	3.09	496	1767	248	b
5. Oklahoma, U.S.A.	*Andropogon gerardi* (C₄) *Panicum virgatum* (C₄)	15.0	674	512	286	5.48	365	1043	256	b
SUBTROPICAL GRASSLANDS										
6. Queensland, Australia	*Thyridolepis mitchelliana* (C₃) *Monachather paradoxa* (C₃)	22.9	664	559	122	2.74	26	181	60	c
7. Queensland, Australia	*Cenchrus ciliaris* (C₄)	22.9	647	616	154	4.29	38	560	165	c
8. Queensland, Australia	*Dichantium sericeum* (C₄)	23.6	610		180	2.21		418		d
9. Queensland, Australia	*Astrebla lappacea* (C₄)	21.4	605	556	185	3.95	60	510		e

[a] After Williams and Markley, 1973; Sims *et al.*, 1978; Boutton *et al.*, 1980; Christie, 1975; and Jacobsen, 1981. [b] Sims and Coupland (1979). [c] Christie (1978). [d] Jacobsen (1981). [e] Christie (1979, 1981).

compared with C_3 pathway plants (Downes, 1969). Such a feature would appear to be an important adaptation for plant growth in low, variable rainfall environments. However, of greater significance in any analysis of yield in such climates would be to examine the effects of an increasing water deficit on photosynthesis for co-occurring C_3 and C_4 grasses. Although C_4 grasses generally tend to occur in hotter, more arid regions, perennial C_3 grasses may be present, for example, *Thyridolepis mitchelliana* in semi-arid Australia and *Agropyron smithii* in semi-arid North America. Very few studies have been undertaken where the effects of an increasing water deficit on the photosynthetic response of co-occurring semi-arid or arid C_3 and C_4 grasses have been compared in the field.

Kemp and Williams (1980) examined the effects of short-term water deficit on leaf gas exchange on the co-occurring North American grassland species *Agropyron smithii* (a C_3 species) and *Bouteloua gracilis* (C_4). Under conditions of non-limiting water supply, net photosynthetic rate of *A. smithii* was limited more by residual (internal) than by stomatal conductance. In contrast, photosynthetic rate of *B. gracilis* was limited mainly by stomatal conductance. Rates of net photosynthesis in both species were very sensitive to leaf water potential and decreased exponentially as potential declined. The decrease in net photosynthesis resulted from pronounced decreases in both stomatal and residual conduction. Kemp and Williams (1980) concluded that the photosynthetic mechanism of *B. gracilis* is not more resistant to low water potential than is the mechanism of *A. smithii*. Moreover, the similarities in these species in their physiological responses to water deficit suggests that seasonal moisture gradients are not as important a parameter along which niche separation has occurred, as are seasonal temperature gradients.

From the studies of Christie and Detling (1982) and Kemp and Williams (1980), the central role of temperature on the growth and competitive ability of these grass species is clear. But as Kemp and Williams (1980), and Ode *et al.* (1980) have suggested, photosynthetic adaptation to temperature and seasonal separation of production activity of C_3 and C_4 grasses may be a strategy for ecosystem utilization through reduction of interspecific competition.

III. ANALYSIS OF PRODUCTIVITY: GRASSLAND COMMUNITIES

A. Net Primary Production

The most appropriate method for determining net primary production of a grassland community is based on the following expression of Milner and Hughes (1968)

$$NPP = \triangle W + L_D + L_C \qquad (13.1)$$

where NPP is the net primary production rate for any given vegetation compartment; $\triangle W$, the change in biomass of the specified compartment over a time interval (t_1-t_2); L_D, the plant losses through death and shedding during (t_1-t_2); and L_C, the plant losses by consumer organisms during (t_1-t_2).

Consumption by herbivores can be considerable (e.g. 10%–40%, Ebersohn and Lee, 1972) in native grasslands. In order to characterize accurately the potential biomass production of any grassland, sequential measurements of W should be made in the absence of grazing or alternatively, modifying the harvest method to take into account consumption. In addition, a sequential sampling programme over time must be related to soil water supply to ensure that available water is not a limiting factor. An example of such an approach is given for a native grassland. Sampling frequency is closely related to available soil water during the summer growing season (Harvests 0–6) and the winter growing season (Harvests 8–11) (Fig. 13.4). Regression analysis of the sequential changes in herbage biomass over time allows short-term growth rates to be derived. These rates provide a more precise means for comparing very different grassland systems, as variation in length of growing season, and hence yield, is offset. A comparison of net primary production for above-ground and below-ground compartments of temperate and subtropical grassland found in very different climatic zones and soils, is given in Table 13.1. Annual green-shoot biomass production varied from 101 g m^{-2} y^{-1} to 286 g m^{-2} y^{-1}. There were only small differences in the overall yield response between the temperate and subtropical communities. A similar trend in short-term rate of maximum green-shoot biomass production also existed. These short-term values of 1.9 g m^{-2} d^{-1}– 5.5 g m^{-2} d^{-1} were much less than those recorded for a fertilized subtropical *Digitaria decumbens* sown pasture of 19.3 g m^{-2} d^{-1} (Bryan and Sharpe, 1965). More striking differences between the temperate and subtropical grasslands existed in annual root biomass production and root and litter biomass than in net primary production for the temperate grasslands. Ratios of underground biomass to green-shoot biomass ranged from 2 to 16 and 2 to 4 for temperate and subtropical grasslands respectively.

B. Mineral Nutrition

Mineral absorption rates by a grassland may be derived following Williams (1948) as

$$U = \frac{1}{W_R} \cdot \frac{dC_N}{dt} \qquad (13.2)$$

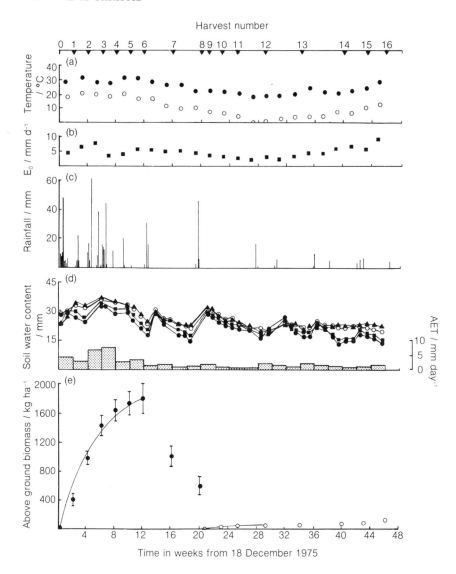

Fig. 13.4 Interrelationships between harvest sequence, climatic conditions, soil water content, actual evapotranspiration and above-ground live biomass production at the study site. (a) Mean fortnightly maximum (●) and minimum (○) temperatures. (b) Mean fortnightly pan evaporation rate, E_0 (■). (c) Daily rainfall. (d) Sequential changes in water content of the 0 to 10 cm (●), 10 to 20 cm (■), 20 to 30 cm (○), and 30 to 40 cm (▲) bands of the soil profile and actual evapotranspiration rate, E (:::). (e) Changes in above-ground biomass, including continuous non-limiting periods of soil water for growth over summer (●) and winter (○), respectively; ± one standard error of the mean biomass weight at each harvest over summer is indicated (from Christie, 1981).

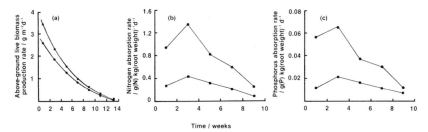

Fig. 13.5 Sequential changes in above-ground green biomass production rate and nutrient absorption rate over a continuous summer growing season for a C_4 mulga (●) and a C_3 buffel (■) grassland. (a) Above-ground biomass production rate. (b) Nitrogen absorption rate. (c) Phosphorus absorption rate (after Christie 1978, 1979).

where U is the absorption rate for mineral nutrient; W_R is the grassland community root biomass; and C_N is the absolute nutrient content of the total herbage biomass on time (t).

In a C_3 mulga grassland and a C_4 buffel grassland on infertile red earths in a semi-arid climate, seasonal decreases in above-ground green biomass production were related to the decline in nitrogen and phosphorus absorption (Fig. 13.5). The relative differences in growth rate and phosphorus absorption rate for these communities were similar to those recorded in single plant studies (Fig. 13.2): the C_3 mulga grassland had a higher phosphorus absorption rate but a lower green biomass production rate compared with the C_4 buffel grassland. Although soil water was non-limiting, the uptake of nitrogen and phosphorus began to decline after eight weeks' continuous growth. This suggests that the upper limit to grassland yield is set in part by these nutrients. In a similar study of a C_4 Mitchell grassland on a moderately fertile, grey cracking clay, the reduction in above-ground green biomass was related to a decline in nitrogen uptake only, phosphorus uptake remaining linear over the twelve week continuous growing season (Christie, 1981).

Over the first eight weeks of the summer growing season when the level of soil water was non-limiting and nitrogen uptake linear for each of the above three grasslands, the relationship between above-ground rate of green biomass production and green-shoot total nitrogen concentration was examined. The decline in biomass production rate over time was associated with a declining tissue nitrogen concentration (Fig. 13.6). Furthermore, under similar environmental conditions both C_4 grassland communities had higher biomass production rates at lower tissue nitrogen concentrations than the C_3 grassland. However in order to understand clearly the significance to yield determination of grassland differences in nitrogen use in this example the fertility of the soil on which the grasslands naturally occur should be considered; also, the phosphorus:nitrogen balance in the green shoot biomass

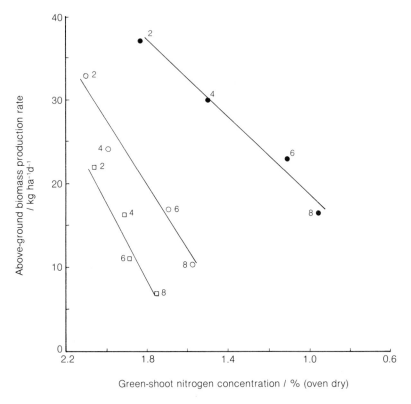

Fig. 13.6 Interrelationships between daily above-ground live biomass production rate and green shoot nitrogen concentration over an eight week continuous summer growing season for a C$_4$ Mitchell grassland (●), a C$_3$ *Thyridolepis* grassland (□) and a C$_4$ *Cenchrus* grassland (○) growing in semi-arid Australia. The numbers in italics refer to the length of the growing period in weeks (from Christie, 1981).

should be considered. Other nutrients affect nitrogen balance in the green-shoot biomass and affect nitrogen uptake mainly through their effects on plant growth and development. Interactions resulting from the phosphorus: nitrogen ratio are the best example of this effect (Cole *et al.,* 1963). Tentative standards (Penning de Vries *et al.,* 1978) for plant growth in relation to the phosphorus:nitrogen ratio suggest that nitrogen level is the major limitation on biomass production for the Mitchell grassland system, but that phosphorus level is the major limitation for the mulga grassland. This analysis tends to be supported by the inherent soil nutrient supply. Soil nitrogen is deficient and phosphorus adequate in the Mitchell grassland, whereas soil phosphorus is grossly deficient in the mulga grassland (Table 13.2; Christie, 1981). These results, although site specific, suggest that one

Table 13.2

Soil fertility, green-shoot biomass nutrient relationships for two native Australian grasslands

Grassland Community Type	Acid extractable soil phosphorus (0cm–30cm) ppm	Total soil nitrogen (0cm–30cm) ppm	Mean P:N ratio in green shoot biomass (growing period 84 day) $g(P).g(N)^{-1}$	Efficiency of nutrient use	
				g (dry matter) $g(P)^{-1}$	g (dry matter) $g(N)^{-1}$
Mulga (C$_3$)	3	410	0.05	1250	64
Mitchell (C$_4$)	55	720	0.15	525	114

adaptive mechanism to account for yield and distribution of these two major grasslands is associated with their efficiencies of absorption and utilization of major nutrients. The distinction between the two grasslands was very clear in this respect.

C. Water Use Efficiency

A feature of natural grasslands is that they are found in environments where water deficits, of varying degrees, occur. The ratio of total rainfall to actual evapotranspiration is around one, a value commonly recorded for semi-arid grasslands (see Table 13.1). This value reflects the close relationship between water and plant growth in this environment.

Because C_4 plants have higher water use efficiencies than C_3 plants, it is sometimes postulated that C_4 plants may have yield advantages over C_3 plants in low, variable rainfall environments (Laetsch, 1969). One approach for estimating water use efficiency for grassland communities is based on measured net primary production and evapotranspiration rates. This approach was used in a comparative study, in a semi-arid environment, for a C_4 buffel grassland and a C_3 mulga grassland, growing on a similar soil over a twelve-week continuous summer growing season and an eight-week continuous winter growing period. Derived values for water use efficiency for the C_3 grassland were about 60% that of the mean value for the C_4 grassland over the summer growing season. Values were similar for both communities for the winter growing season (Christie, 1978). Although differences in efficiency of water use exist, annual production of the C_4 grassland is often not much higher than that of the C_3 grassland. Part of the explanation could be that soil phosphorus and nitrogen limit yield on this soil. In addition, in this environment actual evapotranspiration rates are very high. The 0–40 cm zone of the red earth (sandy clay loam in texture) on which the study was undertaken stores 63 mm at 0.02 MPa and 28 mm at −1.5 MPa potential. Around 81% of the precipitation events that occur in this environment are less than 20 mm (90% are less than 35 mm) so that most precipitation events would not penetrate beyond the 0–40 cm zone. This zone contains most of the grassland root biomass and major nutrients and is considered to be the zone most important for primary production. Over a continuous drying cycle in summer (mean E_0 10 mm d^{-1}) the time taken for the 35 mm of available soil water to be lost by evaporation varied from 14 to 17 days for a bare soil site and a high-ground cover site, respectively. Consequently, advantages in water use efficiency by the C_4 grassland may be offset to some extent for short-term growth, as evaporation is a major component of water loss when E_0 is high and because small precipitation events predominate. On the other hand it might be possible for C_3 grasses to compensate for their lower values of water use efficiency by possessing other strategies for tolerating stress, such as local differences in the vertical pattern of root growth and elongation, and nutrient and water uptake. Noy-Meir (1973) suggests that the adaptive value of higher water use efficiency in C_4 plants may not be a higher net primary production but rather in the capacity of the plant to grow in periods and latitudes with high E_0, where they have the advantages of high saturation of light and flexible temperature acclimation (Caldwell *et al.*, 1972).

D. Integration of Environmental Factors and Grassland Productivity

Characterization of the potential net primary production of any site is one of the fundamental issues in grassland management. Year to year variability may be great so that annual estimates may be quite different from the long-term balance. An accurate simulation model provides the conversion of lengthy field experiments into environment–grassland dynamics quickly and economically. Based on field data that quantified a number of grassland primary production rate processes and soil and water balance, Hughes and Christie (1983) have developed an empirical model for net primary production for an extensively grazed C_3 semi-arid mulga-grass dominant community. Model predictions for soil water balance and herbage biomass balance are in good agreement with independently collected data.

The model works on a time period of one day and requires inputs of only evaporation and rainfall. Daily values for above-ground growth rate are calculated as a function of soil water potential, root distribution, the amount of growth that has occurred during the season and a thermal index for growth. Grasslands have a characteristic seasonal pattern of production in relation to temperature when water is non-limiting (but see Fig. 13.4). Soil nutrient supply sets the upper limit of live biomass production in extended growing seasons. An example of the application of this model in semi-arid Australia is the analysis of historic climatic records. The derived pattern for annual variability in above-ground live biomass production for a grassland (Fig. 13.7) is the essential starting point for livestock management investigations in this environment.

IV. CONCLUSIONS

While considerable research into the environmental and edaphic factors controlling herbaceous biomass production of grasslands has led to a better understanding of the processes involved, there remains a number of areas requiring closer investigation. Firstly, comparative differences in carbon

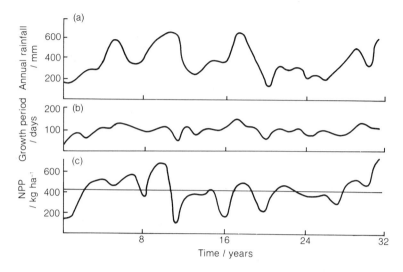

Fig. 13.7 Interrelationships between (a) Total summer rainfall; (b) length of the summer growing season (water balance output); and (c) green herbage biomass produced over this period. Analysis based on historic climatic records for Charleville, Australia (after Christie and Hughes, 1983).

allocation patterns between above-ground and below-ground compartments for C_3 and C_4 grasses should characterize species differences in root growth and spatial distribution, laterally and vertically down the soil profile. Zones of nutrient and water absorption need to be more clearly defined. Ideally an examination of these broad relationships should also be undertaken when environment resources are limiting and studied in relation to defoliation frequency and intensity.

As C_3 and C_4 pathway grasses co-occur in semi-arid environments and because of inherent differences in optimal temperatures for photosynthesis and growth, temperature would be a major factor conditioning yield responses to variable water supply. Because seeds and seedlings would be exposed to supra-optimal temperatures for growth in this environment, at some stage, the physiological responses of C_3 and C_4 grasses to high temperatures could impose a further limit to growth and metabolism. Prolonged exposure to supra-optimal temperatures also leads to heat-induced water stress which results in injury due to desiccation (Levitt, 1972).

Nutrient fluxes, apart from nitrogen, in grassland systems need to be better quantified. Considerable differences exist between native grasslands in the size of the organic compartments and inherent soil fertility. The mechanisms stabilizing organic carbon, nitrogen, phosphorus and sulfur are not necessarily common to all four elements and the mechanisms and pathways of mobilization are specific to the organic material containing the various elements (McGill and Cole, 1981). The balance between mineralization rate, mineral nutrient absorption rate, microbial use and live biomass production rate needs to be explored on a short-term basis and at different times of the year.

Finally, management systems for native grassland should aim at establishing a balance between living vegetation, dead vegetation, other organisms and the soil in order to ensure the system remains productive and stable in the long term (Tomanek, 1969). However, the proportion of the plant energy which can can be transferred, annually, through the grazing and detritus pathways has not not been quantified for many grassland systems. The desirable level of herbage utilization through animal consumption must aim for a balance between near-optimal animal production and grassland ecosystem stability. This issue remains a major challenge for ecologists concerned with the management of natural grasslands grazed by domestic livestock.

REFERENCES

Boutton, T. W., Harrison, A. T., and Smith, B. N. (1980). Distribution of biomass of species differing in photosynthetic pathway along an altitudinal transect in south-eastern Wyoming Grassland. *Oecologia* **45**, 287–298.

Brown, R. H. (1978). A difference in N use efficiency by C_3 and C_4 plants and its implications in adaptation and evolution. *Crop Science* **18**, 93–98.

Bryan, W. W., and Sharpe, J. P. (1965). The effects of urea and cutting treatments on the production of pangola grass in south-eastern Queensland. *Australian Journal of Experimental Agriculture and Animal Husbandry* **5**, 433–441.

Caldwell, M. M., Moore, R. T., White, R. S. & De Puit, E. J. (1972). Gas exchange of great basin shrubs. *Desert Biome Research Memorandum*, 20.

Christie, E. K. (1975). Physiological responses of semiarid grasses IV. Photosynthetic rates of *Thyridolepis mitchelliana* and *Cenchrus ciliaris* leaves. *Australian Journal of Agriculture Research* **26**, 459–466.

Christie, E. K. (1978). Ecosystem processes in semiarid grasslands I. Primary production and water use of two communities possessing different photosynthetic pathways. *Australian Journal of Agricultural Research* **29**, 773–787.

Christie, E. K. (1979). Ecosystem processes in semiarid grasslands II. Litter production, decomposition and nutrient dynamics. *Australian Journal of Agricultural Research* **30**, 29–42.

Christie, E. K. (1981). Biomass and nutrient dynamics in a C_4 semiarid Australian grassland community. *Journal of Applied Ecology* **18**, 907–918.

Christie, E. K. and Detling, J. K. (1982). Analysis of interference between C_3 and C_4 North American grasses in relation to temperature and soil nitrogen supply. *Ecology* **63**, 1277–1284.

Christie, E. K., and Hughes, P. G. (1983). Interrelationships between net primary production, ground-storey condition class and grazing capacity of the *Acacia aneura* rangelands of semiarid Australia. *Agricultural Systems* **12**, 191–211.

Christie, E. K. and Moorby, J. (1975). Physiological responses of semiarid grasses. I. The influence of phosphorus supply on growth and phosphorus absorption. *Australian Journal of Agricultural Research* **26**, 423–436.

Cole, C. V., Grunes, D. L., Porter, L. K., and Olsen, S. R. (1963). The effects of nitrogen on short-term phosphorus absorption and translocation in corn *(Zea mays)*. *Proceedings of the Soil Science Society of America* **27**, 671–674.

Davidson, J. L., and Milthorpe, F. L. (1966). The effects of defoliation on the carbon balance in *Dactylis glomerata. Annals of Botany* **30**, 186–198.

Detling, J. K., Parton, W. J., and Hunt, H. W. (1978). An empirical model for estimating CO_2 exchange of *Bouteloua gracilis* (H.B.K.) Lag. in the shortgrass prairie. *Oecologia* **33**, 137–147.

Doley, D. and Trivett, N. B. A. (1974). Effects of low water potentials on transpiration and photosynthesis in Mitchell grass *(Astrebla lappacea). Australian Journal of Plant Physiology* **1**, 539–550.

Downes, R. W. (1969). Differences in transpiration rate between tropical and temperate grasses under controlled conditions. *Planta* **88**, 261–273.

Ebersohn, J. P., and Lee, G. R. (1972). The impact of sown pastures on cattle numbers in Queensland. *Australian Veterinary Journal* **40**, 217–223.

Hughes, P. G., and Christie, E. K. (1983). Ecosystem processes in semiarid grasslands III. A simulation model for net primary production. *Australian Journal of Agricultural Research*, **34**.

Jacobsen, C. N. (1981). A review of the species of Dichantium native to Australia with special reference to their occurrence in Queensland. *Tropical Grasslands* **15**, 84–95.

Kemp, P. R., and Williams, G. J. (1980). A physiological basis for niche separation between *Agropyron smithii* (C_3) and *Bouteloua gracilis* (C_4). *Ecology* **61**, 846–858.

Laetsch, W. M. (1969). Relationship between chloroplast structure and photosynthetic carbon-fixation pathways. *Science Progress Oxford* **57**, 323–351.

Lauenroth, W. K. (1979). Grassland primary production: North American grasslands in perspective. *In* "Perspectives in Grassland Ecology, Ecological Studies" Vol. 32 (N. R. French, ed.), pp. 3–24. Springer-Verlag, New York.

Levitt, J. (1972). "Responses of Plants to Environmental Stresses" Academic Press, New York.

Lieth, H. (1975). Modelling the primary productivity of the world. *In* "Primary Productivity of the Biosphere, Ecological Studies" Vol. 14 (H. Leith and R. H. Whittaker, eds.), pp. 185–202. Springer-Verlag, New York.

McGill, W. B., and Cole, C. V. (1981). Comparitive aspects of cycling of organic C, N, S and P through soil organic matter during pedogenesis, *Geoderma* **26**, 267–286.

McGill, W. B., Shields, J. A., and Paul, E. A. (1975). Relation between carbon and nitrogen turnover of soil organic fractions of microbial origin. *Soil Biology and Biochemistry* **7**, 57–63.

May, L. H. (1960). The utilization of carbohydrate reserves in pasture plants after defoliation. *Herbage Abstracts* **30**, 239–245.

Milner, C., and Hughes, R. E. (1968). "Methods for the Measurement of the Primary Production of Grasslands". Blackwell, Oxford.

Noy-Meir, E. (1973). Desert ecosystems: Environment and producers. *Annual Review of Ecology and Systematics* **4**, 25–51.

Ode, D. J., Tieszen, L. L., and Lernan, J. C. (1980). The seasonal contribution of C_3 and C_4 plant species to primary production in a mixed pasture. *Ecology* **61**, 1304–1311.

Penning de Vries, F. W. T., Krul, J. M., and Van Keulen (1978). Productivity of Sahelian range-lands in relation to the availability of nitrogen and phosphorus from the soil. Proceedings Conference Nitrogen Cycling in West African Ecosystems, Ibadan, Nigeria.

Pimentel, D. W., Dritschilo, W., Krummel, T. and Kutzman, T. (1976). Energy and land constraints in food protein production. *Science*, **190**, 754–761.

Shantz, H. L. (1954). The place of grasslands in the earth's cover of vegetation. *Ecology* **35**, 142–145.

Sims, P. L., Singh, J. S., and Lauenroth, W. K. (1978). The structure and function of ten western north American grasslands. I. Abiotic and vegetational characteristics. *Journal of Ecology* **66**, 251–285.

Sims, P. L., and Coupland., R. L. (1979). Producers. *In* "Grassland Ecosystems of the World" (R. T. Coupland, ed.), pp. 49–72. Cambridge University Press, Cambridge.

Tomanek, G. W. (1969). Dynamics of mulch layer. *In* "The Grassland Ecosystem: A Preliminary Synthesis" (R. L. Dix and R. G. Beidleman, ed.), pp. 225–254. Colorado State University, Fort Collins.

Trlica, M. J. (1977). Effects of frequency and intensity of defoliation on primary producers of arid and semiarid rangelands. *In* "The Impact of Herbivores on Arid and Semiarid Rangelands" (Proceedings of the Second United States/Australia Rangeland Panel) pp. 27–56. Australian Rangelands Society, Perth.

Williams. G. J., and Markley, J. L. (1973). The photosynthetic pathway type of North American shortgrass prairie species and some ecological implications. *Photosynthetica* 7, 262–270.

Williams. R. F. (1948). The effects of phosphorus supply on the rates of intake of phosphorus and nitrogen and upon certain aspects of phosphorus metabolism in gramineous plants. *Australian Journal of Scientific Research Series A* 1, 311–361.

Wilson, J. R., and Haydock, K. P. (1971). The comparitive response of tropical and temperate grasses to varying levels of nitrogen and phosphorus nutrition. *Australian Journal of Agricultural Research* 22, 573–587.

Woodmansee, R. G. (1978). Additions and losses of nitrogen in grassland ecosystems. *Bioscience* 28, 448–453.

CHAPTER 14

Fruit Crops

K. A. OLSSON, P. R. CARY and D. W. TURNER

I. INTRODUCTION

This chapter examines the environmental, edaphic and physiological factors influencing the yield of three fruit crops: (1) the peach (*Prunus persica*), a deciduous tree grown widely in temperate regions; (2) citrus (*Citrus* sp.), an evergreen tree grown in a wide range of environmentc between the temperate and tropical regions; and (3) banana (*Musa* sp.), an evergreen giant herb, grown mainly in the tropics. The productivity of these crops is influenced

219

by their genetic potential (cultivar) and their response to environment. During our studies on these crops within the Australian context we have interpreted some of the data to show the influence of various factors on yield. These data are the arrangement of the leaf surface in relation to productivity and the importance of water and nutrition.

II. PEACH

Peach (*Prunus persica* (L.) Batsch) is a day-neutral deciduous perennial. The limits of commercial production are determined mainly by winter temperature. Although there is considerable variation between cultivars most have a requirement for winter chilling of 700–1200 h below 7°C to break bud dormancy. Peach is moderately susceptible to frost, however, and the young fruits are damaged by air temperatures below freezing (Yadava and Doud, 1980). Peach trees are sensitive to oxygen deficiency in the root environment (Rowe and Beardsell, 1973) so that they are mostly grown on soils with adequate profile drainage.

A. Canopy Structure and Radiation

Peach tree canopies vary depending on plant spacing and training but most approximate inverted cones. The leaf canopy develops rapidly in spring during shoot elongation and in mature trees a leaf area index (L) of about 10 is reached by mid-summer. At Tatura, Victoria (Australia) the trees are spaced at 5.5 × 5.5 m and reach 4.8 m in height. The individual tree canopies merge to approximate a horizontal leaf layer at heights above 3 m; about 80% of the leaf area is contained in this layer.

For an orchard at Tatura mean daily values of the short-wave albedo vary from 0.18 in early spring to 0.14 at maximum L in mid-summer. Albedo values are considerably lower than for most herbaceous crops and result in greater radiation absorption within the elevated canopy. The leaf canopy absorbs about 70% of incoming short-wave radiation, Q_s, at maximum L. Daytime totals of net radiation, Q_n, ranged from 60% to 70% of Q_s from October to March with an average seasonal value of 64%. The relationship between Q_n and Q_s (Wm^{-2}) at Tatura is

$$Q_{nc} = 0.77 \, Q_s - 76 \qquad (14.1)$$

At maximum L the leaf canopy absorbs 84% of Q_n. The leaf layer above 3 m absorbs about 60% of the net radiation total for the canopy, Q_{nc}, while the leaf layers between 3 and 1.5 m and below 1.5 m absorb 33% and 7% respectively (Olsson, 1977).

B. Transpiration and Photosynthesis

The proportion of Q_{nc} utilized by the canopy in transpiration increases with increasing L to about 0.8 by mid-December (Fig. 14.1) corresponding to an L of about 6. Thereafter, transpiration generally exceeds 0.8 Q_{nc} and approximates potential rates (cf. Monteith, 1965) over successive irrigation cycles where trees are irrigated to minimize reductions in fruit growth rates (Cockroft, 1963). Corresponding changes in the mean daytime canopy resistance (r_c), estimated from changes in stomatal conductance and leaf area with height are small (range 30–45 s m⁻¹) and vary little with time over the irrigation cycle.

Diurnal and seasonal patterns of stomatal conductance and leaf water potential (ψ) have been described for the Tatura orchard by Olsson (1977), Chalmers and Wilson (1978) and Chalmers et al. (1983).

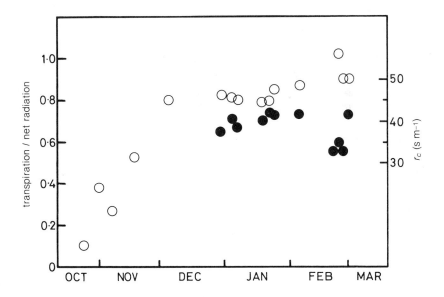

Fig. 14.1 The seasonal course of canopy transpiration expressed as the fraction of daytime net radiation absorbed by the canopy (○) and mean daily canopy resistance (●) for individual days in the period January–March 1974 (from Chalmers et al., 1983).

Peach leaves are hypostomatous. Diurnal and spatial conductance patterns of abaxial stomates, of groups of leaves representing canopy layers have been described by Olsson (1977) and Chalmers *et al.* (1983). Maximum daily values of stomatal conductance (g_s) of 0.4–0.6 cm s^{-1} occur in the upper leaf layers early in the day (before 1000 h) and decrease steadily until stomates close (*c.* 0.02 cm s^{-1}) after sunset. Daily maximum and mean values of g_s are lower in the deeper layers of the canopy despite generally higher leaf ψ_1, suggesting a response to the decreased radiation levels. This is important in radiation partitioning within the canopy (Olsson, 1977).

Leaf ψ_1 starts to decrease in all levels of the canopy soon after sunrise, and reaches minimum values around noon or early afternoon. Leaf ψ_1 then increases during the afternoon to maximum values soon after sunset. Minimum daily leaf ψ_1 decreases with increasing height in the canopy and may reach –2.5 MPa in the uppermost leaf layer towards the end of a drying cycle. The distribution of leaf ψ_1 reflects both resistances to water flow within the crop and energy partitioning within the canopy. Whereas minimum and daily mean leaf ψ decrease during the drying cycle, the maximum changes little (Olsson, 1977). Daily minimum leaf ψ_1 is lower in trees with fruit and during rapid fruit growth (Chalmers and Wilson, 1978) accompanying the higher transpiration rates. The fruit are more sensitive to stress during periods of rapid growth than at other times.

Field observations on peach leaves (Olsson, 1977) suggest that g_s is little influenced by leaf ψ_1 higher than about –1.2 MPa in early January during dry weight stage II of fruit growth (cf. Fig. 14.2). A seasonal trend to a lower leaf ψ_1 at the commencement of stomatal closure is apparent after mid-January.

In addition to the influence of environmental factors that produce variation in rates of transpiration and photosynthesis with height in the canopy (cf. Milthorpe and Moorby, 1979), rates of these processes can be regulated by the assimilate requirements of sinks, possibly by influencing stomatal conductance (Gifford and Evans, 1981). In a mature peach tree the fruit crop accounts for up to 70% of the annual dryweight increment (Chalmers and van den Ende, 1975a). Peach fruit have a double-sigmoid growth curve for fresh and dryweight (Fig. 14.2), comprising two periods of increasing growth rate (DW I and DW III) separated by a period (DW II) of decreasing growth rate (Chalmers and van den Ende, 1975b). Diurnal and seasonal patterns of transpiration and photosynthesis of a peach tree are closely related to the changing demand for assimilates resulting from the changes in the growth rate of the fruit and by fruit removal at harvest (Chalmers *et al.,* 1975; Chalmers *et al.,* 1983).

From mid-January (DW III) to harvest, mean daily values of g_s increased at all canopy levels and are seen as lower values of r_c for the period in Fig.

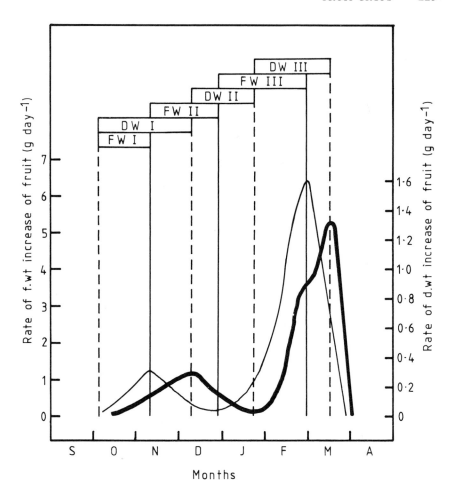

Fig. 14.2 Daily rate of fresh weight (-) and dry weight (■) increase of Golden Queen peach fruit throughout the growing season. The solid lines on the figure are the transitions between fresh weight stages and the broken lines are the transitions between dry weight stages (from Chalmers and van den Ende, 1975a).

14.1 and higher rates of transpiration per unit of radiation absorbed by the canopy. About 60% of the final fruit size is made up during this period and water supply should be unrestricted.

Rates of photosynthesis in leaves are influenced by their proximity to the fruit (Chalmers *et al.*, 1975). Although the leaf area distribution within the canopy can be readily manipulated, fruit distribution is more difficult since it is largely under hormonal control.

C. Modern Orchard Design and Control

Conventional peach orchards may take more than ten years to produce their maximum fruit yields and various strategies have been explored to increase early yields. Modern intensive systems of fruit production usually comprise smaller trees at close spacings that will crop heavily in their early years. In Figure 14.3 experimental (canning) yields of peach trees are compared for several systems over a five year period from planting. Clearly higher total yield is achieved in close-planted orchards than in conventional systems.

Close-planted orchards must achieve a stable balance between vegetative vigour and fruit production to sustain continued high production. More vegetative growth than needed to produce future crops may reduce yield by shading excessively within the canopy and by competing with the fruit for assimilates (Proebsting, 1958; Chalmers and van den Ende, 1975b). In young vigorous peach trees the harvest index is low (Chalmers and van den Ende, 1975a). Management for controlling the growth and cropping of young peach trees includes the use of dwarfing rootstocks and the use of cultural practices that control vegetative vigour and optimize the geometry of the leaf canopy.

For example, the Tatura Trellis is a cropping system which combines close tree spacing with a highly controlled leaf canopy. The trees are grown in parallel rows, usually 6 m apart and spaced at 1 m intervals along the row. Each tree has two limbs; these are trained at 60°–70° to the horizontal and reach 3.5 m height. The leaf canopy is trained intensively on a wired-trellis frame to achieve a low leaf to fruit ratio and a uniform distribution of radiation over most of the leaf area. The trees are grown rapidly to fill their allotted spaces. Vegetative growth is then controlled by crop loading and summer and winter pruning.

At specific stages of fruit growth in young peach trees Chalmers *et al.* (1981) and Mitchell and Chalmers (1982) controlled vegetative vigour in a bed planting in favour of increased fruit yield by inducing root competition and reducing irrigation.

D. Soil Environment

The growth and production of peach trees is greatly influenced by the variation in physical and hydraulic properties of the soil (e.g. Gras, 1962; Cockroft and Wallbrink, 1966b). The recognition of functional equilibria between the growth and activity of root and shoot systems (e.g. Richards, 1977; Thornley, 1977), emphasizes the importance of the soil environment in controlling crop productivity.

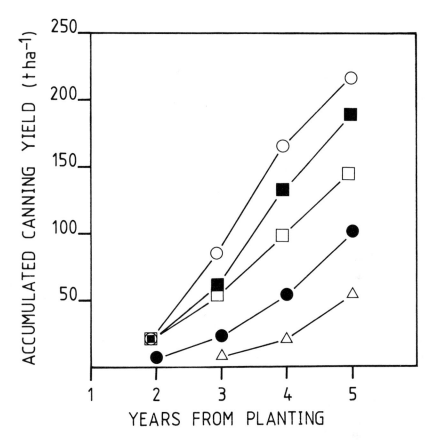

Fig. 14.3 Experimental (canning) yields of Golden Queen peach trees under several systems and tree spacings at Tatura over a five year period from planting. Tatura Trellis, 2222 (trees) ha[-1] (○); 1668 (trees) ha[-1] (■). Regulated bed planting, 4166 (trees) ha[-1] (□); 2500 (trees) ha[-1] (●). Conventional vase canopy, 330 (trees) ha[-1] (△).

Root growth and distribution vary with cultivar, age and plant spacing, and are modified by soil properties and soil and irrigation management. On the shallow, clay-pan soils of south-eastern Australia, few roots penetrate below 1 m (Cockroft and Wallbrink, 1966a). The highest root concentrations occur in the surface soil where root length per unit volume of soil is commonly 2–4 cm[-2]. Root concentration decreases with depth to values less than 0.2 cm[-2] below 0.9 m. Total root length beneath unit area of ground surface is in the range 60–110 cm[-1]. These shallow root systems predispose the crop to water deficits and contrast with those of peach on deep soils where individual roots may reach 5 m depth (Proebsting, 1943). As a consequence irrigation is less frequent on deep soils (Veihmeyer and Hendrickson, 1938).

On shallow soils root distribution may be influenced by soil management practices such as hilling, zero-tillage and mulching. These practices permit roots to fully explore the limited volume of surface soil (Cockroft and Tisdall, 1978).

In mature peach trees, new (white) roots are not produced continuously throughout the growing season as a tendency towards autumn and spring peaks of new root production is observed (Cockroft and Olsson, 1972). The start of root growth in spring usually precedes bud burst. At most times, however, white roots make up only a small proportion of the suberized and woody root system. The periodicity in deciduous orchards has been attributed largely to fluctuations in soil temperature and water content (Rogers, 1939; Kramer and Kozlowski, 1960; Cockroft and Olsson, 1972), but physiological factors relating to the distribution of assimilates and growth-regulating substances also exert considerable control over the timing of growth (Priestley 1970; Vaadia and Itai, 1969).

With adequate levels of soil nutrients the principal factors in the soil environment that influence the growth and function of roots are: soil-water potential, oxygen supply, temperature and mechanical resistance to root penetration. Richards and Cockroft (1974, 1975) considered that soil-water potentials higher than -40 kPa were necessary for maximum growth rates in peach roots. The supply of oxygen to roots depends on diffusion through the air-filled pore space; this in turn varies with water content (Currie, 1962; Greenwood, 1975). Nightingale (1935) showed that the optimum temperature for peach roots was $18°C$ whereas few roots grew at temperatures below $6°C$. Mechanical (penetrometer) resistances greater than 2 MPa seriously restrict rates of root elongation (Greacen, 1981). Mechanical resistance increases with increasing bulky density and with decreasing water content. For soil of a given bulky density, mechanical resistance can be controlled by maintaining high water contents.

Soil water is central to the control of productivity by affecting the plant water balance and dependent processes and also supplies of oxygen, nutrients and the levels of temperature and mechanical resistance in the root environment. Water is thus a key factor in the root environment through which the growth and function of roots can be controlled (Eavis and Payne, 1969; Richards and Cockroft, 1974). Intensive systems of soil and irrigation management based on these criteria have been described by Cockroft and Tisdall (1978) and Olsson and Cockroft (1980).

III. CITRUS

The productivity of a citrus orchard depends on many factors such as location, climate, nutrition, soil type and management practices. Some factors, such as scion–rootstock combination, pre-determine the potential productivity of an orchard even before trees are planted. Therefore, we shall examine those factors affecting productivity that can be manipulated to our advantage. These include the choice of scion–rootstock combination, planting density and soil management practice.

A. Tree Productivity

According to Phillips (1978), "The citrus grove of the future should be thought of as a unit of plant foliage, manipulated to provide the greatest fruit-bearing surface possible and trained to facilitate mechanization of production and harvesting operations in an assembly-line fashion". A similar concept of the potential of citrus productivity is held by Wheaton *et al.* (1978) who have suggested that citrus plantings should be planned to maximize economic returns by interrelating the biological and management aspects. They emphasized that the design of a citrus orchard should be optimized by taking into account interrelationships between light, canopy volume and productivity for the whole life of the orchard rather than just at maturity. They also mentioned that the amount of canopy, or fruit-bearing volume, per unit of land is a useful concept in any design aimed at maximizing productivity. Hence for orange trees in Florida Wheaton *et al.* (1978) calculated the bearing volume per tree for several tree spacings, assuming that the bearing volume occupied a space no deeper than 0.9 m inside the canopy. They then derived a bearing volume index per unit of tree area and assuming a production efficiency of 5.34 kg of fruit per m^3 of bearing volume (Savage, 1966), yield potentials at tree densities up to 600 trees ha^{-1} were calculated. These calculations indicated that at a density of 600 trees ha^{-1} a yield potential of 100 t ha^{-1} could be reached soon after trees were 20 years old, whereas at the lower, more conventional, density of 172 trees ha^{-1} a yield potential of only 50 t ha^{-1} could be reached when trees were 30 years old. Furthermore, these trees could not reach their maximum yield potential until they were 40–50 years old.

Already there is evidence to indicate that a yield potential of 100 t ha^{-1} y^{-1} can be achieved for 20-year-old citrus trees at only two to three times the traditional densities (Castle, 1978; Tucker and Wheaton, 1978). High average annual yields can also be obtained from young (less than ten-year-old) trees, as shown by Hutton (1980) and Hutton and Cullis (1983) for Valencia orange on *Poncirus trifoliata* rootstock inoculated with a mild strain of the exocortis dwarfing virus. For four-year-old to six-year-old trees at densities of 833, 1667 and 5000 trees ha^{-1} he obtained average annual yields of 22.5, 48.0 and 68.4 t ha^{-1}, respectively. However, achieving high average annual yields at high tree densities is only possible by ensuring the control of pest and diseases, and an adequate supply of water and nutrients. Also tree canopies must be pruned regularly to maintain the greatest possible volume of fruit-bearing foliage. However, one major hazard could arise from the use of closely planted viral-dwarfed trees; the virus could be transmitted mechanically from infected to non-infected trees by pruning tools. Although mechanical transmission of the dwarfing exocortis virus has not been reported in Australia (Bacon, 1980; Fraser, 1980), it has been observed in California (Wutscher and Shull, 1975). Thus Bitters *et al.* (1977) consider that in the long term, planting healthy naturally dwarf trees (using dwarfing root-stocks and interstocks) is a more practical approach to high-density plantings than attempts to artificially dwarf trees by the introduction of a growth-retarding disease, such as exocortis, which may be unpredictable in its response.

B. Soil Environment and Root Growth

Some research workers have suggested that the growth pattern of citrus is governed by variations in soil moisture content or by the rhythm of irrigation cycles (Marloth, 1950; Monselise, 1947). Others have produced evidence of alternate root and shoot growth cycles that are governed by seasonal changes and by temperature (Cameron, 1939; Cameron and Schroeder, 1945; Crider, 1927; Reed and MacDougal, 1937). Cary (1970) and Liebig and Chapman (1963) showed that 25°C is a near-optimum soil temperature for growth of citrus roots.

Irrigation strategy should take cycles of root and shoot growth into account. When root and shoot growth is most active it may be desirable to increase either the frequency or intensity of irrigation or both. There are critical periods for the tree's water requirements (Bain, 1958; Sauer, 1953): one just before and another just after the beginning of spring growth and blossoming. In addition, water applications must be carefully supervised until flowering and fruit set have been completed and all fruitlets are larger than about 15 mm in diameter. Adequate water is important for preventing excessive fruit-drop in early summer (Bain, 1949). Even in winter when trees are relatively inactive, special applications of water often are needed to prevent fruit desiccation by frost.

Soil management depends on many factors such as location, slope and soil type. On sloping ground, cover cropping may be essential to prevent soil erosion. On level ground, however, zero-tillage herbicide weed control has been used for up to 28 years without reducing tree growth and yield (Cary, 1968; Cary and Weerts, 1977). Indeed, zero tillage, provided water and nutrients were adequate, gave the highest average annual yields (Fig. 14.4).

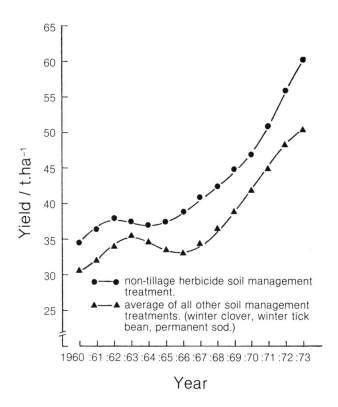

Fig. 14.4 Experimental yields of oranges, based on centred four-year moving averages from a long-term field experiment at Griffith, NSW. Fertilizer was applied annually in midwinter at rates of 100 kg (N) ha^{-1} and 37 kg (P) ha^{-1} (from Carey, 1968 and Cary and Weerts, 1977).

In Californian citrus orchards zero tillage has produced many benefits with little evidence of adverse effects (Jordan and Russell, 1978) whereas in Brazil better yields were obtained by using leguminous cover crops or a dead mulch to control weeds rather than herbicide sprays (Rodriguez *et al.*, 1964).

In California Eaks and Dawson (1979) have evaluated the effects of vegetative ground cover and bare soil on the rind colour of Valencia oranges. In the fruit rind, the caretenoid content was higher and the chlorophyll content lower from ground-cover plots than in fruit from bare-ground plots. Regreening of fruit was also significantly less on ground-cover plots. Improved colour and reduction of regreening were attributed to temperatures under ground cover at 10 cm depth being considerably (5°C) lower than under bare soil. Similarly in Australia the root zone of a Valencia orange orchard warmed up more quickly in spring and cooled down more quickly in autumn when a zero-tillage herbicide treatment was used than under tillage or permanent sod (Weerts and Cary, 1980). The greater heat conductance of the soil receiving zero tillage was of particular benefit to orange trees in spring and early summer (Cary and Evans, 1972) by facilitating nutrient uptake.

In orchards on hillsides, where the use of a zero-tillage and herbicide treatment is likely to result in excessive water run off and soil erosion, some low-growing perennial legumes and clovers can be used. They have the additional advantage that they may contribute about 75% of the trees' nitrogen requirements (Cary, 1968). Moreover, minimum tillage can still be practised in citrus orchards with winter-grown leguminous cover crops. Desiccation of the cover crop using a suitable herbicide in the late spring can form a summer mulch that protects the soil from excessively high temperatures and reduces water loss.

IV. BANANA

A. Growth and Development

In the banana the ontogenetic sequence of vegetative growth, flowering and fruit growth is not seasonal as in citrus or stone fruits. The plant consists of an underground rhizome (commonly called a corm) that is surmounted by a growing point. Lateral buds together with the parent corm produce suckers that form the stool or mat. The stool consists of single axis plants representing up to three visible generations. The first generation is the plant crop. Suckers from the plant crop form the second generation or first ratoon crop. This gives rise to the second ratoon crop and so the sequence is repeated for, say, 50 or more generations. Each growing point produces between 30 and 50 foliate leaves before it becomes reproductive. Suckers start growing after the parent has produced about 12 leaves (Turner, 1972). New suckers may develop from then until the inflorescence (bunch) emerges from the plant. In tropical environments 40–50 leaves per axis are produced annually and one crop cycle may be as short as nine months. In the subtropics only 24–28 leaves are produced annually (Turner, 1971)—the cycle being extended to 1.5–2 years. The harvesting of a plant crop may be spread over 3–4 months. Because ratoon crops commence growth over a broad spread of the parents' life, their harvest period is more extended. In general, once the harvest commences the plantation enters a phase of almost continuous fruit production, although the yield shows large seasonal variation. Fruit maturing at different times of the year reflects its growing conditions. Considerable variation in fresh weight of the harvested bunches can occur from month to month.

B. Generation of Leaf Canopy

Individual shoots produce sequentially larger leaves. The leaf area therefore changes with ontogeny and is usually greatest at bunch emergence (flowering). This is most noticeable in the plant crop, but less so in ratoons where individual shoots of all ages are present. Depending on initial planting density and arrangement the canopy changes from single isolated plants to rows to continuous cover. In ratoon crops of Giant Cavendish varieties at densities of 1800 plants ha^{-1} the radiation profile can be described by

$$Q = Q_o \exp(-kL) \qquad (14.2)$$

where Q is the flux on any plane, Q_o the flux at the top of the canopy, L the leaf area index, and k contains a transmission component and a geometrical component, Θ, where $\Theta = 0.5\pi(L + 1)^{-1}$ and Θ is the mean midrib angle in each layer. This analysis indicates that about 90% of the radiation is intercepted when $L = 4$. There seems little value in increasing L beyond this limit in the field.

Leaf area index changes from 1.0 to 2.5 when plants are half grown to 4.4–4.8 at bunch emergence (Bonhomme and Ganry, 1976). In ratoon crops L is a function of the rate of appearance of new leaves, leaf size and the rate of senescence of old leaves. It can vary from 2.5 to 4.5 in New South Wales (Turner, 1972) and from 0.8 to 4.8 in Israel (Lahav and Kalmar, 1981), the low values occurring in spring and the high values in mid-autumn. Large seasonal changes in L have implications for water use and the application of irrigation in relation to environmental demand and available soil water.

A simple method of changing L is to increase plant density. In ratoon crops this can be done by allowing more than one sucker to develop on each parent plant. Whereas increasing the density can be expected to increase L it also increases the number of fruiting sites, and a direct relationship between L and yield may be expected. This seems true up to 1800–2200 plants ha^{-1}, but in several experiments higher densities reduce bunch weight, slow maturation, produce fruit with soft pulp, increase the incidence of leaf diseases (Phillips, 1970), reduce fruit length (Hord and Gross, 1981) and prevent some plants from becoming reproductive (Robinson and Singh, 1974). In these experiments L was not measured so we can only infer a relationship between leaf area and yield. On individual plants the number of fruit on the bunch is related to the total area of the third to the fifth last leaves produced before flowering. The fresh weight of these fruit at harvest is related to the environment and the leaf area duration of the last three leaves produced before flowering (Turner, 1980). Cultural practices can then be used to modify yield if they can influence L within 1–2 months before flowering and the leaf area duration after flowering. In this context nutrition, irrigation and disease control are important.

C. Production of Fruit

A single axis of banana contains from 4 kg to 16 kg dry weight at maturity, depending on variety and growing conditions. About 30% of this dry matter is allocated to the mature bunch. Productivity will be a function not only of the total amount of dry matter generated but also a function of the dry matter allocated to the fruit.

The nutrients nitrogen (N) and potassium (K) have been shown to be important for the growth and productivity of the banana under many field situations. There is a direct relationship between N and growth. If environmental conditions are favourable the addition of N produces more growth. On the other hand if growth is limited by unfavourable weather the addition of N has little effect; most of it not absorbed by the plant (Martin-Prével and Montagut, 1966). The uptake of K is influenced by the concentration in the medium and the growth of the plant (Turner and Barkus, 1981) and although its supply influences the total production of dry matter it has more profound effects upon its distribution. Turner and Barkus (1980) found that where low K supply halved total dry-matter production, fruit was reduced most (about 80%) and roots least (no change, Table 14.1). Fruit quality can be influenced by a greater K supply than the amount that influences growth

Table 14.1

Dry matter produced by banana tops and various plant parts in relation to potassium supply. (After Turner and Barkus, 1980).

	Potassium supply[a]			LSD (0.01)
	K/10	K/5	K1	
Plant part	Dry weight g/plant			
Fruit	732	1208	3491	405
Stalk	42	57	126	12
Pseudostem	1030	1207	1839	231
Corm	447	524	551	92
Leaves	300	358	639	161
Suckers	109	142	209	85
Whole plant over 3 crop cycles				
Trash (dead leaves)	3740	3835	4857	508
Roots	2715	2329	2335	712
Total plant (kg)	14.4	16.7	27.8	3.5

[a] K1, K/5, K/10 are standard, one fifth and one tenth of standard nutrient supply respectively, where K1 is 7.3 gK/plant/week.

and yield. Vadivel and Shanmugavelu (1978) observed an increased sugar–acid ratio in the fruit as K supply was increased.

Irrigation is practised in arid areas and can have considerable effects on yield (e.g. 100% increase) even in areas of high rainfall, where irrigation is supplementary. Although the relationship between yield and water applied is well established the influence of treatments on the components of yield, such as fruit size, fruit per bunch, and time between crops, is less well known. Indeed very few attempts have been made to measure water use by bananas (Ghavami, 1972; Turner, 1979), although several investigators have examined the relationship between plant response and soil water potential. In general, no response occurs down to –40 kPa, whereas between –40 and –200 kPa a variety of responses occur, probably depending on environmental demand. Below –200 kPa a reduction in growth may be expected in all banana growing climates.

D. Soil Environment and Root Growth

Bananas are successfully grown on a wide range of soils although little experimentation has been done to define the conditions necessary for high yields. The root system of the banana is not inherently shallow but its depth is a function of edaphic factors. A shallow water table or a gravelly layer results in a shallow root system; deep water tables allow roots to penetrate to 1.5 m and beyond. Irrigation methods influence root distribution—under drip irrigation the root system is less extensive than where flood or sprinkler irrigation is practised (Lahav and Turner, 1983). The banana has cord roots (main, secondary and tertiary) that are thick and suberized (2.4, 0.5 and 0.2 g m^{-1} per unit length) and hair roots that arise from any class of cord root. Hair roots are short (2–10 cm) and have an abundance of root hairs. The partitioning of dry matter within the root system appears to be 1:1 cord to hair roots (Turner and Barkus, 1981). Elongation rates of cord roots range up to nearly 3 cm d^{-1} at temperatures of about 28°C (Riopel and Steeves, 1964; Swarbrick, 1964; Lassoudiere, 1971).

Root densities are about 1 cm cm^{-3} which are low compared with other crops and may, in part, account for the sensitivity of banana to low supplies of water and nutrients.

V. CONCLUSIONS

Because photosynthesis is an important component of yield the arrangement of the leaf canopy to maximize production has been a major consideration in horticultural crops. In the three species we have examined the most successful manipulation has been with the peach, especially with regard to the Tatura Trellis. In citrus the focus has been on kg (fruit) m^{-3} of bearing canopy volume and the arrangement of this on the land surface. Dwarf trees at high densities are seen as a way of obtaining high production. With bananas high densities produce high yields and increase the number of fruiting sites ha^{-1}. Yields of about 100 t (fresh fruit) ha^{-1} y^{-1} for peaches and citrus are comparable with 60–70 t ha^{-1} y^{-1} for bananas. In the latter case the fruit is about 20% dry matter and the plant needs to re-establish itself in each crop cycle, because it has no woody framework. Hence about 30%–40% of the total dry matter is allocated to the fruit compared with as much as 70% in a mature peach tree (Chalmers and van den Ende, 1975a).

High yields can only be obtained with adequate supplies of water and nutrients. Our level of understanding of the influence of water on processes that control yields varies from crop to crop. In peaches and citrus the picture is better understood than for bananas where the influence of water supply on the components of yield has yet to be elucidated. Nutrient supply influences the partitioning of assimilates between vegetative and reproductive growth. This is especially important for the bananas where only 30%–40% of the dry matter is partitioned to fruit.

For citrus we have highlighted the soil management approach to influencing productivity and this is important for peach culture, especially in relation to water supply and plant density. By contrast, no cultivation takes place within an established banana plantation, hence any effort to modify soil conditions needs to be done before the bananas are planted.

We have drawn attention to some environmental factors that influence the productivity of peaches, citrus and bananas. We have seen how an understanding of the processes involved can lead to more effective use of environmental resources in producing high yielding crops. This understanding needs not only to be expanded, but to be integrated to ensure its best use in producing fruit crops of high quality.

REFERENCES

Bacon, P. E. (1980). Efects of dwarfing inoculations on the growth and productivity of Valencia oranges. *Journal of Horticultural Science* **55**, 49–55.

Bain, F. M. (1949). Citrus and climate. *California Citrograph* **34**, 382.

Bain, F. M. (1958). Morphological, anatomical and physiological changes in the developing fruit of the Valencia orange, Citrus sinensis (L.) Osbeck. *Australian Journal of Botany* **6**, 1–24.

Bitters, W. P., Cole, D. A., and McCarty, C. D. (1977). Citrus relatives are not irrelevant as dwarfing stocks or interstocks for citrus. *Proceedings of the International Society of Citriculture* **2**, 561–566.

Bonhomme, R., and Ganry, J. (1976). Mesure de l'indice foliare du banananier par photographies hemispheriques faites *in situ*. *Fruits* **31**, 421–425.

Cameron, S. H. (1939). Quantitative relationships between leaf, branch and root systems of the Valencia orange tree. *Proceedings of American Society for Horticultural Science* **37**, 125–126.

Cameron, S. H., and Schroeder, C. A. (1945). Cambial activity and starch cycle in bearing orange trees. *Proceedings of American Society for Horticultural Science* **46**, 55–59.

Cary, P. R. (1968). The effects of tillage, non-tillage and nitrogen on yield and fruit composition of citrus. *Journal of Horticultural Science* **43**, 299–315.

Cary, P. R. (1970). Growth, yield and fruit composition of "Washington Navel" orange cuttings, as affected by root temperature, nutrient supply and crop load. *Horticultural Research* **10**, 20–33.

Cary, P. R., and Evans, G. N. (1972). Long-term effects of soil management treatments on soil physical conditions in a factorial citrus experiment. *Journal of Horticultural Science* **47**, 81–91.

Cary, P. R., and Weerts, P. G. J. (1977). Crop management factors affecting growth, yield and fruit composition of citrus. *Proceedings of the International Society of Citriculture* **1**, 39–43.

Castle, W. S. (1978). Controlling citrus tree size with rootstocks and viruses for higher density plantings. *Proceedings of the Florida State Horticultural Society* **91**, 46–50.

Chalmers, D. J., Canterford, R. L., Jerie, P. H., Jones, T. R., and Ugalde, T. D. (1975). Photosynthesis in relation to growth and distribution of fruit in peach trees. *Australian Journal of Plant Physiology* **2**, 635–645.

Chalmers, D. J., Mitchell, P. D. and van Heek, L. (1981). Control of peach tree growth and productivity by regulated water supply, tree density and summer pruning. *Journal of the American Society of Horticultural Science* **106**, 307–312.

Chalmers, D. J., and van den Ende, B. (1975a). Productivity of peach trees: Factors affecting dry-weight distribution during tree growth. *Annals of Botany* **39**, 423–432.

Chalmers, D. J., and van den Ende, B. (1975b). A reappraisal of growth and development of peach fruit. *Australian Journal of Plant Physiology* **2**, 623–634.

Chalmers, D. J., and Wilson, I. B. (1978). Productivity of peach trees: Tree growth and water stress in relation to fruit growth and assimilate demand. *Annals of Botany* **42**, 285–294.

Chalmers, D. J., Olsson, K. A., and Jones, T. R. (1983). Water relations of peach trees and orchards. *In* "Water Deficits and Plant Growth". (T. T. Kozlowski, ed.), Vol. 7 pp. 197–232. Academic Press, New York.

Cockroft, B. (1963). Timing of irrigation in Goulburn Valley Orchards. *Journal of Agriculture, Victoria* **61**, 492–495.

Cockroft, B., and Olsson, K. A. (1972). Pattern of new root production in peach trees under irrigation. *Australian Journal of Agricultural Research* **23**, 1021–1025.

Cockroft, B., and Tisdall, J. M. (1978). Soil management, soil structure and root activity. *In* "Modification of Soil Structure". (W. W. Emerson, R. D. Bond and A. R. Dexter, eds). pp. 387–391. Wiley, Chichester.

Cockroft, B., and Wallbrink, J. C. (1966a). Root distribution of orchard trees. *Australian Journal of Agricultural Research* **17**, 49–54.

Cockroft, B., and Wallbrink, J. C. (1966b). Soil properties and tree vigour in the Goulburn Valley. *Australian Journal of Experimental Agriculture and Animal Husbandry* **6**, 204–208.

Crider, F. J. (1927). Root growth of citrus trees with practical applications. *Citrus Leaves* **7**, 1–3, 27–30.

Currie, J. A. (1962). The importance of aeration in providing the right conditions for plant growth. *Journal of the Science of Food and Agriculture* **13**, 380–385.

Eaks, I. L., and Dawson, A. J. (1979). The effect of vegetative ground cover and ethylene degreening of "Valencia" rind pigments. *Journal of the American Society for Horticultural Science* **104**, 105–109.

Eavis, B. W., and Payne, D. (1969). Soil physical conditions and root growth. *In* "Root Growth" (W. J. Whittington, ed.) pp. 315–338. Butterworths, London.

Fraser, L. R. (1980). Recognition and control of citrus virus diseases in Australia. *Proceedings of the International Society of Citriculture,* 1978, 178–181. Sydney, N.S.W.

Ghavami, M. (1972). Determining the water needs of the banana plant. *Transactions American Society Agricultural Engineers* **16**, 598–600.

Gifford, R. M., and Evans, L. T. (1981). Photosynthesis, carbon partitioning and yield. *Annual Review of Plant Physiology* **32**, 485–509.

Gras, R. (1962). Some relationships between the physical properties of soil and the growth of peach trees in the Rhone Valley between Vienne and Valence. *Annals of Agronomy* **13**, 141–174.

Greacen, E. L. (1981). Physical properties and water relations. *In* "Red-brown Earths of Australia", (J. M. Oades, D. G. Lewis and K. Norrish, eds.), pp. 83–96. Waite Agricultural Research Institute and CSIRO Division of Soils, Adelaide.

Greenwood, D. J. (1975). Measurement of soil aeration. *In* "Soil Physical Conditions and Crop Production". pp. 261–272. Technical Bulletin 29. Ministry of Agriculture, Fisheries and Food. H.M.S.O., London.

Hord, H. H. V., and Gross, R. A. (1981). Population control in banana plantations. *Fruits* **36**, 83–85.

Hutton, R. J. (1980). High density citrus plantings. *Farmers' Newsletter,* No. **148**, 18–19. Griffith, New South Wales.

Hutton, R. J. and Cullis, B. R. (1983). Tree spacing effects on productivity of high density dwarf orange trees. *Proceedings of the International Society of Citriculture, 1981.* Tokyo, Japan, **1**, 186–190.

Jordan, L. S., and Russell, R. C. (1978). Repeated application of soil-residual herbicides and yield and quality of "Valencia" oranges. *HortScience* **13**, 544–545.

Kramer, P. J., and Kozlowski, T. T. (1960). "Physiology of Trees". McGraw-Hill, New York.

Lahav, E., and Kalmar, D. (1981). Shortening the irrigation interval as a means of saving water in a banana plantation. *Australian Journal Agricultural Research* **31**, 465–477.

Lahav, E., and Turner, D. W. (1983). Fertilizing for high yields—banana pp. 1–62. International Potash Institute, Bulletin No. 7, Berne, Switzerland.

Lassoudiere, A. (1971). La croissance des racins du bananier. *Fruits* **26**, 501–512.

Liebig, G. F., and Chapman, H. D. (1963). The effect of variable root temperatures on the behavior of young navel orange trees in a green-house. *Proceedings of American Society for Horticultural Science* **82**, 204–209.

Marloth, R. H. (1950). Citrus growth studies. 1. Periodicity of root-growth and top growth in nursery seedlngs and buildings. *Journal of Horticultural Science* **25**, 50–59.

Martin-Prével, P., and Montogut, G. (1966). Essais sol-plants sur bananier 8. Dynamique de l'azote la croissance et le development vegetal. *Fruits* **21**, 283–294.

Milthorpe, F. L., and Moorby, J. (1979). "An Introduction to Crop Physiology" 2nd edn. Cambridge University Press, Cambridge.

Mitchell, P. D., and Chalmers, D. J. (1982). The effect of reduced water supply on peach tree growth and yield. *Journal of the American Society of Horticultural Science* **107**, 853–856.

Monselise, S. P. (1947). The growth of citrus roots and shoots under different cultural conditions. *Palestine Journal of Botany,* Rehovot Series **6**, 43–54.

Monteith, J. L. (1965). Evaporation and environment. *Symposium of the Society for Experimental Biology* **19**, 205–234.

Nightingale, G. T. (1935). Effects of temperature on growth, anatomy and metabolism of apple and peach roots. *Botanical Gazette* **96**, 581–639.

Olsson, K. A. (1977). Physical aspects of the water relations of an irrigated peach orchard. Ph.D. Thesis, Macquarie University, Sydney.

Olsson, K. A., and Cockroft, B. (1980). Soil management for high productivity in horticulture. Proceedings Australian Agronomy Conference, Lawes, Qld, Australia, April 1980, pp. 30–39.

Phillips, C. A. (1970). Spacing and density of plantings. Proceedings of the first meeting of the Association for co-operation in banana research in the Caribbean and Tropical America (ACORBAT) St Lucia, Windward Is. 1970, 4–6.

Phillips, R. L. (1978). Tree size control. Hedging and topping citrus in high-density plantings. *Proceedings of the Florida State Horticultural Society* **91,** 43–46.

Priestley, C. A. (1970). Carbohydrate storage and utilization. *In* "Physiology of Tree Crops" (L. C. Luckwill and C. V. Cutting, eds.), pp. 113–127. Academic Press, London.

Proebsting, E. L. (1943). Root distribution of some deciduous fruit trees in a Californian orchard. *Proceedings of the American Society of Horticultural Science* **43,** 1–4.

Proebsting, E. L. Jr. (1958). A quantitative evaluation of the effect of fruiting on growth of Elberta peach trees. *Proceedings of the American Society of Horticultural Science 71,* 103–109.

Reed, H. S. and MacDougal, D. T. (1937). Periodicity in the growth of the orange tree. *Growth* **1,** 371–373.

Richards, D. (1977). Root-shoot interactions: a functional equilibrium for water uptake in peach Prunus persica (L.) Batsch. *Annals of Botany* **41,** 279–281.

Richards, D., and Cockroft, B. (1974). Soil physical properties and root concentrations in an irrigated peach orchard. *Australian Journal of Experimental Agriculture and Animal Husbandry* **14,** 103–107.

Richards, D., and Cockroft, B. (1975). The effect of soil water on root production of peach trees in summer. *Australian Journal of Agricultural Research* **26,** 173–180.

Riopel, J. L., and Steeves, T. A. (1964). Studies on the roots of Musa acuminata cv. "Gros Michel". 1. The anatomy and development of main roots. *Annals of Botany* **28,** 475–479.

Robinson, J. B. D., and Singh, J. M. (1974). Effect of spacing on banana yield in Fiji. *Fiji Agricultural Journal* **36,** 1–5.

Rodriguez, O., Moreira, S., and Rossing, C. (1964). Estudo de nove practicas de cultivo do solo em pomar citrico no planalta Paulista. *Anais do Seminario brasileiro dè herbicidas e ervas daninhas* 257–258.

Rogers, W. S. (1939). Root studies VIII. Apple root growth in relation to rootstock, soil, seasonal and climatic factors. *Journal of Pomology* **17,** 99–130.

Rowe, R. N., and Beardsell, D. V. (1973). Waterlogging of fruit trees. *Horticultural Abstracts* **43,** 533–548.

Sauer, M. R. (1953). Fruit sizes: need for adequate soil moisture in February-May shown by Valencia measurements at Merbein. *Citrus News* **28,** 138–141.

Savage, Z. (1966). Citrus yield per tree by age. *University of Florida Agricultural Extension Service.* Economic Series 66–3.

Swarbrick, J. T. (1964). The growth and root distribution of some temporary shade plants for cocoa. *Tropical Agriculture (Trinidad)* **41,** 313–315.

Thornley, J. H. M. (1977). Root: shoot interactions. *Symposium of the Society for Experimental Biology* **31,** 367–389.

Tucker, D. P. H., and Wheaton, T. A. (1978). Spacing trends in higher citrus planting densities. *Proceedings of the Florida State Horticultural Society* **91,** 36–40.

Turner, D. W. (1971). The effects of climate on rate of banana leaf production. *Tropical Agriculture (Trinidad)* **48,** 238–287.

Turner, D. W. (1972). Banana plant growth. 2. Dry matter production, leaf area and growth analysis. *Australian Journal Experimental Agriculture and Animal Husbandry* **12,** 216–224.

Turner, D. W. (1979). Growth and mineral nutrition of the banana—an integrated approach. Ph.D. Thesis, Macquarie University, Australia.

Turner, D.W. (1980). Some factors related to yield components of bananas, in relation to sampling to assess nutrient status. *Fruit* **35,** 19–23.

Turner, D. W., and Barkus, B. (1980). Plant growth and dry matter production of the "Williams" banana in relation to supply of potassium, magnesium and manganese in sand culture. *Scientia Hoticulturae* **12**, 27–45.

Turner, D. W., and Barkus, B. (1981). Some factors affecting the apparent root transfer coefficient of banana plants (cv. "Williams"). *Fruit* **36**, 607–613.

Vaadia, Y., and Itai, C. (1969). Interrelationships of growth with reference to the distribution of growth substances. *In* "Root Growth" (W. J. Whittington, ed.), pp. 65–79. Butterworths, London.

Vadivel, E., and Shanmugavelu, K. G. (1978). Effect of increasing rates of potash on the quality of banana cv. "Robusta". *Potash Review* **24**, 8.

Veihmeyer, F. J., and Hendrickson, A. H. (1938). Soil moisture as an indication of root distribution in deciduous orchards. *Plant Physiology* **13**, 169–177.

Weerts, P. G. J., and Cary, P. R. (1980). Effects of soil management on soil temperatures in an orange orchard. *Proceedings of the International Society of Citriculture*, 1978, 221–226.

Wheaton, T. A., Castle, W. A., Tucker, D. P. H., and Whitney, J.D. (1978). Higher density plantings for Florida citrus-concepts. *Proceedings of the Florida State Horticultural Society* **91**, 27–33.

Wutscher, K. H., and Shull, A. V. (1975). Machine-hedging of citrus trees and transmission of exocortis and xyloporosis viruses. *Plant Disease Reporter* **59**, 368–369.

Yadava, U. L., and Doud, S. L. (1980). The short life and replant problems of deciduous fruit trees. *Horticultural Review* **2**, 1–116.

CHAPTER 15

Protected Crops

S. W. BURRAGE and P. NEWTON

I. INTRODUCTION

The production of plants within protective structures is usually divided into two phases, a short propagation period when high densities are used, followed by a much longer period from transplanting to cropping when densities are lower.

Continuous introduction of new, useful ideas and techniques based on results from research and development by state and industrially financed establishments, plus observations and experiments by growers, has given a steady improvement in productivity of protected crops. Even greater productivity would be possible if existing knowledge were fully exploited.

Production depends, however, not only on the knowledge of the plant's reaction to environmental or nutritional factors, but also on the economics

239

of applying these treatments. For example, it is possible to improve the yield of glasshouse tomatoes by increasing carbon dioxide levels from 1000 to 1400 μl l^{-1} (Calvert and Slack, 1971). In practice this is not carried out as the cost of maintaining the higher levels of carbon dioxide outweighs the value of the increased yield. Production under protective structures depends as much on the economics of the introduction of a technique as on the improvement of our knowledge of the crop physiology. The areas under investigation are influenced by those economics.

There will always be new ways in which further improvements can be made. In the past, ideas that have improved yield have come forward from growers. The crop scientist has the responsibility of examining these ideas and putting them on a sound scientific basis if we are to exploit them fully.

II. PROPAGATION

An increasing number of plants grown under protection are from seeds of F_1 hybrids. These give higher yields of more uniform quality than the varieties they have replaced. However, their cost is usually a much larger proportion of the overall value of the crop. It therefore becomes increasingly important to ensure a high percentage germination and minimum time to seedling emergence. In this way as many plants as possible are produced at the right time from each gram of seed. In recent years this has resulted in an increase in the use of sophisticated equipment by the propagator; for example, controlled environments for seed chitting (Gray and Salter, 1980), seed germination (Gould and Newton, 1969) and seedling growth (Newton, 1966a,b; Electricity Council, 1972).

A wide range of crop species and varieties that spend part or all their growing period in a protected environment, fruit, vegetable and flowering plants, can be successfully germinated in aerated water and appropriate plant growth substances can be added to ensure high germination percentages of "difficult" varieties; for example, celery (Thomas *et al.*, 1975). The possibility of improving yields by treating seeds with growth substances during chitting needs detailed investigation.

Plants grown with protection are usually transplanted. Therefore it is relatively easy to produce the required crop density compared with crop production that depends on direct seed sowing. However, the production of large numbers of uniform seedlings, even from chitted seed, requires seeds that will withstand extremes of water availability during germination and good control of the soil and aerial environments.

An alternative to more reliable seeds could be treatment of seeds and compost with chemicals; for example, peroxides to provide oxygen to developing embryos (Batch, 1981). The addition of calcium cellulose xanthate (Greenwood and Rowse, 1981) or microgranular polyacrylamides (P. Newton, unpublished) to compost aids the provision of appropriate amounts of water and air during the critical period between seed sowing and irreversible recommencement of embryo growth.

The accurate control of timing of seedling production when all the radiation necessary for photosynthesis is supplied by artificial light (normally fluorescent lamps; Newton, 1982a) is exploited by many growers and specialist propagators in northern temperate latitudes. With large but relatively simple plant growth rooms, carbon dioxide (CO_2) enrichment is inexpensive and can reduce running costs by reducing propagating periods, or photosynthetically active radiation (PAR) levels can be reduced without a reduced growth rate. However, at the moment CO_2 enrichment is not widely used because growers are reluctant to purchase monitoring and control equipment.

The methods used to produce seedlings in controlled conditions can give reproducible results. In practice, however, there is a lack of precision because it is difficult to ensure that the required growing conditions are reproduced accurately. For example, there are inconsistencies with light-measuring instruments in the type, size, mounting method, nature of the covering material over the sensor, cosine correction and spectral response that influence the reading given by the instrument (Newton, 1982b). There may be more agreement today in the units used to measure light (McLaren, 1980; Langhans and Tibbitts, 1981), but it is equally important to get agreement about which instruments are used.

A problem associated with using discharge lamps to supplement daylight or fluorescent lamps as the total light source is the effect of spectral composition on morphogenesis. High levels of far-red radiation within leaf canopies as a result of high leaf area indices can lead to excessive stem and petiole elongation. The addition of far-red radiation by means of incandescent lamps; for example, only aggravates this problem. The lack of far-red radiation in the light produced by fluorescent lamps does not delay flower initiation, and providing plant density is not too high, gives sturdier stems and petioles that facilitate easier handling (Mahasen, 1978).

The rate of dry matter production by seedlings grown under fluorescent lamps is, as far as it is possible to assess, similar to that of plants grown in natural light (Griffin, 1976). If, however, "night break" treatments are used when seedlings are grown with fluorescent lamps, greatly improved growth rates have been found, when compared with growth rates of plants receiving an uninterruped dark period (Griffin *et al.*, 1977). Night breaks are common in commercial installations where, for reasons of economy, lamps are lit continually. It is difficult on a large scale to ensure that plants that should receive a dark period are adequately screened, and to ensure they are moved by hand or automatically from one group of plants to another and back every 12 hours.

The improvement in growth rate associated with night break treatment was found during an investigation to assess the efficiency of different types of fluorescent lamp (Griffin and Newton 1974; Griffin, 1976). The lamp with the highest proportion of PAR (Philips TL33) considered by Gaastra (1970) to be the most suitable for use in controlled environments, did not give the largest amount of plant material per unit amount of light.

The formative effects of light in different parts of the spectrum can obviously be important. A detailed investigation is needed in order to understand the effects of the interaction between spectral composition of light and energy level in the "light" and "dark" period on different varieties of crop plants. Then it will be possible to assess the potential value of interrupting the "dark" period when seedlings are radiated with some or all of the light from artificial light sources.

The value of fluorescent lamps purpose-made for plant growth will need to be considered, particularly for low-energy night break. Several types of fluorescent lights are available; their spectral composition has been adjusted by using appropriate phosphors inside the glass tube to give, for example, light richer in red and far-red than the widely used domestic lamps, or light more similar to natural light; that is, light from a cloudy sky, not direct radiation from the sun. These purpose-made lamps are more expensive than those that are mass produced, and are used almost entirely to illuminate plants grown indoors by enthusiastic amateur gardeners, particularly in parts of North America with very cold winters.

The cost of plants raised in artificial light might be reduced if growth rates can be increased by gibberellin treatment. The improvement in dry matter production by tomato seedlings treated wih gibberellins has been established (Selman and Ahmed, 1962; Mohamed and Newton, 1978), but treatments that could be used on a commercial scale have not been developed. Seedlings that are easier to handle could be produced by treatment with an appropriate plant growth regulator; for example, when excessive stem and petiole elongation occurs as a result of growing plants at high density, or when light has high far-red content. Hypocotyl elongation is also often a problem, even when fluorescent lamps are used, and could presumably be controlled by applying a growth regulator.

Environmental conditions during seed germination and early seedling growth can have lasting effects on subsequent plant performance (Ashby and Wangerman, 1950): the smaller the seedling, the larger the effect on seedling relative growth rate (Milthorpe, 1956). Relatively poor growing conditions, particularly low soil and air temperatures and small amounts of solar radiation, delay the time from seed sowing to development of a complete leaf cover. Effects of fertilizer treatment are usually associated with higher seedling growth rates (Austin, 1964).

Even seed size can be important. Some seed merchants sell "graded" seeds; that is, the relatively small seeds have been removed by sieving. This can be important when seeds are accidentally or deliberately sown with a deep soil covering, or for crops where there is little or no competition; for example, lettuce. However, nearly all the available published data concerned with seed size and yield show that once seedlings begin to compete, any initial advantage of plants grown from large seeds, or seedlings raised in artificial conditions, is lost. A reason for this comes from results of Greenwood et al. (1977) which showed that once the dry matter content of stands of a wide range of vegetables grown with ample fertilizer and soil water had reached 2 tons ha^{-1}, the rate of dry matter production was similar for all species. This is not surprising if the amount of dry matter accumulation is related to radiation absorbed by a crop canopy (e.g. Monteith, 1977).

Where crops are grown for long periods in glass or plastic structures with footpaths between them to facilitate access the relationship between the amount of light reaching the canopy and dry matter production may not follow the same pattern as the canopies grown by Greenwood et al. (1981) and Monteith (1977). Irrespective of this, there must presumably be an optimum pattern of leaf arrangement in relation to the relatively small amount of light reaching the plants in the important period after transplanting until the development of a dense canopy during the poor light period from late autumn to early spring.

It is possible that by using controlled conditions during propagation (i.e. artificial illumination in an insulated building), seedlings can be produced that will perform best under the worst light conditions. With better conditions than the worst, higher yields would automatically occur if sufficient flower buds had been initiated to allow high fruit numbers.

Suitable propagating conditions to guarantee a given quality of *Chrysanthemum* flower have been proposed (Hassan and Newton, 1975), and outlined for early cauliflower production (Merza, 1982). Likewise, light and temperature conditions for *preventing* flowering of chinese cabbage when seedlings are transferred to relatively low temperatures have been established (P. Newton, unpublished). Fluctuating light conditions in a glasshouse with or without supplementary illumination make it difficult to produce the most suitable type of transplant at the right time. It can be cheaper raising transplants in controlled conditions where timing is much more accurate.

III. CROPPING

A. General Design Considerations

The main elements controlling crop production from planting out are (1) crop response to environment; (2) control of the environment; and (3) the economics of the production process. The major limiting factor to crop production in protective structures in Northern Europe is light, particularly during the winter when the growing time may be tripled. For example, the time from planting out to harvest for glasshouse lettuce takes four weeks in July and twelve weeks in November. The low density of plants generally precludes the use of supplementary light. It is critical, therefore, to the production process that optimum use is made of available light. This depends on the glasshouse structure, orientation, environment, management and plant genotypes.

Numerous investigations have been carried out into glasshouse design and orientation to achieve optimum transmission (Lawrence, 1948; Edwards, 1968; Harnett, 1975; Kurata and Tachibana, 1978). Reduction of support structure to glass and orientation in an east–west direction has improved winter transmission from 50% to 75% of available light, but a high level of maintenance and cleaning is necessary to maintain this level of transmission.

Because of the sharp rise in oil prices of 1973 and the increases that have followed, energy costs can now be more than 40% of the overall production costs of long-season tomato crops in the UK. This has caused a reassessment of glasshouse design where energy-saving is given a higher priority. The tall "widespan" structures of the 60s and early 70s with their high-light transmission characteristics have been replaced by the low-profile "Venlo" structure reducing thermal transmittance values and heat loss. Heat loss has been further decreased by sealing the laps between glass panes to reduce the air exchange from inside to outside the structure. Double glazing with fibreglass or polythene has been tried, but light losses of up to 14% can occur and thereby offset the value of the energy saved in the overall cost. New plastic glazing materials have been introduced, notably polycarbonate, which can provide a more effective sealing of the structure, thereby minimizing conduction and convection losses and effectively reducing "U" value from 5.5 to 2.7. Such energy savings are still, however, at the cost of light transmission because of the increased number of interfaces formed by condensation on non-wettable surfaces. These losses are partly offset by the reduced structure required to support these lighter materials. The flexible nature of the plastic materials allows also the use of curved structures more suited to light transmission. A new generation of structures are emerging, where both high transmission and heat retention characteristics are maintained.

An alternative energy-saving technique is the use of internal screens.

These "thermal screens" are similar to the blackout type of screen used to control photoperiodism in crops, and are pulled over in the evening to reduce convective and radiative heat losses at night. The shading caused by these structures reduces light available to the crop, but design has reduced this to an acceptable level of around 2%.

The more effective sealing of glasshouses and reduction of cold surfaces where condensation can take place has altered the pattern of water vapour exchange and crops experience longer periods of high humidity. The effect of this on plant growth has yet to be assessed, but there is some evidence that it is likely to have an adverse effect on crop production. Acock (1971, 1981) found that plants grown constantly at 95% relative humidity had reduced stem extension and fasciation of the leaf tissues. High humidity is rapidly becoming an area where further investigation is required to establish its importance in plant production.

The crop scientist has for long been interested in the underlying factors controlling crop production. Budyko (1966) commented on the complexity of the interrelationship between productivity and the levels of radiation and temperature. In protected cultivation, our ability to manipulate CO_2 levels also adds to the complexity. Gaastra (1959), Bierhuizen and Slatyer (1964), Brun and Cooper (1967), and Enoch and Hurd (1977) have investigated the interaction between photosynthesis, CO_2, temperature and radiation in isolated plants. For example, Enoch and Hurd (1977) found a 60% increase in net photosynthesis when CO_2 levels were increased four-fold and that temperature also had a marked effect. They aimed to obtain the optimal combination of CO_2 and temperature for prevailing light levels. They found that optimal temperatures (T_{opt}) were almost a linear function of light intensity (Q_s). Doubling CO_2 levels $[CO_2]$ increased T_{opt} by similar amounts at all light intensities. From their observations on carnations they derived an empirical eqn. for optimum temperature

$$T_{opt} = -6.47 + 2.34 \ln [CO_2] + 0.0319 Q_s \qquad (15.1)$$

where T_{opt} is expressed in °C, $[CO_2]$ in $\mu l \, l^{-1}$ CO_2 and Q_s as W m^{-2} PAR. There are, of course, problems in applying these isolated plant results to production of a glasshouse crop. The plant canopy is a collection of leaves each adapted to a different light environment and photosynthesis per unit leaf area has been found to vary with previous light environments (Charles-Edwards and Ludwig, 1974). In a crop canopy we may anticipate that a range of photosynthetic potentials will occur, and the overall crop production model becomes complex (Acock et al., 1978). Production also depends on environmental effects on the translocation of accumulated organic solutes and the growth and development of different plant organs; in lettuce we harvest the leaves, in tomatoes, the fruit.

B. Control systems for Aerial Environments

Whereas the crop scientist can give a better understanding of the underlying mechanisms controlling crop performance, the application of this information to crop production is made possible by the ability of the glasshouse engineer to implement appropriate controls to provide the optimum conditions. Warren-Wilson (1972) identified three basic control systems (Fig. 15.1). In the open loop, inputs are fed in at a constant rate and assume a constant level in the controlled system with no modifying factors, such as adding CO_2 to a glasshouse structure. The closed-loop system is where information feedback results in secondary control; for example, as in a thermostat, or, in a more sophisticated form, the Solar-stat (Winspear and Morris, 1965) and the light-modulated temperature controller developed by Bowman and Weaving (1970), where glasshouse temperature is related to the prevailing levels of radiation. Finally, there is the adaptive control where the

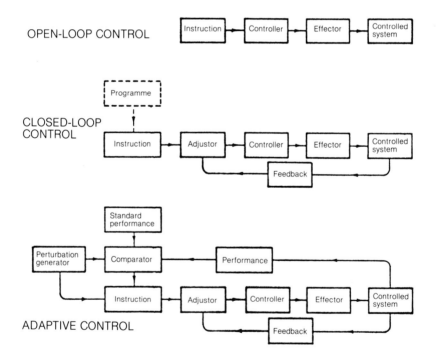

Fig. 15.1 Three types of control system (from Warren-Wilson, 1972).

system makes small changes in response to performance in relation to a standard; if the performance is improved then the standard is upgraded. Control of this type requires considerable sophistication in the models used to develop control algorithms. A high level of environmental control is required to implement these models.

The recent introduction of microcomputer-based control systems into the glasshouse industry in Northern Europe (van de Vooren, 1975; Weaving and Hoxey, 1980; Burrage, 1981) was mainly stimulated by the need to improve the efficiency in the use of energy. Their introduction has, in the first instance, replaced analogue controllers and they therefore fall into the category of sophisticated closed-loop controllers. Microcomputer-based control systems are, however, able to cope with many interrelationships between plant growth and the environment, and with the appropriate software make more precise control decisions. Given the models and the correct software, the opportunity to develop adaptive control techniques comes closer to reality. Unduik ten Cate and van de Vooren (1977) and Bot and Dixhoorn (1978) developed some of the earlier adaptive models maximizing crop yield to environmental factors. Seginer (1980) carried this one stage further, introducing the economic aspects into the model where "optimum" is defined as maximizing the function "income minus cost of operation". Under these criteria, optimum production may not coincide with optimum photosynthesis (Fig. 15.2). Diagrams analagous to Figure 15.2 will form the framework for set point algorithms in the future. The main elements are, as listed by Seginer, (1) an economic objective function; (2) feed forward crop response and its effect on harvest time; and (3) optimization of the various environmental control systems to meet required conditions. The task is still daunting but the opportunities of achieving it have been improved by the rapid developments taking place in microcomputer technology.

C. Control systems for Root Environment

While considerable advances have been made in the monitoring and control of the aerial plant environment, similar advances have been made in the control of the root environment. Traditional growing in the soil has been replaced in many instances by soil-less media; for example, peat and rockwool, following the need to avoid the build-up of pests and diseases in the rooting media. Cooper (1975) introduced the concept of nutrient film technique (NFT) to the production of glasshouse crops. Hydroponic culture of plants had been available commercially for some years (Hoagland and Aron, 1950) and thin film hydroponics had been used for experimental purposes previously (Rudd-Jones and Winsor, 1978).

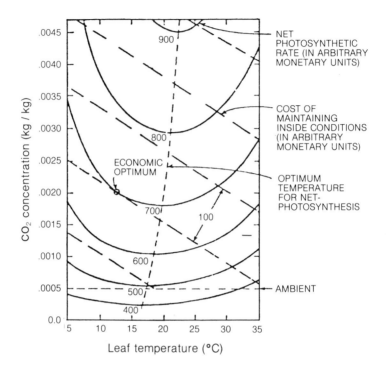

Fig. 15.2 Maximizing the objective function "Income minus operation costs" in the CO_2-temperature plane. Solar radiation is assumed to be constant (from Seginer, 1980).

The concept of NFT is simple: a nutrient solution is circulated in a shallow stream over the roots of growing plants. The solution contains all the nutrients required by the plants and aeration is provided by the motion of the fluid and the large surface area of the solution in contact with the air. Plant roots often grow partially above the surface of the solution. The stream of solution is generally retained in a plastic trough with direct radiation excluded from the roots but it is possible to grow some plants in open troughs with the roots exposed to sunlight; for example, lettuce. In NFT production of plants, the root environment comes more closely under the control of the operator. The plants are continuously supplied with water, removing the necessity for irrigation scheduling.

The composition of the solutions in general use are similar to those originally developed by Hoagland and Aron (1950) with minor modifications (Cooper, 1975). They provide all the basic nutrients required by the plants, but problems can arise where conductivity controllers are used to monitor solution status because of the accumulation of ions such as sulphates and chlorides which are added with the major nutrients but absorbed only slowly. Varley and Burrage (1981) overcame this difficulty by eliminating most of the unwanted ions from the nutrient mix. The pH of the solution is controlled at pH 6 to ensure phosphates and micronutrients, particularly iron, remain in solution.

When first introduced it was thought that NFT would allow the manipulation of the nutrient solution composition and concentration to obtain optimum yields. However, Burrage and Varley (1980) found no response in total fresh weight or dry weight accumulation in lettuce grown over conductivities ranging from 1100 to 3500 μmhos and obtained satisfactory growth up to 5000 μmhos. Massey and Winsor (1978, 1981) found no difference in yield and quality in tomatoes grown in several concentrations of nitrogen and manganese. The picture emerging from NFT culture for many of the nutrients, provided adequate and non-toxic levels are supplied, is that plants will produce a similar yield over a wide range of nutrient concentrations. These conclusions are supported by more fundamental work which has extended the observations to much lower concentrations (Moorby and Besford, in press).

Improvements in quality of a tomato crop (Adams and Winsor, 1977) and resistance to marginal leaf scorch in lettuce (Burrage and Varley, 1980) have been found in plants grown in NFT. The ability to manipulate the root environment has also led to potential energy-saving techniques; for example, solution rather than aerial heating (Moorby, 1980). NFT provides the capability for closer control over the crop environment, with the opportunity to manipulate aerial and root environments independently.

It should be possible with NFT to produce higher crop yields per unit of nutrient applied when compared with soil or peat. For example, Khudheir (1982) has shown that a gram of N in solution may give from 339 to 1236 g fresh weight of tomato. At the moment the technique for optimizing fresh weight production per unit of applied nutrient is not known.

IV. CONCLUSIONS

Considerable advances have been made in environmental monitoring and control of the protected environment. Crops can be grown under more highly controlled conditions than at any previous time. The findings of the crop scientist can be applied with a level of precision previously unattainable. There is a need for more basic information on crop growth, environmental control and the economics of production. In future, the development of algorithms taking into consideration all the aspects of crop production, like that of Seginer (1980) will be necessary if we are to achieve optimum production from protected cultivation.

REFERENCES

Acock, B. (1971). Tomato plant/water relations. 1970 Annual Report of Glasshouse Crops Research Institute (Littlehampton, U.K.) p. 70.

Acock, B. (1981). Analysing and predicting the response of the glasshouse crop to environmental manipulation. *In* "Opportunities for Increasing Crop Yield" (R. G. Hurd, P. V. Biscoe and C. Dennis, eds.) pp. 131–148. Pitman,

Acock, B., Charles-Edwards, D. A., Fitter, D. J., Hand, D. W., Ludwig, L. J., Warren-Wilson, J., and Withers, A. C. (1978). The contribution of leaves from different levels within a tomato crop to net photosynthesis: An experimental examination of two canopy models. *Journal of Experimental Botany* **29**, 815–827.

Adams, P., and Winsor, G. W. (1977). Further studies in the composition and quality of tomato fruit. 1976 Annual Report of Glasshouse Crops Research Institute (Littlehampton, U.K.) pp. 133–138.

Ashby, E., and Wangerman, E. (1950). Studies in the morphogenesis of leaves. IV. Further observations on area, cell size, and cell number of leaves of *Ipomoea* in relation to their position on the shoot. *New Phytologist* **49**, 23–35.

Austin, R. B. (1964). A study of the growth and yield of red beet from a long term manurial experiment. *Annals of Botany* **28**, 637–646.

Batch, J. J. (1981). Recent developments in crop regulators for cereal crops. *In* "Aspects of Crop Growth" pp. 80–94. Ministry of Agriculture Fisheries and Food Reference Book 341. Her Majesty's Stationery Office, London.

Bierhuizen, J. F., and Slatyer, R. O. (1964). Photosynthesis of cotton leaves under a range of environmental conditions in relation to internal and external diffusive resistances. *Australian Journal of Biological Sciences*, **17**, 348–359.

Bot, G. P. A., and Dixhoorn, J. J. (1978). Dynamic modelling of greenhouse climate using a microcomputer. *Acta Horticulturae* **76**, 113–120.

Bowman, G. E., and Weaving, G. S. (1970). A light modulated greenhouse controller. *Journal of Agricultural Engineering Research* **15**, 255–264.

Brun, W. A., and Cooper, R. L. (1967). Effect of light intensity and carbon dioxide concentration on photosynthetic rate of soybean. *Crop Science* **7**, 451–454.

Budyko, M. I. (1966). Solar radiation and the use of it by plants. UNESCO Symposium on Methods in Agroclimatology, p. 392. UNESCO, Paris.

Burrage, S. W. (1981). A personal computer for glasshouse environmental control. *The Agricultural Engineer* **36**, 35–38.

Burrage, S. W., and Varley, M. J. (1980). Water relations of lettuce grown in nutrient film culture. *Acta Horticulturae* **98**, 79–86.

Calvert, A., and Slack, G. (1971). Effects of carbon dioxide concentration on glasshouse tomatoes. 1970 Annual Report of Glasshouse Crops Research Institute (Littlehampton, U.K.) p. 66.

Charles-Edward, D. A., and Ludwig, L. J. (1974). A model for leaf photosynthesis by C_3 plants. *Annals of Botany* **38**, 921–930.

Cooper, A. J. (1975). Crop production in recirculating nutrient solution. *Scientia Horticulturae* **3**, 251–258.

Edwards, R. I. (1968). Transmission of solar radiation in glasshouses. *Acta Horticulturae* **6**, 47–50.

Electricity Council (1972). Growing Rooms. Grow Electric Handbook No. 1. Electricity Council, London.

Enoch, H. Z., and Hurd, R. G. (1977). Effect of light intensity, carbon dioxide concentration and leaf temperature on gas exchange of spray carnation plants. *Journal of Experimental Botany* **28**, 84–95.

Gaastra, P. (1959). Photosynthesis of crop plants as influenced by light, carbon dioxide, temperature and stomatal diffusive resistance. *Mededelingen van de Landbouwhoogeschool te Wageningen* **59**, 1–68.

Gaastra, P. (1970). Climate rooms as a tool for measuring physiological parameters for models of photosynthetic systems. *In* "Prediction and Measurement of Photosynthetic Productivity" pp. 387–398. Centre for Agricultural Publishing and Documentation (PUDOC), Wageningen.

Gould, C. W., and Newton, P. (1969). Germination room can bring controlled growing a step nearer. *Commercial Grower* No. 3809, 17–18.

Gray, D., and Salter, P. J. (1980). "Fluid Drilling." Grower Books, London.

Greenwood, D. J., Cleaver, T. J., Loquens, S. M. H., and Niendorf, K. B. (1977). Relationship between plant weight and growing period for vegetable crops in the United Kingdom. *Annals of Botany* **41**, 987–997.

Greenwood, D. J., and Rowse, H. R. (1981). Soil, Water and Crop Growth. *In* "Aspects of Crop Growth" pp. 15–33. Ministry of Agriculture. Fisheries and Food Reference Book 341. H.M.S.O., London.

Griffin, P. M. N., and Newton, P. (1974). Determination of commercial growth room treatments. *In* "Symposium on Basic Problems of Protected Vegetable Growing". *Acta Horticulturae,* **39**, 221–230.

Griffin, P. M. N. (1976). The effects of light, temperature and carbon dioxide on plant growth. Ph.D. thesis, University of Manchester.

Griffin, P. M. N., Mahasen, A. H. Mohamed, and Newton, P. (1977). Which are the best fluorescent lamps for salad seedlings? *The Grower* **87**, 429–432.

Harnett, R. F. (1975). Study of glasshouse type and orientation. *Acta Horticulturae* **46**, 209–215.

Hassan, M. R. A., and Newton, P. (1975). New light treatments for spray chrysanthemums. *Commercial Grower,* No. 4181, 484, 486 and 493.

Hoagland, D. R., and Aron, D. I. (1938, revised 1950). The water culture method of growing plants without soil. Circular 347, California Agricultural Experiment Station, Berkeley.

Khudheir, G. A. (1982). Water and nutrient uptake and yield of tomatoes. Ph.D. thesis, University of Manchester.

Kurata, K., and Tachibana, K. (1978). Greenhouse structure design with the optimization technique. *Acta Hoticulturae* **87**, 21–30.

Langhans, R. W., and Tibbitts, T. W. (1981). Guidelines for measuring and reporting the environment for plant studies. *Chronica Horticulturae* **21**, 8–9.

Lawrence, W. J. C. (1948). "Science and the glasshouse." Oliver and Boyd, Edinburgh.

Massey, D. M., and Winsor, G. W. (1978). Responses of tomatoes to nitrogen concentration. 1977 Annual Report of Glasshouse Crops Research Institute (Littlehampton, U.K.), p. 77–78.

Massey, D. M., and Winsor, G. W. (1981). Effects of manganese concentration on the growth and yield of tomatoes. 1980 Annual Report of Glasshouse Crops Research Institute (Littlehampton, U.K.), p. 52.

McLaren, J. S. (1980). The expression of light measurements in relation to crop research. *In* "Seed Production" (P. P. Hebblethwaite, ed.), pp. 663–670. Butterworths, London.

Merza, T. I. T. (1982). Buttoning of cauliflowers. M.Sc. thesis, University of Manchester.

Milthorpe, F. L. (1956). The relative importance of the different stages of leaf growth in determining the resultant area. *In* "The Growth of Leaves" (F. L. Milthorpe, ed.). pp. 141–150. Butterworths, London.

Mahasen, A. H. Mohamed (1978). Plant growth and artificial light: Interaction between total energy and spectral composition. M.Sc. thesis, University of Manchester.

Monteith, J. L. (1977). Climate and the efficiency of crop production in Britain. *Philosophical Transactions of the Royal Society of London, Biological Sciences* **281**, 277–294.

Moorby, J. (1980). Effects on manipulating root and air temperature on tomato growth and the efficient use of energy. *In* "Opportunities for Increasing Crop Yields" (R. G. Hurd, P. V. Biscoe and C. Dennis, eds.) pp. 183–194. Pitman,

Moorby, J. and Besford, R. T. (in press). Effects of mineral nutrients on growth. Encyclopedia of Plant Physiology, 15. Editors A. Läuchli and R. Bieleski, Springer.

Newton, P. (1966a). Growing rooms speed tomato propagation. *Commercial Grower*, No. 3696, p. 801.

Newton, P. (1966b). Light and Growth. *Commercial Grower* No. 3698, 886.

Newton, P. (1982a). Commercial seedling production in the U.K. using fluorescent lamps. *Acta Horticulturae* **128**, 193–196.

Newton, P. (1982b). Light measurements and plant growth. *Acta Horticulturae* **128**, 21–32.

Rudd-Jones, D., and Winsor, G. W. (1978). Environmental control of the root zone: Nutrient film techniques. *Acta Horticulturae* **87**, 185–195.

Seginer, I. (1980). Optimizing greenhouse operations for best aerial environment. *Acta Horticulturae* **108**, 169–178.

Selman, I. W., and Ahmed, E. O. S. (1962). Some effects of far-red radiation and gibberellic acid on the growth of tomato plants. *Annals of Applied Biology* **50**, 479–485.

Thomas, T. H., Palevitch, D., Biddington, N. L., and Austin, R. B. (1975). Growth regulators and the phytochrome-mediated dormancy of celery seeds. *Physiologia Plantarum* **35**, 101–106.

Unduik ten Cate, A. J., and van de Vooren, J. (1977). Digital adaptive control of a glasshouse heating system. 5th International Federation of Automatic Control/International Federation of Information Processing International Conference on Digital Computer Application to Process Control. The Hague.

Varley, M. J., and Burrage, S. W. (1981). New Solution for lettuce. *The Grower* **95**, 24.

van de Vooren, G. S. (1975). A computer for crop research and climate control in glasshouses. *Acta Horticulturae* **51**, 169–174.

Warren Wilson, J. (1972). Control of crop processes. *In* "Crop Processes in Controlled Environments" (A. R. Rees, K. E. Cockshull, D. W. Hand and R. G. Hurd, eds.), pp. 7–30. Academic Press, London.

Weaving, G. S., and Hoxey, R. P. (1980). Monitoring and control of greenhouse environment utilizing a Texas Instruments TMS 990/10 microcomputer. *Acta Horticulturae* **106**, 67–76.

Winspear, K. W., and Morris, L. G. (1965). Automation and control in glasshouses. *Acta Horticulturae* **2**, 61–70.

Mangroves

B. F. CLOUGH

I. INTRODUCTION

The term "mangrove" is used to describe halophytic trees and shrubs that grow in the intertidal zone along protected coasts in warmer parts of the world. The term is also used loosely to describe the entire plant community comprising individual mangrove species. The term "mangrove" will be used when describing individual species, and the terms "tidal forest" or "mangrove community" will be used when referring to an assemblage of mangrove species.

There are now over 50 species recognized as mangroves worldwide, of which more than 30 species have been reported from Australia (Dowling and McDonald, 1982). Most mangroves are dicotyledonous, but two ferns, *Acrostichum* spp., and the palm *Nypa fruticans*, are also regarded as mangroves because they occur only in association with other mangroves in the intertidal zone along tropical coastlines. Although the distribution of several species extends well into temperate latitudes, mangroves tend to be replaced in

CONTROL OF CROP PRODUCTIVITY
ISBN 0 12 548280 9

cooler climates by saltmarsh vegetation composed of predominantly herbaceous species. The best developed mangroves, in terms of tree size and lux-
uriance, are found in humid equatorial climates where with favourable
geomorphic conditions they often form extensive forests comparable in
productivity to their more landward counterparts.

Mangroves are not thought of as a crop in industrialized western nations.
In many developing countries of the Indo-Pacific region, however, they are
a major source of firewood and building materials and are greatly valued
as a forest resource. It is therefore not surprising that mangroves have a long
history of silviculture. The best-documented example is in Malaysia, where
large areas of mangrove forest on the Peninsula Malaysia have been
managed on a sustained yield basis since early this century (Walsh, 1974;
Ong, 1982).

In the last decade mangroves have also become appreciated as a nursery
for a wide variety of organisms, including many species of prawn, fish and
crab of commercial value. For these and other organisms, mangroves appear
to be important as a source of reduced carbon, either as dissolved organic
carbon, or as leaves, wood and other debris that falls from the trees and contributes to detritus-based food chains in bays and estuaries (Heald, 1969;
Odum, 1971). However, the trophic relationships in mangrove ecosystems
are largely unknown and the dependence of commercial fisheries production
on mangroves has yet to be established in quantitative terms.

The most obvious feature that distinguishes mangroves from other trees
is their ability to grow in saline soils, which are often extremely anoxic
because of water logging. In addition, mangroves usually grow in hot
climates where a combination of high irradiance and high air temperatures
can lead to high rates of water loss and salt loading, and often gives rise to
leaf temperatures which may be supraoptimal for many physiological
processes. Thus a substantial part of this chapter is concerned with the
physiological processes that maintain the balance for salt, water and carbon
in mangroves. The rest of the chapter discusses how edaphic and climatic
factors interact with these processes to regulate growth and productivity.

II. SALT BALANCE

Although some species have been reported to grow and reproduce successfully in freshwater (McMillan, 1974; Walsh, 1974), experiments have indicated that seedlings of a number of species grow better in the presence of
sodium chloride (NaCl) (Stern and Voigt, 1959; Connor, 1969; Clarke and
Hannon, 1970; Ball, 1981; Downton, 1982; B. F. Clough, unpublished). All
species so far tested (*Avicennia marina, Aegiceras corniculatum, Rhizophora
mangle* and *R. stylosa*) grow poorly in the absence of NaCl and, with the

exception of *R. mucronata* (Stern and Voigt, 1959), the rate of growth is also much reduced at salinities approaching that of seawater (*c.* 550 mM NaCl). For most of these species, the optimum salinity for growth lies between 100 mM NaCl and 350 mM NaCl. A number of other species also appear to grow best in nutrient culture with moderate salinities. These include *Acanthus ilicifolius, Bruguiera gymnorhiza, B. sexangula, Rhizophora apiculata, Sonneratia caseolaris, Xylocarpus australasicus,* and *X. granatum* (B. F. Clough, unpublished).

Like other halophytes, most species of mangrove contain high salt concentrations in their tissues. At high external salinities, sodium (Na^+) and chloride (Cl^-) ions dominate the ionic composition of the tissue, but potassium (K^+), magnesium (Mg^{2+}) and calcium (Ca^{2+}) are also present in significant concentrations (Atkinson *et al.,* 1967; Downton, 1982). At external NaCl concentrations below about 50 mM, K^+ and, to a lesser degree, Mg^{2+} are the dominant cations in the cell sap (Downton, 1982). It is generally accepted that high concentrations of inorganic ions are required by halophytes in order to maintain intracellular osmotic potentials that are lower than the water potential of the soil. This is the minimum requirement for maintenance of a positive water balance. This view is supported by measurements, made both in the laboratory and in the field, that indicate that osmotic potentials in mangrove tissues are generally 0.5–2 MPa lower than that of the substrate on which they are growing (Scholander, 1968; Chapman, 1976; Downton, 1982).

Most evidence suggests that enzyme activity in the cells of halophytes is generally inhibited by high levels of NaCl (Flowers *et al.,* 1977). Thus it seems likely that most of the NaCl in mangrove tissues is compartmentalized in cell vacuoles. Whereas these ions seem to be mainly responsible for osmoregulation in the vacuole, osmotic adjustment in the cytoplasm of cells in mangroves is probably maintained by compatible organic solutes such as choline and betaine (Wyn *et al.,* 1981; Clough *et al.,* 1982).

Despite the requirement for NaCl for osmoregulation it is also clear that mangroves, like many other halophytes, must control the intake and distribution of Na^+, Cl^- and other ions to avoid ion toxicity. Mechanisms thought to be important in regulating the salt balance of mangroves include the capacity of the roots to discriminate against NaCl, the possession by some species of specialized salt-secreting glands in their leaves, the accumulation of salt in various parts of the plant, and the loss of salt when leaves and other organs are shed.

As with the roots of all higher plants, mangrove roots display a high degree of selectivity in ion uptake and transport to the xylem. Measurements of the composition of the xylem sap of mangroves indicate that mangroves exclude 80%–95% of the NaCl in the solution surrounding the roots (Scholander, 1968 and references therein; Atkinson *et al.,* 1967). In contrast, K^+ is present in the xylem sap at higher concentration than in the solution surrounding the roots (Atkinson *et al.*, 1967). Preferential uptake of K^+ relative to other cations has been observed in a wide range of plant species (Laties, 1969; Rains, 1972). There is some evidence to suggest that exclusion of Na^+, Cl^-, and perhaps some other ions is largely a passive process in mangroves (Scholander, 1968 and references therein). On the basis of this and other evidence it has been suggested that uptake of NaCl occurs mainly via apoplastic pathways in mangrove roots, and that differences among species in the capacity to exclude salt may arise from different degrees of development of the casparian strip in the root endodermis (Clough *et al.*, 1982).

A number of mangroves (*Avicennia, Aegiceras, Aegialitis, Acanthus, Laguncularia ?, Sonneratia ?*) have salt-secreting glands in their leaves. These glands mainly secrete NaCl, apparently via an active process (Atkinson *et al.*, 1967; Cardale and Field, 1975). Various aspects of the structure and function of these glands in mangroves have been reviewed by Saenger (1982) and Clough *et al.* (1982).

Although the roots may exclude as much as 90%–95% of the NaCl in the soil solution, the cumulative effect of the remaining 5%–10% that does gain access could lead to a significant flux of salt to the leaves. Calculations based on measurements of the salt concentration in xylem sap, known rates of transpiration and the concentration of NaCl in mangrove leaves suggest that the amount of salt transported to a leaf, via the xylem, during its life (about one year in *R. stylosa)* is more than 30 times that actually found in the leaf. For those species that have salt secreting glands in their leaves the rate of secretion (*c.* 0.9 μmol (Cl^-) m^{-2} s^{-1}; Atkinson *et al.*, 1967) is sufficient to expel most of the salt reaching the leaves via the xylem (Clough *et al.*, 1982).

For those species without salt-secreting glands (most mangroves) two explanations are possible for the discrepancy between the predicted and the actual amounts of NaCl in the leaves. Firstly, NaCl could be withdrawn from the sap as it moves up the xylem to the leaves. For example, some evidence suggests that salt is withdrawn from the xylem in the roots of some plant species by structurally modified xylem parenchyma cells (Läuchli, 1976). However, this phenomenon has not yet been demonstrated to occur in the stem.

The second possibility is that salt is carried to the leaves at roughly the rate calculated, but that it is subsequently retranslocated along with the export of assimilate to other tissues, as suggested by Clough *et al.* (1982). Since carbon assimilation, water loss and presumably the salt flux to the leaves are to some degree all regulated by stomatal conductance, this would give rise to a more or less uniform distribution of salt within the plant. This explanation is supported by analyses of the salt content of different parts of the plant (Walsh, 1974; Clough and Attiwill, 1975; Spain and Holt, 1980).

III. WATER RELATIONS

Mangroves mostly grow in waterlogged soils where the upper limit to their water potential is set by the osmotic potential of the substrate. Thus, the water status of mangroves is influenced significantly by the salinity of the substrate. In areas subject to regular inundation by seawater the maximum water potential of leaves and other organs might be expected to be about –2.5 MPa. However, many mangrove sediments have a low hydraulic conductivity, are poorly drained, and in such cases the salinity of the interstitial water is usually somewhat higher than that of the overlying tidal water. Moreover, exclusion of salt by the roots during water uptake presumably leads to a gradient of decreasing salt concentration away from the roots. This could mean that mangrove roots normally experience much lower substrate osmotic potentials than would be shown by measuring the osmotic potential of either the bulk interstitial water or the overlying tidal water.

Apart from its effect in salt uptake such a gradient would have a significant effect in the water balance of mangroves. Under conditions favouring rapid transpiration one might expect a two-fold, or perhaps greater, increase in salt concentration at the root surface, and thus the roots could be exposed to an osmotic potential of –3 MPa to –5 MPa. This is not too different from the hydrostatic pressure potentials of –2.5 MPa to –6 MPa commonly measured in the xylem of mangroves (Scholander, 1968; Attiwill and Clough, 1980) and suggests that they could indeed suffer a severe water deficit during periods of high evapotranspiration. This could account for reports (e.g. Joshi et al., 1975) of stomatal closure of mangroves during the daytime. Since the hydrostatic pressure potential in the xylem is about ten times greater than the osmotic potential of xylem sap (c. –0.2 MPa; B. F. Clough, unpublished) it is the main component of the xylem water potential.

Mangroves grow mostly in the lower latitudes where solar radiation and air temperatures are commonly high. This results in leaf temperatures that may be up to 5°C–10°C higher than ambient, giving rise to a large vapour pressure deficit between the leaf and the surrounding air, despite the high humidity normally associated with the mangrove environment. For instance, vapour pressure deficits between mangrove leaves and their environment of up to 5 kPa have been recorded in north Queensland (T. J. Andrews, unpublished). Although large vapour pressure deficits could give rise to very high rates of evaporation from mangrove leaves, the leaves of most species of mangrove have a number of anatomical features that restrict the loss of water vapour. These include thick cuticles, wax coatings, and sunken or otherwise hindered stomata, which in all but a few genera (*Sonneratia, Osbornia, Lumnitzera, Laguncularia*) are found only on the abaxial surface (Macnae, 1968; Walsh, 1974; Saenger, 1982). Thus the maximum values for leaf conductance in mangroves are generally somewhat lower than those for most herbaceous plants; for example, less than 0.13 mol m^{-2} s^{-1}, or, in terms of resistance, greater than 3 s cm^{-1} (Moore *et al.*, 1972, 1973; Attiwill and Clough, 1980; Ball, 1981; Andrews *et al.*, in press). Both high leaf temperatures and high vapour pressure deficits reduce the conductance of the leaf to water vapour (Ball, 1981; Andrews *et al.*, in press). However, reduced leaf conductance apparently does not compensate fully for the increase in vapour pressure deficit between leaf and air, with the result that the water loss from the leaf rises with increasing vapour pressure deficit and leaf temperature, despite the lower conductance of the leaf to water vapour (Ball, 1981). The conductance of the leaf to water vapour also decreases with increasing salinity (Ball, 1981).

These effects are probably a result of changes in leaf water status and emphasize the importance of leaf angle as an adaptation to an environment where temperatures and solar irradiance are usually high. Leaves in exposed positions at the top of the canopy are steeply inclined, often nearly vertically, whereas those in the shade, deep within the canopy, are horizontal. As a result, the incident radiation is shared over a larger photosynthetic surface, while the thermal input per unit leaf area, and hence leaf temperature, is reduced.

Since the water potential of mangroves commonly lies in the range of -2 MPa to -4 MPa there is no doubt that they experience a severe and consistent water deficit. What is not clear, however, is the extent to which this constitutes a physiological "stress". Many, if not most, physiological processes in more conventional crop plants would be severely impaired by such low water potentials. It is possible that the integrity of metabolic processes in plants depends more on the state of hydration of the cytoplasm and cell membranes than on water potential *per se*. Water potential–water content isotherms for mangrove shoots (Scholander *et al.*, 1964; Scholander, 1968) show that large changes in leaf water potential give rise to much smaller changes in tissue water content. In this respect the water relations of mangrove leaves resemble those of conifers and xerophytic trees (Scholander *et al.*, 1964; Connor and Tunstall, 1968) and are unlike those of glycophytes (e.g. Gardner and Ehlig, 1965; Wenkert *et al.*, 1978; Acevedo *et al.*, 1979). This could be an important adaptation for survival of mangroves and other halophytes on substrates with low water potentials.

IV. CARBON METABOLISM

A. Photosynthesis

There is currently some controversy about whether mangroves are C_3 or C_4 plants. On one hand, there have been reports (Joshi *et al.*, 1974; Joshi, 1976) of high phosphoenolpyruvate activity in some species of mangrove from India. These workers also found that aspartate and alanine were heavily labelled after a few seconds exposure to ^{14}C and, on the basis of these results, concluded that the C_4 pathway operated. On the other hand, the $^{13}C/^{12}C$ carbon isotope ratios of a number of Australian mangrove species from a wide range of habitats in north Queensland range from -24.6% to -32.2%. (Andrews et al., in press). These values are typical of C_3 plants and considerably more negative than the range for C_4 plants (Smith and Epstein, 1971). Furthermore, photosynthesis in mangroves typically saturates at one-half to two-thirds full sunlight, has a temperature optimum below 35°C, and mangroves have an easily measurable carbon dioxide (CO_2) compensation point (Moore et al., 1972; Ball, 1981; Andrews et al., in press). These characteristics are also typical of the C_3 photosynthetic pathway. Clearly, the nature of the photosynthetic pathway in mangroves needs to be resolved. It may well turn out that some mangrove genotypes have evolved with the C_4 pathway, whereas others employ the C_3 pathway.

Under normal ambient conditions CO_2 is linearly related to leaf conductance (Andrews et al., in press). Since the maximum leaf conductance is somewhat lower in mangroves than in many other plants, the maximum rate of assimilation is also low by comparison with that of most crop plants. Thus, maximum rates of assimilation for mangroves, both in the laboratory and in the field, appear to be less than 15 μmol (CO_2) $m^{-2} s^{-1}$ (Moore et al., 1972; Moore et al., 1973; Attiwill and Clough, 1980; Ball, 1981; Andrews et al., in press).

The linear relationship betwen CO_2 and leaf conductance implies that the CO_2 partial pressure in the mesophyll remains nearly constant. This has been confirmed by measurements in the field and in the laboratory which show that the CO_2 partial pressure lies in the range of 15 Pa to 18 Pa over a wide range of ambient conditions. These values are low for a C_3 plant and point to a high intrinsic efficiency of water use by mangroves. This efficiency can be expressed as the ratio $(p_a - p_i)/(p_a - \Gamma^*)$, where p_a is the partial pressure of CO_2 outside the leaf, p_i is the intercellular CO_2 partial pressure, and Γ^* is the CO_2 compensation point (Andrews et al., submitted). Based on this index, the mangrove R. stylosa has a water use efficiency of 0.64 (Andrews et al., in press), compared with a range of 0.2 - 0.4 for C_3 plants and 0.7 - 0.8 for C_4 species (Fischer and Turner, 1978).

Studies in southern Florida (Miller, 1972; Moore et al., 1972, 1973) and more recently in Australia (Ball, 1981; Andrews et al., in press) have shown that the rate of assimilation is much reduced at high leaf temperatures. In the few species so far examined the rate of assimilation appears to be relatively unaffected by temperature over the range 17°C–30°C, but falls sharply at temperatures above 30°C, and is close to zero at 40°C. Moore et al. (1972, 1973) observed that the optimum temperature for assimilation by mangroves in southern Florida was subject to seasonal variation. The reduction in assimilation at leaf temperatures above 30°C can be explained partly by a concomitant decrease in leaf conductance, but there is also evidence that the efficiency of the carboxylation reactions is reduced at high leaf temperature (Fig. 16.1). Ball (1981) found a similar effect in *Avicennia marina.*

The response to light of photosynthesis in mangrove leaves (Fig. 16.2) is typical of C_3 plants and is similar to that reported for other trees (e.g. Hesketh and Baker, 1967; Larcher, 1969). Assimilation typically saturates at one-third (Attiwill and Clough, 1980) to one-half full sunlight (Ball, 1981; Andrews *et al.*, in press), but somewhat higher intensities were required to saturate photosynthesis in some species in southern Florida (Moore *et al.*, 1972, 1973). An analysis of a typical light response curve for *Rhizophora apiculata* in north Queensland, such as that shown in Figure 16.2, gives an incident quantum efficiency of about 0.017 and a light compensation point of about 30 μmol (photons) m^{-2} s^{-1} (Andrews *et al.*, in press). Differences in the response to light between sun-adapted and shade-adapted leaves of a particular species might be expected, though Ball (1981) could find no significant differences in gas exchange properties betwen sun-adapted and shade-adapted leaves of *Avicennia marina.*

The effect of the partial pressure of CO_2 on assimilation has so far been studied only in leaves of *Aegiceras corniculatum, Avicennia marina* (Ball, 1981) and *Rhizophora apiculata* (Andrews *et al.*, in press). All three species show a typical C_3 response (Fig. 16.3), with a CO_2 compensation point of around 5 Pa–6 Pa at 25°C. The CO_2 compensation point increases significantly with temperature (Moore *et al.*, 1972; Ball, 1981; Andrews *et al.*,

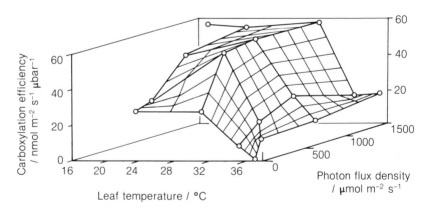

Fig. 16.1 Three-dimensional representation of the effect of leaf temperature and photon flux density (400–700 nm) on the carboxylation efficiency (assimilation rate/intercellular CO_2 partial pressure – compensation CO_2 partial pressure) in a "sun" leaf of *Rhizophora apiculata* (after Andrews *et al.*, in press).

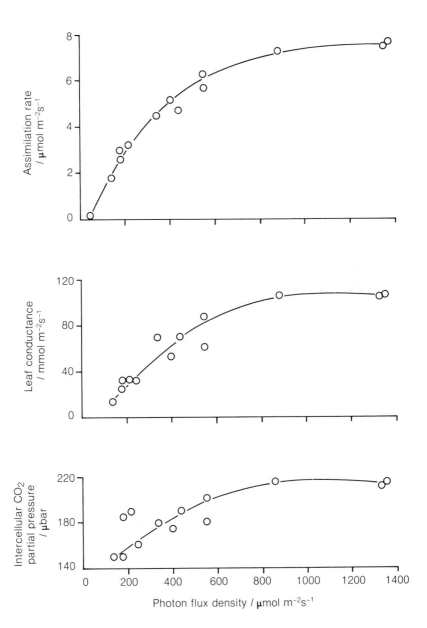

Fig. 16.2 Effect of photon flux density (400–700 nm) on assimilation rate, leaf conductance and intercellular CO_2 partial pressure on a "sun" leaf of *Rhizophora apiculata* (after Andrews *et al.,* in press).

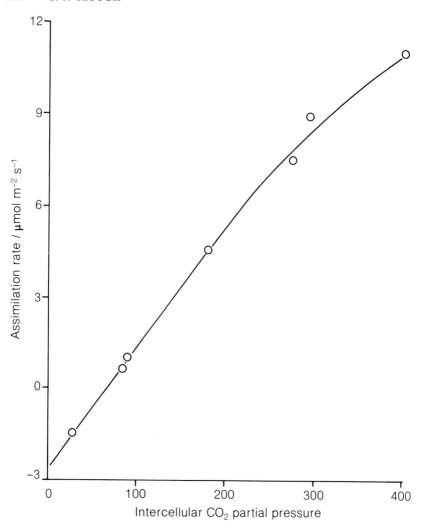

Fig. 16.3 Response of assimilation rate in a "sun" leaf of *Rhizophora apiculata* as a function of the intercellular CO_2 partial pressure at a photon flux density (400–700 nm) of 1400 μmol (photons) m^{-2} s^{-1} (after Andrews *et al.*, in press).

in press), but appears to be unaffected by either salinity or the vapour pressure deficit between leaf and air (Ball, 1981). However, high salinities have been shown to reduce the slope of the initial linear part of the CO_2 response curve in both *Avicennia marina* and *Aegiceras corniculatum* (Ball, 1981). Furthermore, salinity and vapour pressure deficit both affect the shape of the CO_2 response curve at high intercellular CO_2 partial pressures (Ball, 1981). These results suggest that high intracellular salt concentrations inhibit the carboxylation reactions of photosynthesis.

B. Respiration

Rates of respiration in mangrove leaves appear to be similar to that of many other plants, the rates of dark respiration in the leaves of some species of mangrove being commonly of the order 0.7–2 μmol (CO_2) m^{-2} s^{-1} for most species (Moore et al., 1973; Attiwill and Clough, 1980; Ball, 1981). However, Golley et al. (1962) found much higher rates of respiration (c. 10 μmol (CO_2) m^{-2} s^{-1}) in the leaves of *Rhizophora mangle* in Puerto Rico. As expected, rates of leaf respiration in mangroves increase with temperature (Moore et al., 1973).

There is conflicting evidence about the effect of salinity on the respiratory activity of halophytes. In some studies (e.g. Flowers, 1972) dark respiration has been observed to decrease with increasing salinity, whereas others (e.g. Kuramoto and Brest, 1979; Blacquiere and Lambers, 1981) found no effect. In the case of mangroves, Ball (1981) was unable to find any significant effect of salinity on dark respiration in the leaves of *Avicennia marina* and *Aegiceras corniculatum*. This is surprising since the leaves of both species possess salt-secreting glands and secretion of salt by these glands is an active process (Atkinson et al., 1967). On the other hand, Hicks and Burns (1975) found marked differences in the rate of dark respiration in leaves of several species along a gradient of salinity in the field. However, these observations could have resulted from factors other than salinity.

Very little is known about rates of respiration in other organs in mangroves. Though discrimination against NaCl by the roots appears to be largely a passive process (Scholander, 1968), uptake of other ions, such as potassium, most likely involves active transport processes. Furthermore, osmoregulation and compartmentation of ions in the vacuoles of cells in roots and other tissues presumably requires metabolic energy. It therefore seems likely that the respiratory activity of roots and other tissues increases with salinity; this has been reported by Field et al. (1981) for the roots of *Avicennia marina* and *Aegiceras corniculatum*.

V. ENVIRONMENTAL CONTROLS ON GROWTH AND PRODUCTIVITY

Estimates of the net primary productivity of mangrove communities in many parts of the world can be found in Walsh (1974), Lugo and Snedaker (1974), Lugo and Snedaker (1975), Teas (1979), and Clough and Attiwill (1982). Most of these estimates, however, are based on one or more indices such as wood production, leaf production, litter fall, or net leaf photosynthesis. In many cases these parameters have been measured over short periods of time.

No complete data are available for growth rates of whole trees in the field. Nevertheless, it is apparent from the data available that there are considerable regional differences in rates of growth.

Climate and salinity exert a significant influence on the growth and productivity of mangroves through their effect on water status and salt balance. From what we know of the ecophysiology of mangroves, moderate salinities and meteorological conditions that lead to minimal rates of water loss should favour high rates of net primary production and growth. These conditions are most likely to occur in high rainfall areas where substrate salinities are generally much less than in more arid climates. Furthermore, the overcast conditions normally associated with rain reduce the intensity of solar radiation incident on the leaves. This leads to a reduction in leaf temperature and the vapour pressure deficit between the leaf and its aerial environment. This would be advantageous in terms of water loss and salt loading in the shoots. However, it is unlikely that photosynthetic carbon assimilation is much reduced by the lower light levels that accompany overcast conditions, since photosynthesis typically saturates at levels of one-third to two-thirds full sunlight. Indeed, the experimental evidence presently available suggests that carbon assimilation under these conditions will in fact be somewhat higher than in more arid climates because of lower leaf temperatures and higher leaf conductances.

Presently there are not enough comparative data on the growth and primary productivity of mangrove communities from widely different climates to assess the significance of these relationships. However, field observations in northern Australia tend to confirm that given adequate supplies of nutrients, high rainfall mangrove forests (in areas where rainfall exceeds 1500 mm annually; Dowling and McDonald, 1982) have greater floristic diversity, are more luxuriant, and have higher rates of net primary production than those in more arid climates. Thus it seems likely that many of the regional differences in mangrove productivity could be explained in terms of rainfall and salinity. These considerations suggest that a climatic index, perhaps based on the precipitation/evaporation ratio, could be useful when making regional comparisons of mangrove growth and primary productivity.

Local variations in growth rate appear to be related to edaphic factors such as nutrient availability, salinity, and perhaps redox potential. For example, growth rates are generally slower near the upper tidal limits than in lower areas where tidal inundation is more frequent and of longer duration. Recent work has shown that in some areas the reduction in growth with increasing elevation is strongly correlated with decreasing levels of available soil phosphorus and increasing salinity (Boto and Wellington, 1983; Boto and Wellington, in press).

VI. CONCLUSIONS

In recent years considerable progress has been made towards understanding some of the physiological characteristics that enable mangroves to grow in highly saline, anaerobic soils and in climatic conditions that often are not optimal for the growth of C_3 plants. An important part of their success in coping with this environment lies in their ability to maintain a physiologically acceptable balance for salt and water. The high efficiency of water use by mangroves is an important factor in achieving this balance. Nevertheless, it is clear that mangroves contain salt concentrations and experience water deficits that non-halophytes would find intolerable. The nature of this physiological adaptation to high salt concentration and low water potential remains unknown.

This chapter has dealt mainly with the interactions between the environment and the ecophysiology of mangrove leaves. Very little information is available on interactions between the environment and other physiological processes in mangroves. For example, it is not known how the environment influences the allocation of carbon to different parts of the tree. Although roots generally comprise a much greater proportion of the total plant weight in mangroves than in other tree species (Clough and Attiwill, 1982), no information is available on environmental and endogenous controls of the distribution of assimilate within the plant. Furthermore, it would be useful to know how environmental factors such as salinity, temperature and anaerobiosis affect the partitioning of assimilate between maintenance processes and growth in different organs. Future research on the physiology of mangroves will undoubtedly contribute much to our knowledge of the more general problem of salt tolerance in these and other plants.

REFERENCES

Acevedo, E., Fereres, E., Hsaio, T. C., and Henderson, D. W. (1979). Diurnal growth trends, water potential, and osmotic adjustment of maize and sorghum leaves in the field. *Plant Physiology* **64**, 476–480.

Andrews, T. J., Clough, B. F., and Muller, G. J. (in press). Photosynthetic gas exchange properties and carbon isotope ratios of some mangroves in North Queensland. *In* "Physiology and Management of Mangroves" (H. J. Teas, ed.) Junk, The Hague.

Atkinson, M. R., Findlay, G. P., Hope, A. B., Pitman, M. G., Saddler, H. D. W., and West, K. R. (1967). Salt regulation in the mangroves *Rhizophora mucronata* Lam. and *Aegialitis annulata* R. Br. *Australian Journal of Biological Sciences* **20**, 589–599.

Attiwill, P. M., and Clough, B. F. (1980). Carbon dioxide and water vapour exchange in the white mangrove. *Photosynthetica* **14**, 40–47.

Ball, M. C. (1981). Physiology of Photosynthesis in Two Mangrove Species: Responses to Salinity and Other Enviromental Factors. Ph.D. Thesis, Australian National University.

Blacquiere, T., and Lambers, H. (1981). Growth, photosynthesis and respiration in *Plantago coronopus* as affected by salinity. *Physiologia Plantarum* **51**, 265–268.

Boto, K. G., and Wellington, J. T. (in press). Soil characteristics and nutrient status in a northern Australian mangrove forest. *Estuaries* **7**.

Boto, K. G., and Wellington, J. T. (1983). Phosphorus and nitrogen nutritional status of a northern Australian mangrove forest. *Marine Ecology Progress Series* **11**, 63–69.

Cardale, S., and Field, C. D. (1975). Ion transport in the salt gland of *Aegiceras*. *In* "Proceedings of the International Symposium on Biology and Management of Mangroves" (G. E. Walsh, S. C. Snedaker and H. J. Teas, eds.), Vol. 2, pp. 608–614. University of Florida, Gainesville.

Chapman, V. J. (1976). "Mangrove Vegetation". J. Cramer, Vaduz.

Clarke, L. D., and Hannon, N. J. (1970). The mangrove swamp and salt marsh communities of the Sydney district. III. Plant growth in relation to salinity and waterlogging. Journal of Ecology **58**, 351–369.

Clough, B. F., and Attiwill, P. M. (1975). Nutrient cycling in a community of *Avicennia marina* in a temperate region of Australia. *In* "Proceedings of the International Symposium on Biology and Management of Mangroves" (G. E. Walsh, S. C. Snedaker and H. J. Teas. eds.), Vol. 1, pp. 137–146. University of Florida, Gainesville.

Clough, B. F., and Attiwill, P. M. (1982). Primary productivity of mangroves. *In* "Mangrove Ecosystems in Australia: Structure, Function and Management" (B. F. Clough, ed.), pp. 213–222. Australian National University Press, Canberra.

Clough, B. F., Andrews, T. J., and Cowan, I. R. (1982). Physiological processes in mangroves. *In* "Mangrove Ecosystems in Australia: Structure, Function and Management" (B. F. Clough, ed.), pp. 193–210. Australian National University Press, Canberra.

Connor, D. J. (1969). Growth of the grey mangrove *(Avicennia marina)* in nutrient culture. *Biotropica* **1**, 36–40.

Connor, D. J., and Tunstall, B. R. (1968). Tissue water relations for brigalow and mulga. *Australian Journal of Botany* **16**, 487–490.

Dowling, R. M., and McDonald, T. J. (1982). Mangrove communities of Queensland. *In* "Mangrove Ecosystems in Australia: Structure, Function and Management (B. F. Clough, ed.), pp. 79–93. Australian National University Press, Canberra.

Downton, W. J. S. (1982). Growth and osmotic relations of the mangrove *Avicennia marina*, as influenced by salinity. *Australian Journal of Plant Physiology* **9**, 519–528.

Field, C. D., Burchett, M. D., and Pulkownik, A. (1981). Respiration rates in mangrove roots. Abstracts of the XII International Botanical Congress, Sydney, Australia, 21–28 August, 1981, pp. 173.

Flowers, T. J. (1972). Salt tolerance in *Suaeda maritima* (L.) Dum. *Journal of Experimental Botany* **23**, 310–321.

Flowers, T. J., Troke, P. F., and Yeo, A. R. (1977). The mechanism of salt tolerance in halophytes. *Annual Review of Plant Physiology* **28**, 89–121.

Gardner, W. R., and Ehlig, C. F. (1965). Physical aspects of the internal water relations of plant leaves. *Plant Physiology* **40**, 705–710.

Golley, F., Odum, H. T., and Wilson, R. F. (1962). The structure and metabolism of a Puerto Rican mangrove forest in May. *Ecology* **43**, 9–19.

Heald, E. J. (1969). Production of Organic Detritus in a South Florida Estuary. Ph.D. thesis, University of Miami.

Hesketh, J., and Baker, D. (1967). Light and carbon assimilation by plant communities. *Crop Science* **7**, 285–293.

Hicks, D. B., and Burns, L. A. (1975). Mangrove metabolic response to alterations of natural freshwater drainage to Southwestern Florida estuaries. *In* "Proceedings of the International Symposium on Biology and Management of Mangroves (G. E. Walsh, S. C. Snedaker and H. J. Teas, eds.), Vol. 1, pp. 238–255. Gainesville. University of Florida, Gainesville.

Joshi, G. V. (1976). "Studies in Photosynthesis in Saline Conditions" Shivaji University Press, Kolhapur.

Joshi, G. V., Karekar, M. D., Jowda, C. A., and Bhosale, L. (1974). Photosynthetic carbon metabolism and carboxylating enzymes in algae and mangroves under saline conditions. *Photosynthetica* **8**, 51–52.

Joshi, G. V., Bhosale, L., Jamale, B. B., and Karadge, B. A. (1975). Photosynthetic carbon metabolism in mangroves. *In* "Proceedings of the International Symposium on Biology and Management of Mangroves" (G. E. Walsh, S. C. Snedaker, and H. J. Teas, eds.), Vol. II, pp. 579–594. University of Florida, Gainesville.

Joshi, G. V., Jamale, B. B., and Bhosale, L. (1975). Ion regulation in mangroves. *In* "Proceedings of the International Symposium on Biology and Management of Mangroves" (G. E. Walsh, S. C. Snedaker and H. J. Teas, eds.), Vol. II, pp. 595–607. University of Florida, Gainesville.

Kuramoto, R. T., and Brest, D. E. (1979). Physiological response to salinity by four salt marsh plants. *Botanical Gazette.* **140**, 295–298.

Larcher, W. (1969). The effect of environmental and physiological variables on the carbon dioxide gas exchange of trees. *Photosynthetica* **3**, 167–198.

Laties, G. G. (1969). Dual mechanisms of salt uptake in relation to compartmentation and long-distance transport. *Annual Review of Plant Physiology* **20**, 89–116.

Lugo, A. E., and Snedaker, S. C. (1974). The ecology of mangroves. *Annual Review of Ecology and Systematics* **5**, 39–64.

Lugo, A. E., and Snedaker, S. C. (1975). Properties of a mangrove forest in southern Florida. *In* "Proceedings of the International Symposium on Biology and Management of Mangroves" (G. E. Walsh, S. C. Snedaker and H. J. Teas, ed.), Vol. I, pp. 170–212. University of Florida, Gainesville.

Läuchli, A. (1976). Symplasmic transport and ion release to the xylem. *In* "Transport and Transfer Processes in Plants" (I. F. Wardlaw and J. B. Passioura, eds.), pp. 101–112. Academic Press, New York.

Macnae, W. (1968). A general account of the flora of mangrove swamps and forests in the Indo-West-Pacific region. *Advances in Marine Biology* **6**, 73–270.

McMillan, C. (1974). Salt tolerance of mangroves and submerged aquatic plants. *In* "Ecology of Halophytes" R. J. Reimold and W. H. Queen, eds.), pp. 379–390. Academic Press, New York.

Miller, P. C. (1972). Bioclimate, leaf temperature, and primary production in red mangrove canopies in South Florida. *Ecology* **53**, 22–45.

Moore, R. T., Miller, P. C., Albright, D., and Tieszen, L. L. (1972). Comparative gas exchange characteristics of three mangrove species in winter. *Photosynthetica* **6**, 387–393.

Moore, R. T., Miller, P. C., Ehleringer, J., and Lawrence, W. (1973). Seasonal trends in gas exchange characteristics of three mangrove species. *Photosynthetica* **7**, 387–394.

Odum, W. E. (1971). Pathways of energy flow in a South Florida estuary. University of Miami Sea Grant Technical Bulletin. 7. University of Miami, Miami.

Ong, J. E. (1982). Mangroves and aquaculture. *Ambio* **11**, 252–257.

Rains, D. W. (1972). Salt transport by plants in relation to salinity. *Annual Review of Plant Physiology* **23**, 367–388.

Saenger, P. (1982). Morphological, anatomical and reproductive adaptations of Australian mangroves. *In* "Mangrove Ecosystems in Australia: Structure, Function and Management" (B. F. Clough, ed.), pp. 153–191. Australian National University Press, Canberra.

Scholander, P. F. (1968). How mangroves desalinate seawater. *Physiologia Plantarum* **21**, 251–261.

Scholander, P. F., Bradstreet, E. D., Hammel, H. T., and Hemmingsen, E. A. (1966). Sap concentrations in halophytes and some other plants. *Plant Physiology* **41**, 529 –532.

Scholander, P. F., Hammel, H. T., Hemmingsen, E. A., and Bradstreet, E. D. (1964). Hydrostatic pressure and osmotic potential in leaves of mangroves and some other plants. *Proceedings of the National Academy of Science USA* **52**, 119–125.

Scholander, P. F., Hammel, H. T., Hemmingsen, E. A., and Garey, W. (1962). Salt balance in mangroves. *Plant Physiology* **37**, 722–729.

Smith, B. N., and Epstein, S. (1971). Two categories of 13C/12C ratios for higher plants. *Plant Physiology*, **47**, 380–384.

Spain, A. V., and Holt, J. A. (1980). "The Elemental Status of the Foliage and Branchwood of Seven Mangrove Species from Northern Queensland. Commonwealth Scientific and Industrial Research Organization, Australia. Division of Soils, Divisional Report No. 49.

Stern, W. L., and Voigt, B. K. (1959). Effect of salt concentration on growth of red mangrove in culture. *Botanical Gazette* **121**, 36–39.

Teas, H. J. (1979). Silviculture with saline water. *In* "The Biosaline Concept" (A. Hollaender, J. C. Aller, E. Epstein, A. San Pietro and O. R. Zaborsky, eds.), pp. 117–161. Plenum Press, New York.

Walsh, G. E. (1974). Mangroves: A review. *In* "Ecology of Halophytes" (R. J. Reimold and W. H. Queen, eds.), pp. 51–174. Academic Press, New York.

Wenkert, W., Lemon, E. R., and Sinclair, T. R. (1978). Water content-potential relationships in soya bean: Changes in component potentials for mature and immature leaves under field conditions. *Annals of Botany* **42**, 295–307.

Wyn Jones, R. B., and Storey, R. (1981). Betaines. *In* "Physiology and Biochemistry of Drought Resistance in Plants" (L. G. Paleg and D. Aspinall, eds.), pp. 171–204. Academic Press, Sydney.

CHAPTER 17

Trees and Forest Restoration

F. J. BURROWS

I. INTRODUCTION

Here we turn to a different aspect of management of crop productivity, the re-establishment of trees in relatively unmanaged environments, such as land surfaces reclaimed after mining. The subject is topical because there are extensive mining operations throughout the world often near conurbations. In the United States, for example, a total of 500 000 hectares of land was disturbed by surface or strip mining between 1965 and 1974, leaving over 1 000 000 hectare requiring reclamation (Bradshaw and Chadwick, 1980). Furthermore, the rates of mining and revegetation are accelerating. In Western Australia there has been a twenty-fold increase in the rate of land cleared and mined for bauxite since the mining began sixteen years ago (from 12 to 250 hectares annually; Institute of Foresters of Australia, 1980). At Weipa, Queensland, the area regenerated annually has increased from 50 to 500 hectares between 1972 and 1981 (Comalco, 1981). The major ecosystem affected is frequently forest. Different tree species are involved: jarrah (*Eucalyptus marginata*) in Western Australia; blackbutt (*E. pilularis*) and red gum (*Angophora costata*) in coastal ecosystems of New South Wales (Lewis, 1980a); woolybutt (*E. miniata*) and stringybark (*E. tetrodonta*) in Arnhem Land, Northern Australia. Because of the high cost of restoration (about

CONTROL OF CROP PRODUCTIVITY
ISBN 0 12 548280 9

$1300 to $4500 per hectare: Comalco, 1981; Bradshaw and Chadwick, 1980) and the low mineral nutrient requirements of native vegetation, "aftercare" (Bradshaw and Chadwick, 1980) is kept to a minimum. After initial establishment procedures the process is left to natural succession.

Each stage of mining causes serious disruption to the edaphic, biotic and climatic environment. For example, burning the vegetation volatalizes nutrient stocks contained in the standing crop (Raison, 1980); if trees are harvested as timber before burning, nutrients are more effectively removed from the ecosystem. Burning can also destroy soil organic matter. Stockpiling soil can reduce soil mycorrhiza (Lewis, 1980b). Restoration and reconstruction leads to soil compaction and on the surface of the restored mine, growing plants experience severe climatic conditions.

This chapter considers firstly the environment of disturbed, in contrast to undisturbed, coastal dunes and the productivity of indigenous trees during re-establishment in the disturbed ecosystem. I have mostly used climatic data collected from a dune restored after sand mining and data from beneath the adjacent undisturbed forest in the Myall Lakes National Park, 250 kilometres north of Sydney, New South Wales. Conditions on this restored dune were particularly severe; the dune was elevated (about 30 m) and exposed to the ocean (400 m east). At each site measurements were made of incoming radiation, Q_s, net energy, Q_n, and soil heat flux, G (mV signals were recorded on an integrating recorder); temperature, humidity and wind speed were measured at half-hourly intervals (using wet and dry bulb thermometers mounted in a standard Stevenson screen and a cup anemometer 2 m above the ground). Hourly estimates of evaporative demand were calculated using a combination formulae (Monteith, 1973). Data on tree growth was taken from experiments in the Myall Lakes National Park and obervations on trees from a restored bauxite mine in Arnhem Land.

II. SEEDBED ENVIRONMENT

Levels of Q_s, Q_n and G beneath the canopy were consistently below those on the restored dune (Fig. 17.1). At highest insolations in the middle of the day, Q_s and Q_n beneath the canopy were 55% of the levels outside. The levels beneath the canopy were high compared with other canopies but reflected the relative openness of coastal eucalypt forests and the generally low leaf area index of eucalypts (Carbon et al., 1979). At these times G beneath the canopy was only 30% of the value on the restored dune partially because of a layer of leaf litter. Wind speed on the restored dune was about 2.5 times the wind speed beneath the canopy: maximum speeds were 25 km h^{-1} and 10 km h^{-1} (7 and 3 m s^{-1}) for each site respectively. Air temperatures and humidities were slightly higher beneath the canopy and the mean saturation

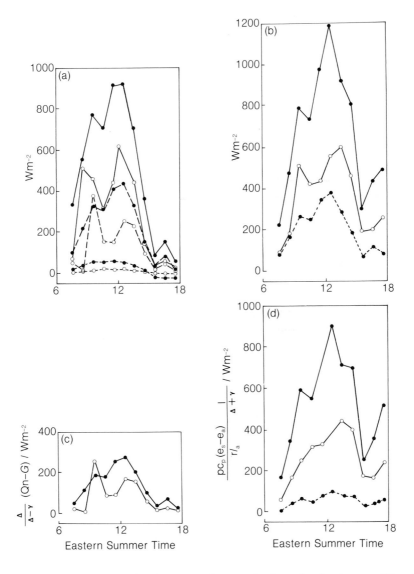

Fig. 17.1 Diurnal variation in the components of evaporation beneath a forest canopy (○) and on an exposed dune restored after sand mining (●); Myall Lakes National Park, NSW, Australia, November 1977. (a) incoming shortwave radiation, Q_s (— — —), net radiation, Q_n (————), and ground heat flux, G (●●●); (b) total evaporative demand; (c) energy component of total evaporative demand; and (d) wind component of total evaporative demand. Values for (b) and (d) over a restored dune have been calculated for aerodynamically smooth (— — —) and rough (----) surfaces. (Δ and γ are slope of the saturation vapour pressure/temperature curve and psychrometric constant, respectively; e_s and e_a the saturation and actual vapour pressure; ρ, air density and C_p specific heat at constant pressure; and r_a the aerodynamic resistance to vapour transfer).

deficit was 8.069 mb and 8.435 mb on the restored dune and in the forest, respectively.

The calculation of evaporative demand depends on assigning a sensible value to surface roughness (see Monteith, 1973 for theory). There is little guide to the value of this parameter beneath forest canopies. The average height of the understorey vegetation was 1.5 m. For a first approximation roughness of the forest site was calculated as if the subcanopy vegetation formed a continuous dominant canopy. In contrast, the surface of the restored dune was smooth and wind polished; that is, of low roughness. However, the undulating topography and the closeness of the dune to the forest edge would increase site roughness. For estimates of evaporative demand above this site I have used two roughness values; one appropriate for a canopy of 0.01 m (about the size of pieces of surface debris) and the other for a canopy of 1.5 m. The former will probably result in an underestimate of evaporative demand. The latter may be closer to the value for a restored dune with some vegetation (although the presence of vegetation will reduce the magnitude of the radiation component). The estimates (Fig. 17.1) suggest daily rates of evaporation of 5.7, 3.2 and 10.8 mm d^{-1} for the forest site and the dune with low and high roughnesses, respectively. Disregarding the estimate for the unrealistically smooth surface, exposed plants were required to supply water at about twice the rate of plants growing beneath the forest canopy.

Soil temperatures were measured at the surface and 1, 10 and 30 cm depth using copper/constantin thermocouples. Maximum surface temperatures measured over four days in spring were 26.1°C in the forest and 46.0°C on the restored dune. Temperatures at all depths below the surface of the dune site were above air temperatures except between 10.00 and 14.00 h at 30 cm. In contrast, the soil temperatures in the forest were 2°C–4°C (depending on depth) below air temperature (maximum 24.0°C) during daylight hours. Higher soil temperature would be expected in summer. During February maximum temperatures at 2 cm were 56°C and 35°C in restored dune and forest sites, respectively. Corresponding air temperatures were 36.5°C and 36.0°C (Dalby, 1977). Surface temperatures would be further elevated.

The bulk density of soil in the forest was below that in the restored dune where bulk density increased with depth (Table 17.1). This reflected both the effects of the heavy earth-moving machinery used to contour the dune during restoration and the reduction in organic matter. The increase in bulk density was accompanied by a marked increase in soil hardness or, strictly, the resistance of soil to penetration. In the undisturbed forest the resistance to penetration was about 2.5 arbitrary units (a.u.) at the surface and at 30 cm. On an adjacent newly restored dune, surface values were the same but hardness at 30 cm was 7.4 a.u. There appeared to be significant changes with time, possibly due to settlement. After three years the surface hardness of the same dune had increased to 6 a.u. and with depth to about 12 a.u.

Table 17.1

Characteristics of soils from restored dune and undisturbed forest sites in Myall Lakes National Park.

Site	Total N ppm	Soluble P ppm	Carbon %	Bulk density g/cm^3	Water holding capacity[a]
Restored dune					
Topsoil (0–30 cm)	900	0.6	1.63	1.52	9.1
Sand (30–60 cm)	–	–	0.5	1.67	4.2
Undisturbed forest					
Topsoil (0–30 cm)	900	0.5	4.9	1.46	15.2

[a] Between Field Capacity and –1.5 MPa in cm^3 cm^{-3}%.

Moisture release curves for soils from the forest and top soil and sand from the restored dune were typical for sands: differences in water holding capacities (Table 17.1) reflected differences in the organic matter content. Soil from the forest contained the greatest amount of organic matter and showed the highest moisture holding capacity.

The mineral content of dune soils is notoriously low yet there are few comprehensive studies of the nutrient stocks in Australian ecosystems that can be used as a guide. Lewis (1978) examined the Myall Lake ecosystem; Westman and Rogers (1977) documented more completely the eucalypt forest on the sand dunes of North Stradbroke Island, Queensland, where, with the exception of nitrogen (N) and sodium (Na), nutrients in the soil were extremely low (Table 17.2). A significant proportion of nutrient stocks were present in the vegetation.

The decreased levels of organic carbon on the restored dune have already been noted. As significant as the loss of carbon is a change in form. In the soils of the undisturbed forest carbon is present mainly as organic carbon. In the restored dune, however, significant quantities appear to be mineral carbon (charcoal) that comes from the burning of vegetation during clearing. Analysis cannot distinguish between the two forms. Mineral carbon is of little use in increasing the soil moisture-holding capacity or in supplying mineral nutrients to plants.

III. PHYSIOLOGICAL RESPONSES TO ENVIRONMENT

Plants growing on restored dunes in the Myall Lakes and on other restored mines, if these data can be taken as a guide, are placed under a range of environmentally induced stresses that are considerably greater than those on plants growing beneath an established tree canopy.

Table 17.2

Nutrient content of vegetation (kg ha⁻¹) and (b) stocks in forest ecosystem as a percentage of total stocks in the vegetation. N. Stradbroke Island. (After Westman and Rogers, 1977.)

	N	P	K	Ca	Mg	Na	Total
(a)							
Trees	413	15.3	171	335	72.2	165.8	1342
Understorey	43.4	2.4	20.8	8.9	5.0	3.6	84.80
Total	456.4	17.7	191.8	343.9	77.2	169.4	1326.8
(b)							
Litter	42.7	38.6	11.8	41.3	44.0	4.6	32.6
Soil							
A horizon	665	24.3	12.3	29.0	55.8	43.7	262
B_1 horizon	3260	119	73.3	62.3	340	489	1380
B_2 horizon	3250	309	53.9	56.3	292	691	1440

Higher levels of Q_s will certainly encourage increased leaf conductances (Burrows and Milthorpe, 1976) and photosynthetic rates in plants on the re-stored dune. Counteracting these effects are the high evaporative demands to which plants are subjected as a result of increased Q_n and wind speed. The water relations of trees on the restoration area are further jeopardized by the reduced moisture-holding capacity of the soil. The amount of water available for evaporation is reduced at the Myall Lakes site—to between 25% and 60% of the amount available in the forest. The size of the reservoir of water available to plants may be reduced still further if root extension is ham-pered by increases in soil compaction. Compaction significantly reduces the number and length of the primary laterals and decreases the tap root length of seedlings. Observations in the Myall Lakes National Park suggest that the roots of at least relatively young plants ($<$ 4 years) growing on restored dunes are confined within the depth of the returned top soil (nominally 30 cm but varying between 0 cm and 200 cm). It is tempting to assume this is the result of the increased soil bulk density; maximum values of 1.67 g cm⁻³ have been recorded at the interface of the sand and top soil (Table 17.1).

With increased evaporative demand and a reduced supply of water to a restricted root system, plants growing on a restored dune will experience water stress well before those growing beneath the forest canopy. Observa-tions of the diurnal changes in leaf conductance and xylem water potential of two species growing in the Myall Lakes National Park illustrate the kinds of response to water stress that might be expected (Fig. 17.2). Measurements were made in late autumn on *Acacia longifolia* (a tall shrub) and *Angophora costata* (a tree) growing on a dune restored for four years. Concurrent

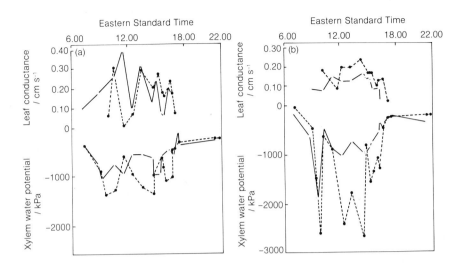

Fig. 17.2 Diurnal variation in leaf conductance (cm s⁻¹) and xylem water potential (kPa) in (a) *Acacia longifolia* adjacent to and (b)*Angophora costata* growing beneath the undisturbed forest canopy (○) and on a dune restored after sand mining (●); Myall Lakes National Park, NSW, Australia, May 1976.

measurements were made on an *Acacia* growing close to the undisturbed forest on a dune previously cleared but not mined and on an *Angophora* sapling beneath the forest canopy. The root environments of these plants were relatively undisturbed compared with those from the dune. The aerial environments of *Acacia* from the two sites were similar but those of *Angophora* differed markedly. Levels of Q_s were 98% and 36% of the levels on the restored dune at each site, respectively.

Mean conductance of *Acacia* from the two sites was not significantly different (0.191 and 0.173 cm s⁻¹ from the forest and restored dune, respectively) reflecting the similar levels of Q_s. In contrast, lower conductances were recorded from *Angophora* growing in the forest (0.112, compared with 0.146 cm s⁻¹ outside); a result of lowered light intensity. Water potentials of plants of both species from the restored dune were significantly lower than from other sites (-808 kPa and -592 kPa for *Acacia* and -1041 kPa and -683 kPa for *Angophora* from dune and forest sites, respectively). A minimum value of -2700 kPa was recorded in exposed *Angophora costata*. Critical levels of water potential (i.e. levels that cause stomatal closure) were not reached (Fig. 17.3). The levels of water potential required to maintain the *same* conductance were lower in plants growing on the dune for both species. Clearly, plants growing in shade or with undisturbed root environments enjoyed better water relations than plants on the restored dune.

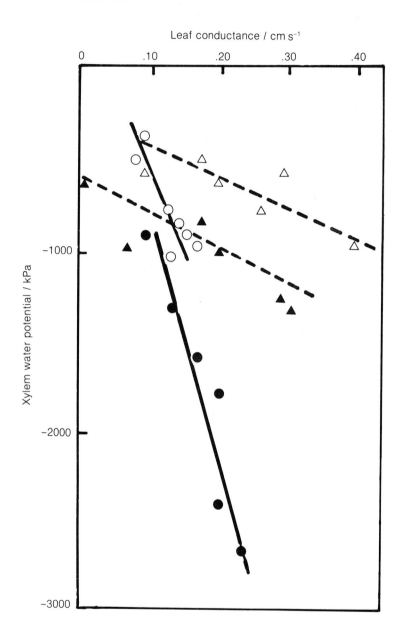

Fig. 17.3 Relationship between leaf conductance and xylem water potential in *Acacia longifolia* (△) adjacent to and *Angophora costata* (○) growing beneath an undisturbed forest canopy (○△) and on a dune restored after sand mining (●▲); Myall Lakes National Park; NSW, Australia, May 1976.

In *Angophora costata* levels of water potential considerably below permanent wilting point (–1500 kPa) were reached without any reduction in conductance. This may indicate a degree of adaptation to severe water stress in this species. With *Acacia longifolia* the minimum potentials were above the permanent wilting point suggesting that this species has available a larger reservoir of water or extracts water more efficiently. Efficient extraction might prove disadvantageous under severe stress. As sandy soils dry out they release moisture completely at high potentials. Thus *A. longifolia* might experience sudden, severe and fatal stress. This species is frequently found moribund or dead indicating this might be the case and suggesting *A. longifolia* has not evolved a mechanism to cope with these conditions.

Mineral nutrients added to the surface of the restored dune in the initial and later stages of revegetation may exacerbate the problems of water stress. Plants growing in low nutrient conditions direct a higher proportion of assimilates into root production than plants well supplied with nutrients. For example, after three weeks 72% of the dry weight of *A. costata* grown in sand watered daily with half-strength Hoagland solution was in above-ground parts whereas only 33% was in these parts for *A. costata* growing in sand from the forest floor at Myall Lakes (Fig. 17.4). If this laboratory response reflects conditions in the field, root systems of plants growing on a well-fertilized restored dune are forced to support a disproportionately large aerial system. The result is an increased drain on the soil water reservoir. Conversely, a low level of nutrients in the restored dune may result in improved water relationships by reducing top relative to root growth.

The effect of added fertilizer has wider implications for reconstruction of dune ecosystems. The return of species not deliberately reintroduced (i.e. those contained in the returned topsoil) is badly affected by increasing the length of time topsoil is in a stockpile, the width of the mining pathway, and the addition of fertilizers (Shelley, 1979).

Most species present before mining that return to the restored site do so within the first year. Shelley re-examined sites on restored dunes studied by Clark (1975) and found little further improvement towards full recovery after the first year on sites restored for up to eight years. She used Percentage Difference (PD) as a measure of dissimilarity between pre-mining and post-mining communities

$$PD = 100 - 200 \Sigma \frac{\min P_{ia},\, P_{ib}}{(P_{ia} + P_{ib})} \qquad (17.1)$$

where P_i is the percentage cover of species i, before, b, and after, a, mining. The measure contains an estimate of abundance, P, and provides valuable supplementary information to qualitative indexes based on species lists (e.g.

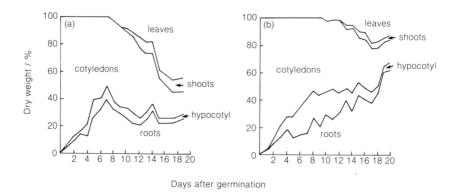

Fig. 17.4 Distribution of dry weight in *Angophora costata* seedlings growing in (a) sand watered daily with half-strength Hoagland solution and (b) unfertilized forest soil from Myall Lakes National Park, NSW, Australia.

Mueller-Dombois and Ellenberg, 1974). Significantly, after six years fertilized sites (PD = 44.5) were more dissimilar to pre-mining conditions than non-fertilized sites (PD = 40.5). Seedlings of many species normally growing on infertile soils are sensitive to high levels of phosphorus (Specht, 1975). The sensitivity may operate through water stress, or directly through toxic effects on the plants (Siddiqi *et al.,* 1976).

The effects of increased solar radiation and wind speed may not be solely as a result of increased evaporative demand. For example, higher levels of solar radiation increased *G* on the restored dune above levels in the forest. This in turn leads to surface and soil temperatures close to the thermal death point of tissues. Such temperatures are likely to cause reductions in shoot and root growth; seedlings growing close to the surface will experience reductions in photosynthesis. However, heat resistance may develop (Kramer and Kozlowski, 1979) that can lead to lignotuber formation in eucalypts (Pryor, 1976). Tropical and subtropical species can withstand higher soil temperatures than temperate species (Russell, 1977). Moreover, elevated soil temperatures may be required to break dormancy of hard seeds or seeds with a high temperature requirement (Mott, 1978).

The main effects of wind speed—other than mechanical damage, such as loss of leaf area—are as a result of increased salt absorption by a leaf surface abraded by sand grains (Grace, 1977) that can lead to death in even salt-tolerant species (Grieve and Pitman, 1978).

IV. TREE GROWTH

Two strategies are used to accelerate the return of trees to disturbed sites. With selected tree species either tube stock is transplanted after some time has elapsed for covering vegetation on the site to re-establish or seed is collected and sown into the restored site with a cover crop. The former strategy has been used widely after sand mining; the latter after bauxite mining (Hinz, 1980a; Hinz and Doettling, 1979). Each method supplements, but does not rely upon, the growth of tree seeds stored in stockpiled topsoil.

Seedlings for transplanting are commonly grown in tubes about 10–15 cm long. These mostly die because the roots are too small to supply adequate water to satisfy high demands. Conditions in the soil described above aggravate the water supply problem. Pilot experiments with *Angophora costata* have shown that larger seedlings survive better: percentage survival was 85, 60 and 30 from transplants having tap root lengths of 60, 40 and 20 cm respectively. Above-ground growth was also related to the size of the original root systems (Fig. 17.5). Under the severe field conditions in the Myall Lakes National Park, leaf area and number initially declined but seedlings with the longest roots recovered their original size after two months and continued to expand leaves and increase leaf number at an almost exponential rate. It was not until after almost six months that surviving seedlings with shorter roots had increased these parameters beyond their size at transplanting (Fig. 17.5). Increase in number and length of shoots was initially much greater in seedlings with larger roots (Fig. 17.5).

The early growth of trees, which is particularly important to quantify in revegetation studies, has received little attention. The relationship between growth and time may be fitted best by complex Richard's functions (Milthorpe and Moorby, 1979) but simpler relationships are often adequate. For example, scanty data of Craciun (1978) suggests that height growth in *E. tetrondonta* may be described by either the logistic or monomolecular growth curves. Mucha (1979) has data that indicate the increase in basal area of *E. miniata* and *E. tetrodonta* is linearly related to time over the entire ontogeny.

Growth of *Eucalyptus* species sown onto a restored bauxite mine at Gove, in Northern Australia was described over four years using three models (Table 17.3). From the fitted curves the time for trees growing on the mine to reach the same size as those in the forest was estimated. The logistic and monomolecular relationships indicated that both species would reach equivalent heights between 8 and 22 years after sowing, respectively: the linear relationship gave an estimate of 16 years for *E. tetrodonta* and 14 years for *E. miniata*. Estimates of the time taken for the basal area of *E. miniata* and *E. tetrodonta* to reach the same values as trees in the forest were 66 and 181 years, respectively. The last estimate is ridiculously long as it reflects the

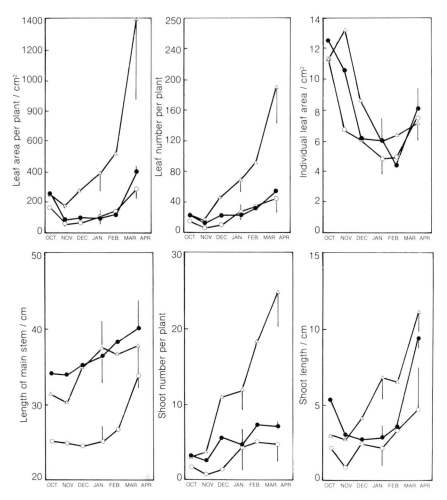

Fig. 17.5 Temporal changes in (a) leaf area, leaf number per plant and mean area of individual leaves and (b) length of main stem, shoot number and mean shoot length of *Angophora costata* seedlings planted as tube stock on a restored dune with tap root lengths of 20 cm (○), 40 cm (●) and 60 cm (△). Vertical bars are standard errors.

very slow increase in girth in this species during early growth. Rate of girth increase may be expected to increase after about five years (Craciun, 1978). Therefore, from the limited evidence available, the time taken for these species to achieve the same stature as trees in the forest is close to 60 years. However, equivalent heights will be reached from 15 to 22 years after restoration.

These estimates from limited data compare well with the age of the mature trees in the forest adjacent to the mine. Tree age there is probably between 25 and 50 y. The upper age limit for indigenous trees is set by the frequency of fire and the effect of cyclones on trees weakened by termite attacks (Mucha, 1979). The lower age limit is set from measurements on a population of thirty *E. miniata* and *E. tetrodonta* to determine the average size of mature trees (> 10 cm diameter at breast height). The mean diameter at breast height (dbh) of these populations, 20 (± 1.3) cm and 16 (± 1.4) cm, respectively, compared well with a much smaller sample aged 25 y measured by Mucha (17.6 cm dbh for both species).

The mortality rate of trees growing from sown seed is higher than for trees planted from tube stock. Hinz (1980b) sowed *Eucalyptus* in a mixture of native species at rates equivalent to 7000 ha^{-1}. (In the undisturbed forest the density of *Eucalyptus* of height > 1 m, representing mature trees and regrowth, was 2100 ha^{-1}). After four years between 70 and 100 trees ha^{-1} had survived—a mortality of 99% between sowing and establishment. To regain pre-mining densities the seed rate would have to be increased 20–30 fold. The increased costs of collecting enough seed would preclude this tactic. Instead, after preliminary planting or sowing further increase in the number of trees will result from germination of seed dropped from either mature trees growing at the edge of the mine site or from sown or transplanted trees when they reach maturity.

Successful germination and growth of seed depends on several factors, many climatic and well studied (e.g. Heydecker, 1973). One not often considered, however, is the presence of understorey vegetation. For example, *E. pilularis* seed germinates best on surfaces devoid of vegetation. Superior stocking (trees > 0.6 m in height) is achieved if the understorey is mechanically cleared rather than burnt, but no stocks are established on uncleared land (Floyd, 1962). Consequently, successful rehabilitation of understorey species may be disadvantageous for increase in the numbers of dominant tree species. Trees reach maturity after a varying number of years. Some eucalypts produce seed at ten years old but usually they don't produce seed until between 25 and 30 y old (e.g. Cremer *et al.*, 1978). Thus whereas trees on the mine edge can contribute towards increasing the population immediately those on the restored site will not do so for a number of years.

The contribution to the spread of trees across the site will depend on the distance seed can be thrown from trees. Floyd (1962) concluded that seed trees of *E. pilularis* were only effective in regeneration within 6 m of the crown when 37 m high and 60 cm dbh. Consequently dispersal rates that rely on seed throw are likely to be slow. The distance to which a seed can be carried by wind

Table 17.3

Equations relating height (h) and basal area (A), with time for *Eucalyptus tetrodonta* and *E. miniata*. Equations were fitted from data collected on 30 trees of each species aged one and four years.

	Attribute	Relationship	Equation	r^2
Eucalyptus tetrodonta	[a]Height	monomolecular	$\ln(1 - h/15) = 0.02 - 0.05t$	0.65
	[a]Height	logistic	$\ln((15 - h)/h) = 4.27 - 0.625$	0.80
	Height	linear	$h = 0.65t - 0.25$	0.68
	Basal area	linear	$A = 1.34t - 1.34$	0.23
Eucalyptus miniata	[a]Height	monomolecular	$\ln(1 - h/14) = 0.03 - 0.07t$	0.70
	[a]Height	logistic	$\ln((14 - h)/h) = 3.95 - 0.64t$	0.77
	Height	linear	$h = 0.82t - 0.28$	0.72
	Basal area	linear	$A = 5.45t - 5.45$	0.33

[a] The values 14 and 15 are maximum heights (m) of the two species in the undisturbed forest.

$$D \simeq V_w h / V_t \qquad (17.2)$$

where D (m) is a function of wind velocity V_w (m s⁻¹) the height h (m) of the tree producing the seed and the terminal velocity of the seed V_t (ms⁻¹) (Cremer, 1977). Seed terminal velocity is simply related to seed weight W (mg) by

$$V_t = 4.27 \, W^{0.27} \qquad (17.3)$$

I have calculated the rate at which trees spread across the site (the advance of the tree front) assuming the monomolecular relationship between height and growth for *E. miniata*

$$\ln (1 - h/14) = 0.03 - 0.07t \qquad (17.4)$$

and seed size for *E. pilularis* (2.16 mg; Cremer, 1977). Given average wind speeds of 2.78 m s⁻¹ (10 km h⁻¹) and assuming *E. miniata* begins producing seed at 6 m height (i.e. 5 years old; probably an underestimate) the resulting advance of the trees is as shown in Figure 17.6a. In these assumed circumstances mature trees on the site edge 14 m high throw seed a maximum distance of 6 m. After germination and growth the resulting trees begin to throw seed when production starts after they have reached a height of 6 m. The distance thrown by these trees is about 2 m. They continue to produce and

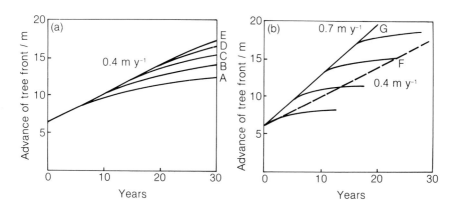

Fig. 17.6 Spread of trees across a restored site from the adjacent undisturbed forest (a) without and (b) with the contribution from canopy development. Lines A to E are the spread of successive suites of trees; lines F and G represent the rate of spread in the absence of, and with allowance for, canopy development, respectively.

throw seed as they grow to full height. Gradually trees spread over the re-stored mine at a rate illustrated by curve A (Fig. 17.6a). The first seed from this suite of trees follows the same time course (curve B) and so on (curves C to E). Only the first dispersal of seed from each suite contributes to the advance of the tree front as subsequent dispersals, although thrown further because of increase in size of the seeding tree, are overtaken by the first seed thrown from newly matured trees of the previous dispersal. This arises simply as a result of selecting the monomolecular curve to describe growth. The overall rate is given by the straight line joining points describing the distance moved on first dispersal. For the conditions assumed the rate is 0.4 m y^{-1}.

These calculations assume dispersal takes place annually from trees maturing at the site of germination. A considerable impetus is given to the rate of advance if account is taken of the rate of spread of the canopy. Canopy width (CW) and dbh are linearly related: for messmate *E. obliqua* (Curtin 1964, 1970)

$$CW = 0.785 + 0.210 \text{ dbh} \qquad (17.5)$$

Estimates of the rate of canopy expansion (the curved lines of Fig. 17.6b) can be made by using the relationship (Curtin, 1970; Kira, 1978).

$$dbh = 0.78 \text{ h} (1 - 0.012/h) \qquad (17.6)$$

The effect of increase in width is to increase the rate of advance of trees growing at the same rate as *E. miniata* to 0.7 m y^{-1}.

Delays in seed trees reaching maturity will reduce these estimates. Furthermore, since seedlings will be growing in the shade of larger trees, growth rates will also be reduced. In young stands of *E. sieberi*, for example, overwood trees exerted a deleterious influence on young stands over a dis-tance of three crown diameters (Incoll, 1979). At the edge of the crown there were reductions of about 14%, 20% and 30% in the height, basal area and volume of trees in the understorey respectively, relative to trees beyond the influence of the crown. Nevertheless, these calculations serve to illustrate an important feature of tree spread over restored landscapes. For trees whose growth–time relationships can be described by the monomolecular growth curve—and this appears to be most trees used for timber production (Kramer and Kozlowski, 1960)—trees advance at a rate determined by first dispersal of seed from newly matured trees. This rate is determined by tree growth rates, age at which seed is produced and characteristics of the seed.

An additional impetus to population dispersal may come from the produc-tion of root suckers, a phenomenon in many species including *Eucalyptus* (Lacey and Whelan, 1976). Vegetative spread is in response to damage or destruction of the parent plant, usually by fire (Lacey, 1974) or wind throw.

Suckering is regarded as undesirable in forest production (Lacey and Whelan, 1976) where considerable lengths are taken to eradicate suckers (Easterbrook, 1951). However, ability to reproduce vegetatively may be desirable for restocking restored areas since it provides a supplementary strategy to seed throw for encouraging tree spread. There is little quantitative evidence as to rates of production and growth and the boundary of suckers produced by a single tree. Suckering was observed at Gove in seven-year-old *Acacia leptocarpa* which had been burnt in the previous year. The maximum distance of suckers from the parent tree was 4.5 m (average 3.0 m); that is, a spread of 0.64 m y^{-1} (average 0.43 m y^{-1}) from sowing. The age at which species are capable of producing vegetative regrowth is unknown.

V. CONCLUSIONS

To some extent the harsh seedbed environment on a restored mine can be ameliorated by spreading brushwood or by planting a cover crop (Bradshaw and Chadwick, 1980). The former method has the added advantage of providing seed and a slowly-released supply of nutrients. Fertilizer application is required to establish a cover crop and often the recommended rates are those necessary to support a productive agro-ecosystem rather than the transient vegetation cover needed for surface stabilization. Fertilizers reduce species diversity (Shelley, 1979) possibly by preventing Clementsian succession through encouraging the growth of exotic weeds and inhibiting native species (Noble and Slatyer, 1981). Root growth of the cover crop may aggravate problems of compaction and restrict growth of trees attempting to establish later (Sands and Bowen, 1978). Consequently, if cover crops are necessary, fertilizer levels could be reduced and tube stock planted in the seed-bed. Alternatively, the seed-bed of soils where compaction has not previously been thought a problem (e.g. sands) could be ripped as is common practice in known hard soils (e.g. bauxite mine restoration, Hinz, 1980b).

Perhaps the greatest challenge to mine restoration is the development of novel techniques to enhance the rates of dispersal of trees and other desirable species that are established in low numbers and often with patchy distribution. Appropriate aftercare technology is required. At present, a *laissez faire* attitude is adopted with perhaps the occasional dusting with fertilizer and the active prevention of fire. The disadvantages of fertilizer use have already been discussed; however, it is possible that localized fertilizer placement may increase the growth rates of selected individuals without deleteriously affecting susceptible species. Fire, at the auspicious time, may play an important role in dispersal through its effect on vegetative reproduction. However these and other techniques need evaluating experimentally before recommendations can be made. Often in unmanaged ecosystems it may be desirable to use techniques that may be agronomically unacceptable but which will more rapidly lead to a truly self-sustaining ecosystem.

ACKNOWLEDGEMENTS

This chapter is based on a talk given to the 1981 North Australian Mine Rehabilitation Workshop (reproduced in the proceedings Ed. D. A. Hinz, Nabalco, Northern Territory). Thanks are due to Nabalco Pty Ltd, Nhulunbuy, Northern Territory, and Mineral Deposits Ltd, Bundall, Queensland, for permission to conduct observations on restored mine sites and to J. Kohen and third year students for assistance with the field observations.

REFERENCES

Comalco (1981). "Regenerating Mined Land at Weipa". Comalco, Melbourne.
Bradshaw, A. D., and Chadwick, M. J. (1980). "The Restoration of Land". Blackwell, London.
Burrows, F. J., and Milthorpe, F. L. (1976). Stomatal conductance in the control of gas exchange. *In* "Water Deficits and Plant Growth" (T. T. Kozlowski, ed.), Vol. IV, pp. 103-152. Academic Press, New York.
Carbon, B. A., Bartle, G. A., and Murray, A. M. (1979). Leaf area index of some eucalypt forests in South-west Australia. *Australian Forest Research* 9, 323-326.
Clark, S. S. (1975). The effect of sand mining on coastal heath vegetation in New South Wales. *In* "Managing Terrestrial Ecosystems" (J. Kikkawa and H. A. Nix, eds.), pp. 1-16. Ecological Society of Australia, Canberra.
Craciun, G. C. J. (1978). *Eucalyptus* trials in the Northern Territory coastal region. *Australian Forest Research* 8, 153-161.
Cremer, K. W. (1977). Distance of seed dispersal in Eucalypts estimated from seed weights. *Australian Forest Research* 7, 225-228.
Cremer, K. W., Cromer, R. N., and Florence, R. G. (1978). Stand establishment. *In* "Eucalypts for Wood Production" (W. E. Hillis and A. G. Brown, eds.), pp. 81-135. Commonwealth Scientific and Industrial Research Organization, Australia.
Curtin, R. A. (1964). Stand density and the relationship of crown width to diameter and height in *Eucalyptus obliqua. Australian Forestry* 28, 91-105.
Curtin, R. A. (1970). Dynamics of tree and crown structure in *Eucalyptus obliqua. Forest Science* 16, 321-328.
Dalby, J. M. (1977). Effect of Top Soil Depth and Type on Root Growth of *Angophora costata.* Honours Thesis, B.A. Macquarie University, Australia.
Easterbrook, B. (1951). "Control of eucalypt seedlings and suckers". Division of Plant Industry, Advisory Leaflet No. 214. Queensland Department of Agriculture, Brisbane.
Floyd, A. G. (1962). "Investigations into the natural regeneration of Blackbutt—*E. pilularis*". Division of Forest Management Research Note No. 10. Forestry Commission of New South Wales, Sydney.
Grace, J. (1977). "Plant Response to Wind". Academic Press, New York,.
Grieve, A. M., and Pitman, M. G. (1978). Salinity damage to Norfolk Island Pines caused by surfactants. III. Evidence for stomatal penetration as the pathway of salt entry to leaves. *Australian Journal of Plant Physiology* 5, 397-413.

Heydecker, W. (1973). "Seed Ecology". Proceedings of the nineteenth Easter School in Agricultural Science, University of Nottingham. Butterworths, London.

Hinz, D. A. (1980a). Return to nature after mining. *Landline*. Newsletter No. 3. Australian Mining Industry Council, Dickson, Australian Capital Territory.

Hinz, D. A. (1980b). Land returned to indigenous forest after bauxite mining. *Quarry, Mine and Pit* **19**, 6-7.

Hinz, D. A. and Doettling, H. P. (1979). Rehabilitation of mined-out bauxite areas and red mud pond surfaces at Gove, N.T., Australia. Paper presented to the 108th American Institute of Mining Engineers Annual Conference, Louisiana, New Orleans.

Incoll, W. D. (1979). Effect of overwood trees on growth of young stands of *Eucalyptus sieberi*. *Australian Forestry* **42**, 110-116.

Institute of Foresters of Australia (1980). Bauxite: A Report on Bauxite Mining in the Darling Range. Institute of Foresters of Australia, Western Division, Nedlands, Western Australia.

Kramer, P. J., and Kozlowski, T. T. (1960). "Physiology of Trees". McGraw Hill, New York.

Kramer, P. J., and Kozlowski, T. T. (1979). "Physiology of Woody Plants". Academic Press, New York.

Kira, T. (1978). Community architecture and organic matter dynamics in tropical lowland rain forests of Southeast Asia with special reference to Pasoh Forest, West Malaysia. *In* "Tropical Trees as Living Systems" (P. B. Tomlinson and M. H. Zimmerman, eds.), pp. 561-590. Cambridge University Press, Cambridge.

Lacey, C. J. (1974). Rhizomes in tropical eucalypts and their role in recovery from fire damage. *Australian Journal of Botany* **22**, 29-38.

Lacey, C. J., and Whelan, P. I. (1976). Observations on the ecological significance of vegetative reproduction in the Katherine-Darwin region of the Northern Territory. *Australian Forestry* **39**, 131-139.

Lewis, J. W. (1978). Ecological Studies of Coastal Forest and its Regeneration After Mining. Ph.D. thesis. University of Queensland, Australia.

Lewis, J. W. (1980a). Environmental aspects of mineral sand mining in Australia. *Minerals and the Environment* **2**, 145-158.

Lewis, J. W. (1980b). Mycorrhizal fungi: How important are they in the rehabilitation of mined land. *Landline*. Newsletter No. 3. Australian Mining Industry Council, Dickson, Australian Capital Territory.

Milthorpe, F. L., and Moorby, J. (1979). "An Introduction to Crop Physiology" 2nd edn. Cambridge University Press, Cambridge.

Monteith, J. L. (1973). "Principles of Environmental Physics." Edward Arnold, London.

Mott, J. J. (1978). Dormancy and germination in five native grass species from savannah woodland communities of the Northern Territory. *Australian Journal of Botany* **26**, 621-631.

Mucha, S. B. (1979). Estimation of tree ages from growth rings of eucalypts in Northern Australia. *Australian Forestry* **42**, 13-16.

Mueller-Dombois, D., and Ellenberg, H. (1974). "Aims and Methods of Vegetation Ecology". Wiley, New York.

Noble, I. R., and Slatyer, R. O. (1981). Concepts and models of succession in vascular plant communities subject to recurrent fire. *In* "Fire and the Australian Biota" (A. M. Gill, R. H. Groves and I. R. Noble, eds.), pp. 311-335. Australian Academy of Science, Canberra.

Pryor, L. D. (1976). "The Biology of Eucalypts". Studies in Biology, No. 61. Edward Arnold, London.

Raison, R. J. (1980). Possible forest site deterioration associated with slash-burning. *Search* **11**, 68-72.

Russell, Scott, R. (1977). "Plant Root Systems". McGraw-Hill, Maidenhead.

Sands, R., and Bowen, G. D. (1978). Compaction of sandy soils in radiata pine forests. II. Effects of compaction on root configuration and growth of radiata pine seedlings. *Australian Forest Research* **8**, 163–170.

Shelley, J. M. (1979). Revegetation Following Sand Mining on the East Coast of New South Wales. Honours thesis, B.A. Macquarie University, Australia.

Siddiqi, M. Y., Myerscough, P. J., and Carolin, R. C. (1976). Studies in the ecology of coastal heath in New South Wales. IV. Seed survival, germination, seedling establishment and early growth in *Banksia serratifolia* Salisb., *B. aspleniifolia* Salisb., and *B. ericifolia* L.F. in relation to fire: temperature and nutritional effects. *Australian Journal of Ecology* **1**, 175–183.

Specht, R. L. (1975). The effect of fertilizers on sclerophyll (heath) vegetation—the problems of revegetation after sand-mining of high dunes. *Search* **6**, 459–461.

Westman, W. E., and Rogers, R. W. (1977). Nutrient stocks in a subtropical eucalypt forest, North Stradbroke Island. *Australian Journal of Ecology* **2**, 447–460.

CHAPTER 18

Modelling Environmental Effects on Crop Productivity

J. M. MORGAN

I. INTRODUCTION

In 1965, F. L. Milthorpe argued the need for a dynamic approach to the quantitative analysis of crop responses to environmental fluctuations. That paper did not aim to "erect an adequate model from which effects of short and long-term climatic changes on crop yields can be assessed and a satisfactory forecast of yields made." However, it presented a framework within which yield responses to changes in the environment could be calculated by describing yield in terms of its components developed at different stages of ontogeny, and by describing overall growth by sigmoid functions. He suggested mathematical relationships between the major environmental variables of light, temperature, water and mineral nutrients. It would seem that

289

CONTROL OF CROP PRODUCTIVITY
ISBN   0 12 548280 9

after 18 y we are still a long way from achieving his goal of an "adequate model". This chapter develops some of these relationships and expresses them in dynamic form to obtain estimates of yield. Temperature and water stress are emphasized.

II. GENERAL CONCEPTS

The most obvious initial problem faced in any attempt to quantify growth is how to define it. Growth may be and is usually understood as a change in the size or number of a biological entity (Richards, 1969). Size may be described by dimensions of length or mass. Although individual cell sizes may vary, there is merit in viewing growth as an increase in cell number, as early physiologists did, even if just to focus attention on growth and as the outcome of two competing influences—exponential increase and factors mitigating against this. These factors include reduction in the supply of substrate in cell cultures or loss of capacity to divide in specialized cells of plant organs. Whether growth is viewed in terms of numbers or mass, both influences must be taken into account. This will be true for considering growth at the primary, cellular, level, at the secondary, organ, level or at the tertiary, plant or community, level. Furthermore, attempts to simulate crop growth where estimates of grain yield are sought will require that calculations are based upon numbers, mass and linear dimensions since seed yield is the product of seed number, individual seed mass per stem and stem number per unit ground area. A determinant of seed yield, water used or transpired, also depends on leaf area.

As Richards (1969) points out, it is easier to quantify and define the exponential processes than to describe the growth-limiting factors. This is more relevant for organisms showing secondary design features, where some cells become specialized and are incapable of further division, than of cultures of single-celled organisms. In the latter, deviations in growth rates from exponential may be explained largely by competition for substrates, assuming constant temperatures. In the former, however, in addition to substrate limitations, design constraints are reflected in individual cell arrangements that characterize various plants. A crop growth model must therefore incorporate these genetic constraints as well as respond to variations in supply factors. At the simplest level, where growth is described as accumulations of carbohydrates fixed by photosynthesis, these constraints are expressed in the parameters of the photosynthesis equation and in the dry-matter partitioning coefficients. In models where growth equations are used to describe dry-matter accumulation, as in the model described below, the genetic constraints are expressed in the parameters of the growth equations. One of the major difficulties in modelling plant or crop growth lies in the equivocal nature of observations of individual processes. The interlocking nature of the feedback mechanisms makes it almost impossible to measure processes in isolation. This not only affects the establishment of functional relationships but also makes testing difficult. Yet unless growth is described in terms of individual processes, it is impossible to properly allow for the contribution of factors such as stored assimilates, to, for example, grain growth in wheat. This dilemma, however, remains unresolved in the following analysis—largely because of absence of information on the control of storage and utilization of reserve carbohydrates.

A convenient approach to the quantitative analysis of the effects of water stress on growth would be to define it in terms of active growth processes and the negative factors of reduced assimilate supply or senescence. This approach may apply generally at either cellular, whole plant or crop level. If we wish to analyse the effects of water stress on the growth and grain yield of wheat, then, bearing in mind the inherent difficulty of predicting responses that are more than one heirarchy level apart (de Wit, 1969), it should be possible to predict variation in biomass or grain yield in terms of growth of the major plant parts. Yield per unit area, for example, is the product of seed number and seed size. Seed number may be further divided into stem number per unit area and seed number per stem. So a logical choice of state variables would be seed weight, stem weight and number, ear size or weight (which determines seed number). We must also add leaf area or weight and root size, which control rate of water use, and hence affect the other state variables. Naturally, in order to compute the changes in each of these, and with variations in moisture stress, appropriate relationships between environmental parameters and the rates of change of the state variables, and feedback relationships must be included. As the model will involve numerical integration of these rates of change, appropriate time steps must be chosen to give the required sensitivity. In a crop model a one day time step should accommodate moisture fluctuations in the soil and their effects on growth rates.

III. GROWTH IN RESPONSE TO ENVIRONMENTAL FACTORS

The growth and development of a plant is a composite of the growth of individual parts, the timing of initiation and numbers initiated, and the subsequent timing and rate of senescence (Milthorpe and Moorby, 1979). Water deficits may affect overall growth by affecting each of these processes (Begg and Turner, 1976). Therefore, to quantify the effects of water deficit on plant growth, we must define the appropriate functional relationships between water stress and each of these processes.

A. Individual Organ Development

Except for primordial regions, changes in weight or cell number with time may be described by a sigmoid relationship, which is usually represented mathematically by either a polynomial (e.g. Milthorpe and Moorby, 1979) or by one of a family of logistic relationships (Richards, 1969). The use of a logistic relationship enables the rate of growth to be defined in terms of physiologically meaningful parameters. Ideally, the growth of an individual plant part may be described in terms of cell division or increase in number and cell expansion. Most evidence suggests that cell number is less affected initially by moisture stress than individual cell weight (Begg and Turner, 1976). Thus for any time period, the change in cell number will be given by

$$\frac{dN}{dt} = RpN \tag{18.1}$$

where N is cell number, p is the proportion that remains meristematic, and R is the growth rate of the meristematic cells (Richards, 1969). The total number of cells in the mature organ N_f, tends to be stable or a characteristic of the genotype. The shape of the growth equation is determined by the value of p. For meristematic regions where all cells are dividing, $p = 1$, whereas for a simple logistic, $p = N_f-N/N_f$. Forms for other growth equations are given by Richards (1969). Here, as a first approximation, the simple logistic is used. The number of non-dividing cells is $(1-p)N$ and their rate of increase is

$$\frac{d(1-p)N}{dt} = spN \tag{18.2}$$

where s is the relative growth rate of non-dividing cells (Richards, 1969). The increase in weight of non-meristematic cells will be $spNW_m$ where W_m is the mean final weight achieved by expanding cells (assuming that expansion occurs immediately). This final weight depends on effects of water stress through turgor pressure and assimilate supply from photosynthesis. Increase in weight as a result of cell division should be ultimately affected by assimilate supply also, but unaffected by small changes in water potential. The absence of data providing the relationship needed between water stress and cell size or weight, necessitates the modification of this approach to use more easily measured area or dry weight data. The increase in dry weight, dW/dt will be the sum of the contribution of increase in cell number and size. Reductions in rate as a result of water stress, however, will largely reflect reduced cell expansion and its summative effect on the final size of the organ. Thus, the increase in weight will be

$$\frac{dW}{dt} = pN (sW_n + (r-s)W_m) \qquad (18.3)$$

where W_n is the mean weight of non-meristematic cells and W_m the mean weight of meristematic cells. Any reduction in growth resulting from water stress will be reflected in the maximum size, W_{max}. If individual cells lose the ability to expand further, growth rate of the organ will be reduced by the summated reductions in cell W_{max}. However, evidence from leaf expansion data (Acevedo et al., 1971) of maize suggests that substantial recovery occurs after stress is relieved (Fig. 18.1) such that the reduction in final length is about 20% of the initial reduction in size before the stress was relieved. As a first approximation, a value of 50% accommodates both recovery and non-recovery situations.

Therefore, we describe changes in growth by the eqn

$$dW/dt = kpW \qquad (18.4)$$

In order to avoid establishing a unique relationship for each stage of ontogeny (e.g. Milthorpe, 1965), functional relationships are set up between measures of the environment and the relative growth rate of growing tissue, k, rather than the relative growth rate $(dW/dt)/W$. Ideally, separate relationships should be found for each major plant part, but such information is rarely available for any one species, and generalized forms or estimates must be used.

The relationship for temperature (T), based on various growth chamber experiments (e.g. Friend et al., 1962) is the well-defined bell-shaped curve (Fig. 18.2a) with parameters that vary with genotype. With wheat, for example, maximum growth occurs at about 20°C whereas for sorghum the maximum occurs at about 30°C. A polynomial fit of the relationship for wheat based on the data of Friend et al. (1962) gives

$$k/k_{max} = 0.0974T - 0.001965T^2 - 0.199 \qquad (18.5)$$

The direct effects of water stress on growth are more difficult to quantify, particularly in terms of units of weight and for the major plant parts. The problem is exacerbated as the response may vary with acclimation to stress through osmotic adjustment (Turner, 1979), with different plant parts (Barlow et al., 1977b), between species (Turner, 1979) and within species (Morgan, 1980a). Nevertheless, it is possible to describe a general relationship between expansion growth and water potential based on observations of various species and organs (Boyer, 1970; Barlow et al., 1977a). Almost invariably there is a range of water potentials for which growth is unchanged, followed by a linear or curvilinear decline (Fig. 18.2b). Whereas

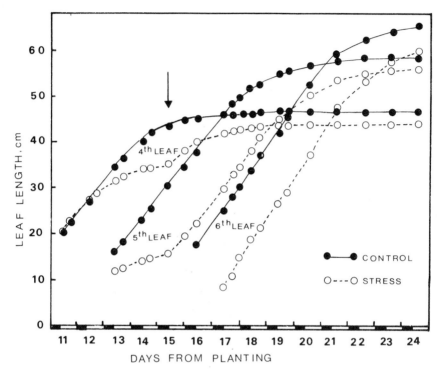

Fig. 18.1 Responses of maize leaf growth to a 2–3 d period of water stress. Arrow indicates time of relief of stress, solid lines are controls and broken lines stressed treatment (after Acevedo *et al.*, 1971).

the initial phase almost certainly reflects the extent of full osmotic adjustment, the following phase of growth decline may or may not reflect partial adjustment, although for most observations reported, the transition from unchecked growth to little or no growth was rapid (e.g. 0.3 MPa of water potential for corn leaves; Barlow *et al.*, 1977a). As a first approximation, it may be assumed that the major source of variation between species and plant parts is the extent of the first phase. To further simplify computations, for shoots the extent of the first phase may be assumed to be about one third of the response of transpiration and photosynthesis to water potential (Barlow *et al.*, 1977a; Boyer, 1970). A similar relationship to the one defined later for transpiration may then be used, where relative transpiration rate is expressed as a function of relative soil water content and potential evaporation, thereby avoiding direct calculations of shoot water potential.

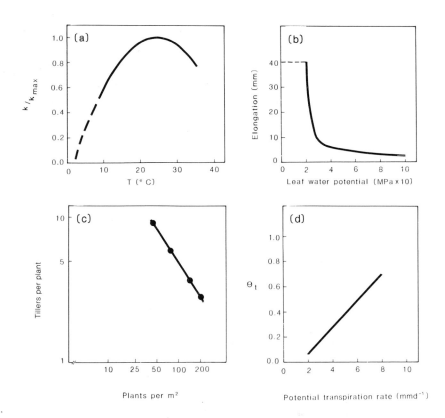

Fig. 18.2 (a) The response to temperature of relative growth rate of the growing portion of a plant part, expressed as a fraction of the maximum of that rate (after Friend *et al.*, 1962); (b) changes in growth rate with leaf water potential (from Boyer, 1970); (c) relationship of ear-bearing tillers per plant with plant establishement (Doyle pers. comm., 1983); (d) a relationship of the relative soil water content at which actual crop transpiration falls below the potential rate θ_t, and the potential transpiration, E_p (after Denmead and Shaw, 1962, but with intercept adjusted for wheat).

B. Meristematic Activity

Although there has been much interest in effects of water on expansion growth, there seems to have been little interest in the effects of water stress on tillering (Begg and Turner, 1976), despite its importance as a determinant of grain yield in most cereals. One can not therefore more than guess at the threshold water potential as being the same as that for cell division; that is, for reduction in transpiration or photosynthesis (Denmead and Shaw, 1962).

The other important determinant of tillering is plant density. Whereas ideally, the effect of changes in plant density should be expressed in terms of reduction of, or competition for assimilates as a result of a decrease in plant photosynthesis, in the present analysis we can do little more than use an empirical relationship (Fig. 18.2c). In wheat the number of ear-bearing tillers per plant is related to plants m^{-2}, in a similar manner to the allometric relationships between plant parts

$$\ln N_T = 5.101 - 0.722 \ln E \qquad (18.6)$$

where N_T is tiller number and E, plants m^{-2}. The slope of this relationship was based on means of three sowing times (data of Doyle pers. comm., 1982), and the intercept was adjusted for non-stressed conditions (Cooper, 1977).

C. Senescence

Although the reduction in leaf area through leaf senescence is one of the more important adaptations to water stress, there is very little information relating senescence of leaves of various ages to measures of water relations either of individual leaves or of the plant or crop (Thomas and Stoddart, 1980; Begg and Turner, 1976; Turner, 1979). Without the effects of nutrition or disease, at least two responses must be accounted for: senescence of lower leaves as a result of ageing and reduced light as the canopy develops, and the senescence as a result of water stress. For wheat, the results of a field experiment involving a number of irrigation treatments (Cooper, 1980) provide an estimate of the time course of senescence under well-watered conditions and the response to soil water stress. Rather than derive a relationship with time, which would depend on temperature, the percentage of senesced leaves is expressed as a function of a measure of physiological age or time—in this instance the proportion of dry weight of the stem that had ceased growing. The stem was chosen because its growth spans most of the ontogeny of the plant, and because its weight was recorded. In estimating this, only weights up to the dough stage were used, because stem weights decline past the dough stage as a result of export of assimilates to the grain. This approximate relationship was

$$D = 0.710 P_s - 0.353 P_s^2 - 0.0036 \quad (r^2 = 0.993) \qquad (18.7)$$

where P_s is the fraction of non-growing stem and D the fraction of senesced leaf. After the dough stage, the leaf area declined rapidly with the onset of maturity about 30 d later. For this period, the average rate of senescence was 2% of the total leaf weight per day.

Senescence as a result of soil water deficit did not occur until anthesis, and then only in the rain-fed treatment. The absence of difference between irrigation treatments until the dough stage suggests that senescence did not commence until the soil water content was about 30% of the water holding capacity, which was assumed about 200 mm. It may be assumed that once this level is reached senescence will occur rapidly, similar to pot-grown plants (Fischer, 1973). If it is assumed that senescence is complete once the soil water store is fully depleted, the relationship between the fraction of senesced leaves (D) and fractional water deficit (Φ) will be 0 at zero available soil water to 1 when soil water is at field capacity.

$$D = 3.33 \ \Phi \qquad (18.8)$$

In contrast to leaves on shoots that survive to bear grain, the development of fertile tillers appears to be sensitive to water deficits. The results of the irrigation experiment by Cooper (1980) indicated that reductions in number of fertile tillers as a result of water stress occurred only in the period before anthesis. The effect of water stress was not evident in the most-frequently irrigated treatment, where 93% of tillers were fertile at the dough stage. Allowing for the possibility that fertility may be reduced by factors other than water stress, the threshold of soil water deficit for tiller fertility would be close to 50 mm or about 75% of the water-holding capacity. At deficits greater than this, the proportion of fertile tillers (N_t) was reduced at the rate of 0.39% per mm of soil water deficit. When related to fractional soil water deficit

$$N_t = 1.2 - 0.8 \ \Phi \qquad (18.9)$$

$(0.2 < \Phi < 0.7)$ and

$$N_t = 2.88 - 3.20 \ \Phi \qquad (18.10)$$

for $0.7 < \Phi < 1.0$, with a minimum tiller number of one.

A further way in which yield may be reduced by water stress is through reduced seed set per shoot (Bingham, 1966; Morgan, 1980b). This will only occur when the upper leaves are near wilting, when seed set may be reduced rapidly by as much as 50%. If stress is subsequently relieved, compensation may occur through increased seed size (e.g. Wright, 1981). Seed set will only be affected in this way during meiosis, between 5 and 15 d before anthesis (Fischer, 1973). We therefore use a relationship similar to Fischer (1973) but related to the soil water deficit and potential evaporation parameters discussed below (Fig. 18.2). For the maximum number of seeds per ear, S_n, an empirical relationship with ear weight, W_e, is used.

$$S_n = 7.754 + 0.042 \ W_e \qquad (18.11)$$

D. Ontogenetical Events

In this model where the growth of major plant parts is calculated, it is possible to sequence ontogenetical events such as jointing, meiosis and anthesis
by using the relative development of existing plant parts; that is,
physiological timing, rather than the more usual day degrees technique.
Jointing, which coincides with the early stages of stem growth, becomes
evident in wheat when the stem dry weight is about 25% of the maximum
and continues to about 50% of maximum weight. Meiosis usually coincides
with the full emergence of the last or flag leaf; that is, when the leaf weight
has reached its maximum and anthesis usually occurs about 14 days after
this. The timing of ear initiation, which occurs after the initiation of the last
leaf, depends on the number of leaves initiated, and on environmental factors which affect this process, mainly temperature. For spring wheat grown
at day/night temperatures of 20°C/15°C, with a total of seven leaves on the
main stem, the period from seed germination to ear initiation is about 15
d (Williams, 1966). At the same plastochrone time, this would be about 21
d for the more usual nine leaves per stem. To describe the effects of temperature on this time interval, the temperature response curve for growth, which
was described above, may be used. Primary tillers are also initiated during
this time interval, with a spacing of about four days (Rawson, 1971).

IV. ENVIRONMENTAL RELATIONSHIPS

The growth responses that form the set of functional relationships used to
calculate changes in the state variables of the model described above involve
changes in the state of soil water available to the plant and the state of water
in the atmosphere. In the following section functional relationships describing changes in these variables are defined including feedbacks with the variables defining the state of the plant. For computation of soil water, the
effective rooting zone, taken as 1.2 m, is divided into four layers of 30 cm
each. The water content on any day will be the water content of the previous
day plus the change in water content, where the change in water content

$$\Delta\Theta = f_i - f_o \tag{18.12}$$

where f_i is the inflow and f_o the outflow (mm). At the surface, the inflow
will be the sum of rainfall and irrigation, less run off. Below the surface, the
inflow of a given layer will be the overflow of the layer above, calculated
as the quantity of water in excess of the field capacity, after evaporative losses
have been accounted for. The outflow from a given layer occurs through
evaporation and drainage.

A. Evaporation

The computations of evaporation follow well-established relationships. At the surface, the two components of soil and plant must be accounted for, whereas below the surface layer only losses as a result of transpiration are computed. Following the procedures outlined by Ritchie (1972) for the calculation of an unstressed crop, potential transpiration, E_p is expressed as a function of potential evaporation, E_o, and leaf area index, L

$$E_p = E_o (-0.21 + 0.7L^{0.5}) \qquad 0.1 \leq L \leq 2.7 \qquad (18.13)$$

The potential evaporation is given by the combination formula of Penman (1963).

$$E_o = \frac{\Delta}{\gamma}Q_n + 0.262 (1 + 0.0061u)(e_s - e_a) (\frac{\Delta}{\gamma} + 1)^{-1} \qquad (18.14)$$

where Δ is the slope of saturation vapour pressure curve at mean air temperature; γ, the psychometric constant; u, the daily wind run at 2 m; e_s, saturation vapour pressure at mean air temperature (mbar); e_a, actual vapour pressure (mbar); Q_n, the daily net radiation (mm equivalents). The net radiation is either measured, or calculated from daily total short-wave radiation (Linacre, 1968). Where L is less than 0.1, it is assumed that crop evaporation is negligible, and losses are due only to soil surface evaporation, E_s. For this an empirical relationship is used (F. L. Milthorpe pers. comm., 1972).

$$E_s = 1.136 \times 10^{-3} (15 + \Psi_s) Q_{ns} \qquad (18.15)$$

where Q_{ns} is the net radiation at the soil surface (mm) and Ψ_s is the water potential of the surface layer (MPa $\times$ 10). The magnitude of Q_{ns} depends on the degree of shading (Ritchie, 1972).

$$Q_{ns} = Q_n \exp - 0.398L \qquad (18.16)$$

During periods of water deficit crop evaporation will fall below the potential rate due to stomatal closure. When leaf water potential reaches a threshold level approximating zero turgor (Turner, 1974), the relative available soil-water content at which this occurs, Θ_t, will depend on the evaporative demand of the atmosphere, or the potential evaporation rate (e.g. Denmead and Shaw, 1962).

$$\Theta_t = 0.106E_p - 0.14 \qquad (E_p > 2) \qquad (18.17)$$

The actual crop evaporation, E_c, will then be

$$E_c = E_p \Theta_r / \Theta_t \qquad \Theta_t \geq \Theta_r \geq 0 \qquad (18.18)$$

where Θ_r is the relative available soil-water content, and equal to the actual total available water divided by the potential available water.

B. Available Water

The volume of soil water available for transpiration depends on the physical parameters of the soil defining the limits of extraction by roots and surface evaporation, and the extent of the root system (Tennant, 1976). Evidence from field measurements suggests that, with the possible exception of the extremity of the root system, wheat plants are able to extract water about equally over the depth of soil penetrated by the roots (Tennant, 1976) which for wheat is about 120 cm, when the deepest roots reach 150–180 cm. However, under well-watered conditions the amount of water extracted may vary with depth, particularly before anthesis (Shalhevet et al., 1976). In order to accommodate this variation, it is assumed that full extraction to a limit of –1.5 MPa (or the known value for a cultivar) may occur to a depth of 120 cm. For each 30 cm layer, it is assumed that extraction will occur according to the pattern reported by Shalhevet et al. (1976) for well-watered spring wheat in which the volume of available water varies with root growth. If a layer dries to wilting point, the evaporation is apportioned to the remaining layers that still contain extractable water and lie within the rooting depth.

V. EVALUATION

The parameters defining the growth of the various plant parts were derived from field measurements made under well-watered conditions (Cooper, 1977) and from growth studies in a controlled environment (Williams, 1966). They approximate cv. Timgalen, an Australian mid-season spring wheat (Table 18.1). Particular features of this cultivar are that a larger than average number of synchronous tillers are produced and smaller than average number of seeds set per ear.

The test data consisted of observations made by Doyle (pers. comm., 1983). Experimental methods and meteorological data are presented in Doyle and Fischer (1979). Briefly, the season produced moderate stress conditions with a crop rainfall of 236 mm. The total available moisture at sowing was 132 mm to a depth of 120 cm. The soil, which was a uniformly textured red earth, had a water-holding capacity of 136 mm. The lower limit of available water

Table 18.1

Growth parameters and initial weights for the various plant parts.
(Data adapted from Cooper, 1977.)

Plant part per Tiller	Initial weight W_i (mg)	Maximum weight W_{max} (mg)	Rate term $k(d^{-1})$
Leaves	0.001	300	0.30
Stems	0.01	950	0.20
Ear	0.001	230	0.45
Grain (single)	1.0	36	0.25

Table 18.2

A comparison of components of grain yield predicted using weather data and the growth model, and field plot observations (Doyle, pers. comm.).

	Grain weight (mg)	Grain number per ear	Ear number per m^2	Grain yield t ha^{-1}
Predicted	32	22	448	3.2
Observed	34	23	405	3.3

was the measured limit of extraction for cv. Timgalen (400 mm) growing in this soil type. The sowing date was 2 July and plant density 100 m^{-2}.

As a summary of model performance, calculated changes in total above-ground dry weight, grain dry weight and leaf area index are presented in Figure 18.3a. Calculated leaf area and total weight were about 10% below measured values. Calculated grain yield (3.2 t ha^{-1}) was within 10% of the measured value although calculated yield components varied both above and below the measured values (Table 18.2).

The sensitivity of the model to variations in water supply was examined using meteorological data from 1974 and 1968 (Doyle, 1972) both with and without rainfall. The data are expressed as the grain yield that correspond to various amounts of cumulative evapotranspiration over crop growth (Fig. 18.3). The actual data presented were derived from measurements made on various spring wheat cultivars grown in different seasons in NSW. Although this type of relationship may not be stable, particularly if applied to environments that are different to those where the data were collected, it does show that the calculated responses correspond reasonably well to those derived from field measurements.

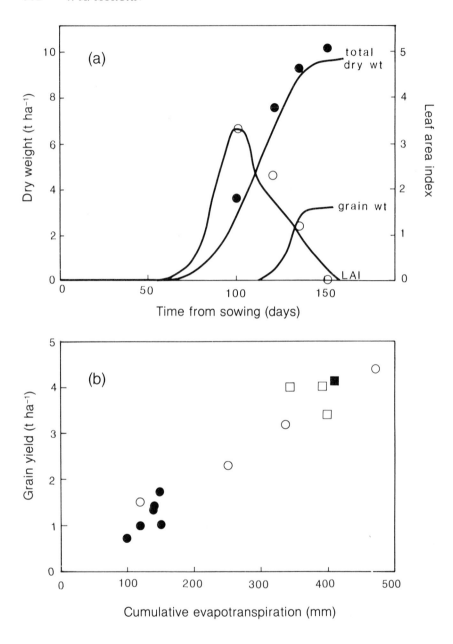

Fig. 18.3 (a) Calculated (lines) and measured (points) values of total above ground dry weight (●), leaf area index (○) and grain weight (measured values from Doyle, pers. comm., 1983); (b) relationship of calculated grain yield to calculated evapotranspiration (○) compared with measured value; Morgan, not previously published (●), Doyle, pers. comm. 1983, (□), Cooper, 1980 (■).

VI. CONCLUSION

Although the results of the calculations are within acceptable limits, they can do little more than provide encouragement for the necessary and more extensive further evaluation. These results do, however, suggest that the overall framework may be a suitable one for analysing growth and yield responses to variations in water and temperature. This, then, highlights the need to better define some of the functions used—which presently are little more than educated guesses. However, whether better defined data are warranted or not depend on the interests of the user and the size of computer available. The present model is a suitable size for use on a microcomputer, so further increases in size could limit utility.

Another deficiency with this model is the failure to account for the effects of redistribution of dry matter, resulting from senescence and mobilization of stored carbohydrates. In order to accommodate these effects, a more complex supply–demand model (e.g. Morgan, 1976) is required. Whereas increases in complexity of this kind may have the potential to produce a more accurate result, they also introduce concepts such as pool or storage sizes, partitioning based on source–sink relationships and proximity to source of supply, for which there is little unambiguous quantitative data to use in developing functional relationships and evaluation. There is also the general problem of how to properly test models with too many uncertain relationships that are integrated with feedback mechanisms. It may seem, then, that the only feasible approach at present is to aim at simpler models for which data are readily available and to accept the consequent limitations.

REFERENCES

Acevedo, E., Hsiao, T. C., and Henderson, D. W. (1971). Immediate and subsequent growth responses of maize leaves to changes in water status. *Plant Physiology* **48**, 631–636.

Barlow, E. W. R., Boersma, L., and Young, J. L. (1977a). Photosynthesis, transpiration, and leaf elongation in corn seedlings at suboptimal soil temperatures. *Agronomy Journal* **69**, 95–100.

Barlow, E. W. R., Munns, R., Scott, N. S., and Reisner, J. (1977b). Water potential, growth and polyribosome content of the stressed wheat apex. *Journal of Experimental Botany* **28**, 909–916.

Begg, J. E., and Turner, N. C. (1976). Crop water deficits. *Advances in Agronomy* **28**, 161–217.

Bingham, J. (1966). Varietal response in wheat to water supply in the field, and male sterility caused by a period of drought in a glasshouse experiment. *Annals of Applied Biology* **57**, 365–377.

Boyer, J. S. (1970). Leaf enlargement and metabolic rates in corn, soybean and sunflower at various leaf water potentials. *Plant Physiology* **46**, 233–235.

Cooper, J. L. (1977). The adaptation of a semi-dwarf wheat in the Murrumbidgee irrigation area. M.Sc. Thesis, The University of Sydney.

Cooper, J. L. (1980). The effect of nitrogen fertilizer and irrigation frequency on a semi-dwarf wheat in south-east Australia. *Australian Journal of Experimental Agricultural and Animal Husbandry* **20**, 359–364.

Denmead, O. T., and Shaw, R. H. (1962). Availability of soil water to plants as affected by soil moisture content and meteorological conditions. *Agronomy Journal* **54**, 385–390.

de Wit, C. T. (1969). Dynamic concepts in biology. *In* "Prediction and Measurement of Photosynthetic Activity" (I. Setlik, ed.), pp. 17–23. PUDOC, Wageningen.

Doyle, A. D. (1972). The influence of superphosphate and time of sowing on the growth, development and nutrient uptake of four wheat cultivars. M.Sc.Agr. Thesis, The University of Sydney.

Doyle, A. D., and Fischer, R. A. (1979). Dry matter accumulation and water use relationships in wheat crops. *Australian Journal of Agricultural Research* **30**, 815–829.

Fischer, R. A. (1973). The effect of water stress at various stages of development on yield processes in wheat. *In* "Plant Response to Climatic Factors." Proceedings of Uppsala Symposium (1970). pp. 233–241. UNESCO, Paris.

Friend, D. J. C., Helson, V. A., and Fisher, J. E. (1962). The rate of dry weight accumulation in Marquis wheat as affected by temperature and light intensity. *Canadian Journal of Botany* **40**, 939–955.

Linacre, E. T. (1968). Estimating the net-radiation flux. *Agricultural Meteorology* **5**, 59–63.

Milthorpe, F. L. (1965). Crop responses in relation to the forecasting of yields. *In* "The Biological Significance of Climate Changes in Britain." (C. G. Johnson and L. P. Smith, eds.), pp. 119–128. Academic Press, New York.

Milthorpe, F. L., and Moorby, J. (1979). "An Introduction to Crop Physiology." Cambridge University Press, Cambridge.

Morgan, J. M. (1976). A simulation model of the growth of the wheat plant. Ph.D. thesis, Macquarie University. Australia.

Morgan, J. M. (1980a). Osmotic adjustment in the spikelets and leaves of wheat. *Journal of Experimental Botany* **31**, 655–665.

Morgan, J. M. (1980b). Possible role of abscisic acid in reducing seed set in water stressed wheat plants. *Nature* **285**, 655–657.

Penman, H. L. (1963). Vegetation and hydrology. *Commonwealth Bureau of Soils Technical Communication* **53**.

Rawson, H. M. (1971). Tillering patterns in wheat with special reference to the shoot at the coleoptile node. *Australian Journal of Biological Sciences* **24**, 829–841.

Ritchie, J. T. (1972). Model for predicting evaporation from a row crop with incomplete cover. *Water Resources Research* **8**, 1204–1213.

Richards, F. J. (1969). The quantitative analysis of growth. *In* "Plant Physiology" (F. C. Steward, ed.), pp. 3–76. Academic Press, New York.

Shalhevet, J., Mantell, A., Bielorai, H., and Shimshi, D. (1976). Wheat. *In* "Irrigation of Field and Orchard Crops under Arid Conditions" pp. 7–15. International Irrigation Information Centre, Bet Dagan, Israel.

Tennant, D. (1976). Wheat root penetration and total available water on a range of soil types. *Australian Journal of Experimental Agriculture and Animal Husbandry* **16**, 570–577.

Thomas, H. and Stoddart, J. L. (1980) Leaf senescence. *Annual Review of Plant Physiology,* **31**, 83–111.

Turner, N. C. (1974). Stomatal response to light and water under field conditions. *Royal Society of New Zealand Bulletin* **12**, 423–432.

Turner, N. C. (1979). Drought resistance and adaptation to water deficits in crop plants. *In* "Stress Physiology in Crop Plants" (C. H. Mussell and R. C. Staples, eds.), pp. 343–372. Wiley, New York.

Williams, R. F. (1966). Physiology of growth in the wheat plant. III. Growth of the primary shoot and inflorescence. *Australian Journal of Biological Sciences* **19**, 949–966.

Wright, G. C. (1981). Adaptation of grain sorghum to drought stress. Ph.D. thesis, University of New England, pp. 47–50. Australia.

Glossary of Main Symbols

I have tried to use symbols in a consistent way throughout this book. This glossary represents the main symbols that have been used.

Note — bar above symbol indicates mean; for example, $\bar{A}$ is average rate of assimilation A

— Δ before symbol indicates change, either an increment or decrement

A   assimilation rate
α transfer coefficient (ion uptake)

β_L angle of leaf inclination

C   concentration of CO_2 at site of carboxylation
C_i intercellular partial pressure of CO_2
C_p specific heat
$_r$ concentration of ion at root–environment interface
C_N concentration of element (nutrient) expressed on whole-plant dry weight basis
$[CO_2]$ concentration of CO_2 in air
CW   (tree) canopy width

d   air density
d_p partitioning of dry weight to photosynthetic apparatus in leaf
d_L partitioning of dry weight to leaves relative to change in whole plant dry weight
dbh   diameter of tree at breast height

D   distance a seed may be dispersed
D   dimensionless reduction factor to quantify nitrogen limitation
Δ slope of saturation vapour pressure/temperature curve

e_a actual vapour pressure
e_s saturated vapour pressure
E   actual crop evaporation rate
E_o potential evaporation rate
E_p potential transpiration rate
E_s actual evaporation from soil surface
E_t density of enzyme sites
ε efficiency with which radiation is used to produce dry matter

F   photorespiration rate

g   conductance. Subscript *l* leaf, s soil.
G   net growth rate (dry weight basis)
G   soil heat flux
γ psychometric constant

I   photosynthetically active radiation

J   potential electron transport rate

k   canopy extinction coefficient
k_C catalytic turnover number of carboxylase
K'_r Michaelis constant for fully activated ribulose bis phosphate
     carboxylase-oxygenase (Rubisco)
K''_r effective Michaelis constant for Rubisco

L   leaf area index
L_C plant loss through consumer organisms
L_D plant loss of biomass through death and shedding

M   concentration of free magnesium
M   maximum tissue concentration of nitrogen

N_T proportion of tillers which are fertile
N   number (of cells)
N_T number (of cells) in mature organ
NAR   net assimilation rate

O   concentration of oxygen at site of Rubisco oxygenation

p   proportion
P   measure of abundance
P weight of photosynthetic apparatus
ψ water potential. Subscript l leaf, r root, s soil

Q   primary acceptor of photosystem II
Q_n net radiation above crop
Q_{nc} net radiation absorbed by the leaf canopy
Q_{ns} net radiation at soil surface
Q_s incoming short wave radiation
Q_{10} increase in rate per 10°C rise in temperature

r   quanta required per electron transported
r_a aerodynamic resistance
r_c canopy resistance
R   radius of root
R_d rate of day respiration
R_M relative growth rate of meristematic cells
R_S relative growth rate of non-dividing cells
RF   rate of grain filling

S_N number of seeds per inflorescence

T   temperature
T_{opt} optimum temperature
T minimum permissible tissue nitrogen concentration
Γ_* CO_2 compensation concentration
Φ fractional soil water deficit (zero at wilting point to 1 at field capacity)
Θ soil water content
Θ_r available soil water relative to potential available soil water

u   daily wind run
U   rate of ion uptake

V_c rate of Rubisco carboxylation
V_o rate of Rubisco oxygenation
V_t terminal velocity (of seed during dispersal)
V_w wind velocity

W   weight. Subscript e ear (inflorescence), l leaf, r root, n non-meristematic cells, m meristematic cells

WU   water use efficiency

Y   yield (protein)

Index